World Forest Biomass
and Primary Production Data

World Forest Biomass and Primary Production Data

Compiled by

M.G.R. Cannell

Natural Environment Research Council,
Institute of Terrestrial Ecology, Penicuik,
Midlothian, Scotland

1982

ACADEMIC PRESS

A Subsidiary of Harcourt Brace Jovanovich, Publishers
London   New York
Paris   San Diego   San Francisco
São Paulo   Sydney   Tokyo   Toronto

ACADEMIC PRESS INC. (LONDON) LTD.
24/28 Oval Road
London NW1

United States Edition published by
ACADEMIC PRESS INC.
111 Fifth Avenue
New York, New York 10003

British Library Cataloguing in Publication Data
Cannell, M.G.R.
 World forest biomass and primary production data.
 1. Biomass energy   2. Forests and forestry
 I. Title
 662'.6   TP360

 ISBN 0-12-158780-0

 LCCCN 82-45028

Printed in Great Britain by
Whitstable Litho Ltd, Whitstable, Kent
from camera-ready copy supplied
by Publication Preparation Service
Penicuik, Scotland

PREFACE

This book has been produced as a reference document for research workers, foresters, ecologists and all those interested in the structure, function and dry matter production of forest stands. It assembles published data abstracted from about 600 papers on the dry biomass of stems, branches and foliage and some components of the net primary production of over 1200 forest stands in 46 countries - where production means the *current* annual increment in standing biomass plus annual litterfall, mortality and consumption. An attempt has been made to include most of the major studies published in the western and Japanese literature up to mid-1981, including the period of the International Biological Programme. All types of woody plant stands are included, ranging from oak woodlands and coniferous plantations to tropical rainforests and mangrove swamps. Values are also given for bamboos, some shrub stands and oil palm, rubber and tea plantations. It would be foolish to claim that the literature has been covered exhaustively, but this book does give a much more comprehensive coverage than hitherto available. Each table has the same simple format, and in most instances footnotes have been added so that readers may avoid some of the pitfalls in interpretation and may decide whether they need to refer to the original publications.

The world literature on forest biomass and production expanded rapidly in the 1960s and 1970s, and in 1981 *Forestry Abstracts* listed about eight new titles per month. Before planning further studies it is important that researchers take stock of the large data base that already exists. This was the main reason for undertaking this compilation. Also, researchers often need to search the literature for particular, comparable or contrasting studies, or to establish the range of some critical variables or relationships. The data in this book have been formatted and indexed with those tasks in mind.

Inevitably some important studies may have been overlooked; if so, I apologise to the authors. For those studies that are included, I take full responsibility for having correctly abstracted and interpreted the published data, because it was not feasible to contact all the authors. However, I am grateful to the following authors for their help in locating the literature and/or for providing additional information: P.B. Alaback, A. Albrektson, T. Ando, P.M. Attiwell, D. Auclair, D.I. Bevege, M. Cantiani, A. Clark, A.W. Cooper, R.N. Cromer, T.R. Crow, J. van der Drift, O.J. Eong, J.F. Franklin, R.H. Gardner, S. Greene, C.C. Grier, D.F. Grigal, D. Hanley, R.K. Hermann, H.Z. Hitchcock, H. Hytteborn, Y. Katazawa, S. Kawanabe, J.P. Kimmins, M. Kimura, T. Kira, H. Klinge, J.N. Long, A.E. Lugo, D.A. MacLean, H.A.T. Madgwick, H.G. Miller, E.W. Mogren, C.D. Monk, L. Olsvig-Whittaker, E. Paavilainen, J. Pardé, D.E. Reichle, R.W. Rogers, G. de las Salas, P.A. Sanchez, T. Satoo, K.P. Singh, G.L. Switzer, Y. Tadaki, E.V.J. Tanner, J. Turner, W.E. Westman, H.E. Young and J. Zavitkovski.

   I would also like to thank Dr Lucy Sheppard for help with the species index.
And, most of all, I thank my wife for planning the layout of the tables and making
many editorial suggestions.

December 1981   Melvin G.R. Cannell
Institute of Terrestrial Ecology
Bush Estate, Penicuik.
Scotland.

C O N T E N T S

viii   Contents

INTRODUCTION

Scope

The forest studies chosen for this book were those published by mid-1981 that provided data on the dry biomass of stems, branches and foliage per hectare. Among
those studies, preference was given to data sets that included information on stand
structure (tree number, height, basal area etc.) and estimates of stand production.
The term 'production' here means the *current* net annual increment of stems, branches,
foliage etc. plus losses due to mortality and litterfall, usually measured over
periods of 1 to 10 years. This book does not include data from studies confined to
estimates of stem biomass, foliage biomass, root biomass, leaf area index or litterfall alone. Readers interested in those aspects will find many more values by consulting *Forestry Abstracts*, appropriate review papers, and forestry stand volume
tables which can be used to estimate merchantable stem biomass and increment. Those
interested in the mineral nutrient content of forest biomass should consult the
subject index and the footnotes in this book, which state whether nutrient contents
were determined.

    The approach used to compile this book was first to search the papers and refer-
ences given in the following books and reviews, many reporting work promoted during
the International Biological Programme: Rodin and Bazilevich (1967), Duvigneaud
(1971), Young (1971), Ellenberg (1971), Art and Marks (1971), Golley and Golley
(1972), Young (1973), Golley and Medina (1975), Lieth and Whittaker (1975), Young
(1976), Shidei and Kira (1977), Lamotte and Bourlière (1978), Lieth (1978), Pardé
(1980) and Reichle (1981). A literature search was then conducted within forestry
and ecological journals, with a bias towards the west European, English-language and
Japanese literature: of the 1200 or so forest stands included, 332 were in the USA
and 384 were in Japan. Most of the data on forests in the USSR were taken from Rodin
and Bazilevich (1967) or DeAngelis *et al*. (1981). No attempt was made to trace all
the titles given in *Forestry Abstracts*, nor to include stands of woody species used
in horticulture. However, all of the 116 studies detailed by DeAngelis *et al*. (1981)
in the Oak Ridge IBP Woodland Data Set were covered, and most of the Japanese
studies which provided English summaries and table legends have been included. Data
are also given on stands of bamboo, some shrub stands, *Hevea* rubber, oil palm and
tea plantations, as well as all kinds of natural and man-made forests. In some instances it was possible to obtain unpublished theses, and a few data were supplied
by correspondence, but for the most part the data given here were abstracted from
the publications cited. Data were supplied by the original authors only if there
were serious discrepancies or omissions, or where it was not clear what was included
in certain values.

Arrangement

The data have been arranged by country, and within each country data on broadleaved species precede those on conifers, each in approximate alphabetical order of the main species in each stand. Non-alphabetical ordering occasionally occurs where a study included similar mixed species stands with different compositions (e.g. on page 123 a stand in Japan in which 52% of the biomass was *Tsuga sieboldii* has been listed with its associated species *Abies firma* because other stands in that study were dominated by *A. firma*). However, where a study included several stands with completely different species compositions (e.g. Whittaker, 1966), the study details have been repeated so that the stands could be placed in their correct alphabetical positions. Readers wishing to locate the work of particular authors should consult the author index.

Footnotes

One major difficulty encountered when arranging these data in a standard format was that not all authors used the same definitions or recognized the same division of components. Some included branches with stems, others included current year's twigs with foliage, understorey shrubs with overstorey trees, roots with above-ground woody parts, and so on. Authors defined the numbers and heights of trees in uneven-aged multi-storeyed stands in different ways, few authors gave estimates of losses from mortality or consumption, some omitted woody and foliage litterfall altogether, while others measured litterfall without estimating woody increments. Rather than list all possible combinations of biomass and production values, it was decided to maintain the simplest possible basic format and to use footnotes to tell readers what was or was not included in particular values. The footnotes are, therefore, an integral part of each table.

Another problem of using a standard format was that studies with very different objectives and of different merit were given equal space. Some studies were based on small samples, or small plots, or used simple assumptions about the ratios between tree parts, while others were based on large samples and detailed analyses. Consequently, where information was available, the footnotes state the number of trees sampled (i.e. cut down and divided), the size of the sample plots, whether roots were excavated, and the general approach used to extrapolate from sample tree values to stand values, recognizing the following four basic methods (see also Pardé, 1980; Ogawa and Kira, 1977).

1. All trees within a sample plot were harvested and weighed.

2. Regression equations were developed between the dimensions of sample trees and the dry weights of their component parts, and those equations were applied to all individuals within sample plots. The dimensions used were normally breast-height diameter D, and tree height H, and this method has been abbreviated in the footnotes by statements such as 'stand values were derived from regressions on D and H'.

3. It was assumed that the ratio of the sum of the biomass of sample trees to the biomass of the stand was equal to the ratio of their respective basal areas, abbreviated in the footnotes as 'stand values were derived by proportional basal area allocation' (a method, and phrase, commonly used by Japanese researchers).

4. The dry weights of trees of about average D and H were multiplied by the numbers of trees per hectare.

However, the footnotes do not describe the many methods used to estimate production values, and readers intending to make detailed comparisons between studies may wish to consult the original publications to see whether differences occurred in the ways in which sample trees were subsampled, in the variance of regression equations used, in the use of corrections for bias after logarithmic transformations, in the number of years over which increments were estimated, and so on. The footnotes on methodology provide only a basic guide to help readers to decide whether particular data sets are likely to be valuable for their purposes.

Guide to users

When consulting the data in this book it is important to bear in mind the following points.

Units. All published values have been converted to metric units, and bearing in mind the errors involved in all estimation procedures, biomass values have usually been rounded to the nearest 0.1 t/ha, and production values to the nearest 0.01 t/ha. Units of t/ha were preferred to g/m² although sample plots rarely exceeded 1 ha. Also, where it might be helpful, values have been given for individual sample plots within stands, rather than the overall stand means, recognizing that values for small plots may sometimes not apply over areas of one hectare.

Literature citations. The first publication cited on each page was the main one consulted. Where appropriate, a list is given of other selected references relating to the particular study, especially if nutrient contents or litterfall were reported separately. Sometimes conflicting data were published in several places, in which case the most recent publication was taken as the most authoritative, or the alternative data sets are given. Journals are cited according to the 4th edition of the World List of Scientific Periodicals.

Location. The locations of the study sites are given by latitude, longitude and altitude, and by country and place name.

Species. The authorities of Latin names are given in the species index. Generic names have been abbreviated only where they occur more than once on the same page. Where possible, for mixed species stands, an estimate is given of the percentage of the stand basal area, biomass or stem volume accounted for by each of the main species. All stands may be assumed to be natural or naturally regenerated (the latter if even-aged), unless they are stated to be plantations. Known recent management treatments, such as thinning and pruning, have been noted.

Site. No attempt has been made to give details of site conditions, but in many instances a brief statement has been included on the soil type. Also, known recent treatments, such as fertilizer applications, have been noted.

Age (years). Stand ages have normally been reckoned by counting growth rings at breast height, or in the case of plantations, from the date of planting. The ages of mature, uneven-aged, natural stands are given as upper values, age ranges, or by the word 'mature'.

Density (trees/ha). Densities include all living trees accounted for in the biomass data. If small-diametered trees were omitted this is noted. For coppiced stands, the density values refer to the number of stems per hectare.

Tree height (m). Unless otherwise stated, these are mean tree heights or height ranges. However, they will often be footnoted as stand top heights or the heights of dominant trees. For multi-storeyed stands the heights are given of each storey where these values were known.

Basal area (m²/ha). Basal areas were usually calculated from diameters at breast height (D) by the authors, but occasionally the compiler has calculated values using published diameter frequency distributions.

Leaf area index. The values given are one-sided leaf areas of broadleaved trees and *projected* leaf areas of conifers. The all-sided needle areas of most pines are about 2.8 times their projected areas, and for most other conifers the ratio is about 2.3. Wherever authors gave the all-sided leaf area indices of conifers these are given in the footnotes. In most instances the leaf area index given is the maximum value obtained during the year.

Biomass (t/ha). The biomass is given of stem wood, stem bark, living branches, foliage and roots (including the stumps) of all living trees, plus the biomass of any understorey shrubs. The biomass of any dead trees, or dead branches, is given in the footnotes. The biomass of ground-layer, herbaceous vegetation is not given. Occasionally very different values were obtained using different methods in which case both estimates are given as explained in the footnotes.

Current Annual Increment (CAI, m^3/ha/yr). These values are the current (or periodic) annual increments in stem volume including the bark.

Net production (t/ha/yr). Data are given for as many as possible of the following components: the current net annual increment in the standing biomass of stem wood, stem bark, branches, fruits, etc. (including all sexual reproductive parts), foliage and roots, plus annual losses of dry matter due to mortality, all kinds of litter-fall, decay, consumption and the turnover of fine roots. However, in most instances some of these components have been pooled or are missing, as is made clear in the footnotes, and in some instances missing values (e.g. mortality or consumption) may have been negligible. *All non-footnoted values refer to increments*: thus stem wood production data presented as 3.12 + 1.00 t/ha/yr mean that the current net annual increment was 3.12 t/ha/yr to which have been added losses of 1.00 t/ha/yr as foot-noted.

Annual stem biomass increments were usually estimated for periods of 1 to 10 years, either by repeated sampling or retrospectively, from measured increases in D, H and/or stem volume (see Pardé, 1980; Ogawa and Kira, 1977). Branch increments were often estimated similarly, or from regressions of branch biomass on branch age.

Annual foliage production in deciduous stands may have been assumed to be equal to the foliage biomass in late summer, in which case early summer leaf fall will have been ignored, unless otherwise stated. Alternatively, annual foliage production may have been assumed to be equal to mean annual foliage litterfall, in which case losses due to translocation, decay and consumption will have been omitted. The foot-notes often indicate the period over which litterfall was measured, but all values in the tables always refer to periods of one year. For evergreen stands annual foliage production may have been derived from (i) regression equations, (ii) annual foliage litterfall, (iii) the biomass of current year's foliage or (iv) foliage biomass divided by mean foliage longevity.

Root production values are by far the least reliable. Recent data on the high turnover of fine roots (e.g. on pages 242, 264, 337 and 341) suggest that most root production values published before about 1975 were underestimates. Most were based on regression equations of root biomass on D, or assumed some other proportionality between the growth of above- and below-ground parts.

References

Art, H.W. and Marks, P.L. (1971). A summary table of biomass and net annual primary production in forest ecosystems of the world. In: "Forest Biomass Studies" (H.E. Young, ed.), pp. 1-32. University of Maine, Orono, U.S.A.

DeAngelis, D.L., Gardner, R.H. and Shugart, H.H. (1981). Productivity of forest ecosystems studied during the IBP: the woodland data set. In: "Dynamic Properties of Forest Ecosystems" (D.E. Reichle, ed.), pp. 567-672. Cambridge University Press, Cambridge, London, New York and Melbourne.

Duvigneaud, P. (ed.) (1971). "Productivity of Forest Ecosystems." UNESCO, Paris.

Ellenberg, H. (ed.) (1971). "Integrated Experimental Ecology." Springer-Verlag, Berlin, Heidelberg and New York.

Golley, F.B. and Medina, E. (eds) (1975). "Tropical Ecological Systems." Springer-Verlag, Berlin, Heidelberg and New York.

Golley, P.M. and Golley, F.B. (eds) (1972). "Tropical Ecology, with an Emphasis on Organic Productivity." Institute of Ecology, University of Georgia, Athens, U.S.A.

Lamotte, M. and Bourlière, F. (eds) (1978). "Problèmes d'Ecologie: Structure et Fonctionnement des Ecosystèmes Terrestres." Masson, Paris.

Lieth, F.H. (ed.) (1978). "Patterns of Primary Production in the Biosphere." Benchmark Paper in Ecology. Vol.8. Dowden, Hutchinson and Ross, Stroudberg, Pennsylvania.

Lieth, F.H. and Whittaker, R.H. (eds) (1975). "Primary Productivity of the Biosphere." Springer-Verlag, Berlin, Heidelberg and New York.

Ogawa, H. and Kira, T. (1977). Methods of estimating forest biomass. In: "Primary Productivity of Japanese Forests." (T. Shidei and T. Kira, eds), pp. 15-25. University of Tokyo Press, Tokyo, Japan.

Pardé, J. (1980). Forest biomass. Forestry abstracts review article. *For. Abstr.* 41, 343-362.

Reichle, D.E. (ed.) (1981). "Dynamic Properties of Forest Ecosystems." Cambridge University Press, Cambridge, London, New York and Melbourne.

Rodin, L.E. and Bazilevich, N.I. (1967). "Production and Mineral Cycling in Terrestrial Vegetation." Oliver and Boyd, Edinburgh and London.

Shidei, T. and Kira, T. (eds) (1977). "Primary Productivity of Japanese Forests." JIBP Synthesis Vol.16. University of Tokyo Press, Tokyo, Japan.

Whittaker, R.H. (1966). Forest dimensions and production in the Great Smoky Mountains. *Ecology* 47, 103-121.

Young, H.E. (ed.) (1971). "Forest Biomass Studies." College of Life Sciences and Agriculture, University of Maine, Orono, U.S.A.

Young, H.E. (ed.) (1973). "IUFRO Biomass Studies." College of Life Sciences and Agriculture, University of Maine, Orono, U.S.A.

Young, H.E. (ed.) (1976). "Oslo Biomass Studies." College of Life Sciences and Agriculture, University of Maine, Orono, U.S.A.

Institute of Terrestrial Ecology,
Bush Estate,
PENICUIK.
Midlothian, Scotland.

Melvin G.R. Cannell

December 1981

Moore, A.W., Russell, J.S. and Coaldrake, J.E. (1967). Dry matter and nutrient content of a subtropical semiarid forest of *Acacia harpophylla* F. Muell. (Brigalow). *Aust. J. Bot.* <u>15</u>, 11–24.

27°17'S 149°45'E   287 m   Australia, Queensland, 11 km SE of Meandarra.

Deep clay soils	*Acacia harpophylla* with a few *Geijera parviflora*

Age (years)	Mature
Trees/ha	1552[a]
Tree height (m)	8 (2 to 15)
Basal area (m²/ha)	
Leaf area index	
Stem volume (m³/ha)	

Dry biomass (t/ha)

Stem wood	$\left.\begin{array}{l}\\\\\end{array}\right\}$ 84.3
Stem bark	
Branches	11.0
Fruits etc.	
Foliage	7.7[b]
Root estimate	15.8[c]

CAI (m³/ha/yr)

Net production (t/ha/yr)

Stem wood	
Stem bark	
Branches	
Fruits etc.	
Foliage	
Root estimate	

All trees were harvested within a 0.04 ha plot including some root stumps; the biomass of the remaining root stumps was derived from a regression on D². There was 52.4 t/ha of standing dead wood and 58.0 t/ha of dead stumps. Nutrient contents were determined.
a. Trees at least 5 cm stem diameter; there were 1297 smaller trees per hectare.
b. Including twigs.
c. Mainly stumps.

Hingston, F.J., Turton, A.G. and Dimmock, G.M. (1979). Nutrient distribution in Karri (*Eucalyptus diversicolor* F. Muell.) ecosystems in southwest Western Australia. *For. Ecol. Manage.* 2, 133-158.

ca. 34°26'S 116°00'E   300-400 m   Australia, Western Australia, near Pemberton.

	Eucalyptus diversicolor Lateritic red earth	*E. diversicolor* and *Eucalyptus calophylla* Yellow podzolic soil
Age (years)	36	Mixed
Trees/ha	440	438
Tree height (m)	30	20-30
Basal area (m²/ha)	26.0	37.5
Leaf area index	2.6	5.0
Stem volume (m³/ha)		
Dry biomass (t/ha)		
Stem wood	183.8	207.9
Stem bark	20.0	35.7
Branches	14.9	31.1
Fruits etc.	0.0	0.5
Foliage	4.5	8.9
Root estimate		
CAI (m³/ha/yr)		
Net production (t/ha/yr)		
Stem wood		
Stem bark		
Branches		
Fruits etc.		
Foliage		
Root estimate		

Six trees of *E. diversicolor* and 3 trees of *E. calophylla* were sampled in winter. Stand values for the above two plots of 0.16 ha were estimated from regressions on basal area per tree. There was 1.8 t/ha of dead wood in the 36-year old stand. Nutrient contents were determined.

Cromer, R.N., Raupach, M., Clarke, A.R.P. and Cameron, J.N. (1975). Eucalypt plantations in Australia. The potential for intensive production and utilization. *Appita* 29, 165-173. (Also, in "Oslo Biomass Studies" 1976, pp. 31-40. College of Life Sciences and Agriculture, University of Maine, Orono, USA).

38°20'S 146°20'E 150 m Australia, Victoria, near Morwell.

Plantation Friable, porous brown-red earth	*Eucalyptus globulus* Fertilizers applied during first 2 years			
	No fertilizer	34 kg/ha N 15 kg/ha P	101 kg/ha N 45 kg/ha P	202 kg/ha N 90 kg/ha P
Age (years)	4	4	4	4
Trees/ha	2196	2196	2196	2196
Tree height (m)	4.3	6.5	7.2	7.8
Basal area (m²/ha)	2.1	5.2	6.5	8.3
Leaf area index				
Stem volume (m³/ha)				
Dry biomass (t/ha) Stem wood	3.4	8.6	12.3	16.3
Stem bark	0.7	1.9	2.6	3.3
Branches	0.8	2.0	2.4	3.9
Fruits etc.				
Foliage	1.1	2.5	3.3	5.2
Root estimate				
CAI (m³/ha/yr)				
Net production (t/ha/yr) Stem wood	1.56	3.39	4.79	6.53
Stem bark	0.31	0.73	0.98	1.27
Branches	0.46[a]	0.81[a]	0.93[a]	1.84[a]
Fruits etc.				
Foliage	0.79[b]	2.10[b]	2.85[b]	4.02[b]
Root estimate				

Twelve trees were sampled per treatment at ages 2 and 4. Stand biomass values for four 159m² plots per treatment were derived from regressions on individual tree basal area. Increments refer to the period from ages 2 to 4 years. Nutrient contents were determined.
a. Including accumulated dead branches, assuming no branch litterfall.
b. Mean foliage biomass between ages 2 and 4 years.

Cromer, R.N., Williams, E. and Tompkins, D. (1980). Biomass and nutrient uptake in fertilized *E. globulus*. "Proceedings IUFRO Symp. and Workshop on Genetic Improvement and Productivity of Fast-growing Trees." Sao Pedro, Sao Paulo, Brazil.

38°20'S   146°20'E   150 m   Australia, Victoria, near Morwell.

Plantation. Friable, porous brown-red earth	*Eucalyptus globulus* Fertilizers applied during first 2 years			
	No fertilizer	34 kg/ha N 15 kg/ha P	101 kg/ha N 45 kg/ha P	202 kg/ha N 90 kg/ha P
Age (years)	9.5	9.5	9.5	9.5
Trees/ha	2196	2196	2196	2196
Tree height (m)				
Basal area (m²/ha)				
Leaf area index				
Stem volume (m³/ha)	34.9	76.5	96.2	109.6
Dry biomass (t/ha) Stem wood	19.2	40.2	51.6	58.4
Stem bark	4.7	9.0	11.0	11.4
Branches	2.6	3.9	5.0	5.5
Fruits etc.				
Foliage	4.0	5.1	6.7	6.6
Root estimate				
CAI (m³/ha/yr)				
Net production (t/ha/yr) Stem wood	2.9	5.7	7.1	7.7
Stem bark	0.7	1.3	1.5	1.5
Branches	0.5^a	0.5^a	0.7^a	$>0.5^a$
Fruits etc.				
Foliage	2.6^b	3.8^b	5.0^b	ca.5.9^b
Root estimate				

Twelve trees were sampled per treatment. Stand biomass values for four 159 m² plots per treatment were derived from regressions on individual tree basal area. Increments refer to the period from ages 4.0 to 9.5 years. There was 0.7, 0.9, 1.0 and 1.4 t/ha of dead branches in columns left to right. Nutrient contents were determined.
a. Excluding branch mortality.
b. Mean foliage biomass between ages 4.0 and 9.5 years.

Bradstock, R. (1981). Biomass in an age series of *Eucalyptus grandis* plantations. *Aust. For. Res.* (in press).

Turner, J. and Lambert, M.J. (1981). Nitrogen cycling within a 27-year-old *Eucalyptus grandis* stand. In: "Managing Nitrogen Economies of Natural and Man-made Ecosystems" (F.J. Hingston, ed.). CSIRO Division of Land Resources Management, Mandura, Western Australia.

30°20'N   153°00'E   10-100 m   Australia, New South Wales, near Coffs Harbour.

Eucalyptus grandis

Plantations   Fertilizers applied to all stands

			Lower Permian sediments			Upper Permian granodiorite		
Age (years)		2	5	6	16	27	10	12	15
Trees/ha		996	961	810	756	790	762	830	1219
Tree height (m)									
Basal area (m²/ha)		4.9	12.3	7.5	23.3	30.4	13.1	22.5	31.5
Leaf area index									
Stem volume (m³/ha)									
Dry biomass (t/ha)	Stem wood	6.3	30.4	16.0	137.4	328.8	60.6	147.2	131.0
	Stem bark	2.1	6.1	3.6	26.5	38.2	11.1	29.9	18.5
	Branches	6.0	11.2	5.9	17.8	20.8	8.5	14.8	11.4
	Fruits etc.								
	Foliage	3.9	4.5	2.0	5.7	6.2	4.0	4.8	3.8
	Root estimate								
CAI (m³/ha/yr)									
Net production (t/ha/yr)	Stem wood					15.31^{a}			
	Stem bark					1.06^{a}			
	Branches					0.31^{a}			
	Fruits etc.								
	Foliage					0.01^{a}			
	Root estimate								

Four to six trees were sampled per stand. Biomass values for 2 plots of about 400m² in each of the 8 stands were derived from regressions on D. Understoreys of *Acacia* spp. *et al.* weighed 2.7, 17.0, 15.6, 11.4, 4.2, 13.7, 7.1 and 23.9 t/ha in columns left to right. Nutrient contents were determined.

a. Preliminary increment values, excluding all litterfall and mortality. Understorey increment was estimated to be 2.8 t/ha/yr excluding litterfall.

Hingston, F.J., Dimmock, G.M. and Turton, A.G. (1981). Nutrient distribution in a Jarrah (*Eucalyptus marginata* Donn ex SM.) ecosystem in southwest Western Australia. *For. Ecol. Manage.* 3, 183-207.

32°45'S 116°00'E   --   Australia, Western Australia, Dwellingup.

Lateritic sandy gravels	*E. marginata* $(74\%)^a$ and *Eucalyptus calophylla* with understorey shrubs.

Burned 4 years previously

Age (years)	60
Trees/ha	356 + 203
Tree height (m)	25
Basal area (m²/ha)	25.5 + 8.9
Leaf area index	2.7 + 1.0
Stem volume (m³/ha)	

Dry biomass (t/ha)

Stem wood	148.3 + 37.9 ⎫
Stem bark	27.8 + 8.9 ⎪
Branches	25.1 + 6.2 ⎬ $+ 4.5^b$
Fruits etc.	0.8 + 0.2 ⎪
Foliage	4.1 + 1.7 ⎭
Root estimate	

CAI (m³/ha/yr)

Net production (t/ha/yr)

Stem wood	
Stem bark	
Branches	⎫
Fruits etc.	⎬ ca. 2.7^c
Foliage	⎭
Root estimate	

Ten trees of each of the two *Eucalyptus* species were sampled in autumn (May). Stand values for a 0.36 ha plot were derived from regressions on D. Values are given above for *E. marginata* plus *E. calophylla* There was 1.4 t/ha of dead wood on each of the two *Eucalyptus* species. Nutrient contents were determined.
a. Percentage of total basal area.
b. Understorey shrubs.
c. Total litterfall only.

Stewart, H.T.L., Flinn, D.W. and Aeberli, B.C. (1979). Above-ground biomass of a mixed Eucalypt forest in eastern Victoria. *Aust. J. Bot.* 27, 725-740.

Harrington, G. (1979). Estimation of above-ground biomass of trees and shrubs in a *Eucalyptus populnea* F. Muell. woodland by regression of mass on trunk diameter and plant height. *Aust. J. Bot.* 27, 135-143.

Australia	37°25'S 149°33'E 350 m Victoria, 10 km N of Genoa	30°55'S 146°30'E 100-200 m New South Wales, near Coolabah
At Genoa:	*Eucalyptus muellerana* (39%)[a]	*Eucalyptus populnea* (54%)[b]
loamy soils	*Eucalyptus sieberii* (27%)[a]	*Eucalyptus intertexta* (25%)[b]
overlying mottled yellow clays	*Eucalyptus agglomerata* (19%)[a] et al.	with understorey shrubs

Age (years)	to over 100	
Trees/ha	123	$36 + 4804^c$
Tree height (m)	28-40	
Basal area (m²/ha)	30.1	
Leaf area index		
Stem volume (m³/ha)		

Dry biomass (t/ha)	Stem wood	195.6	⎫	⎫
	Stem bark	51.3	⎬ +1.9^c	⎬ 42.20 + 9.15^c
	Branches	72.9^d		
	Fruits etc.			
	Foliage	5.6	⎭	1.03 + 2.34^c
	Root estimate			

CAI (m³/ha/yr)		
Net production (t/ha/yr)	Stem wood	
	Stem bark	
	Branches	
	Fruits etc.	
	Foliage	
	Root estimate	

At Genoa, Stewart *et al.* (1979) sampled 31 trees in the autumn and derived stand values for 17 plots of 0.1 ha from regressions on D. At Coolabah, Harrington (1979) sampled 20 trees and derived stand values for 85 plots of 0.2 ha from regressions on D and H.
a. Percentage of the total basal area.
b. Percentage of the total biomass.
c. Understorey shrubs.
d. Comprised of 12.7 t/ha branch bark, 6.4 t/ha twigs and 53.8 t/ha branch wood.

Attiwell, P.M. (1979). Nutrient cycling in a *Eucalyptus obliqua* (L'Hérit.) forest. III Growth, biomass, and net primary productivity. *Aust. J. Bot.* 27, 439-458.

Attiwell, P.M., Guthrie, H.B. and Leuning, R. (1978). *Aust. J. Bot.* 26, 79-91.

Attiwell, P.M. (1966). *Ecology* 47, 795-804.

Attiwell, P.M. (1981). In: "Dynamic Properties of Forest Ecosystems" (D.E. Reichle, ed.). p.573. Cambridge University Press, Cambridge, London, New York.

37°25'S 145°10'E   545 m   Australia, Victoria, Mt Disappointment.

Red, friable porous
krasnozems pH 5.2-5.9   *Eucalyptus obliqua*

	a	b	c	
Age (years)	43.7	50.7	60.7	66.2
Trees/ha	914	865	655	568
Tree height (m)	22-29	22-29	25-32	25-32
Basal area (m²/ha)	49.5	57.6	62.7	64.8
Leaf area index	3.3	4.2	4.9	5.4

Stem volume (m³/ha)

Dry biomass (t/ha)

	a	b	c	
Stem wood	185.3	227.8	263.4	282.5
Stem bark	37.7	44.0	47.4	49.5
Branches	13.8	19.3	25.3	29.8
Fruits etc.				
Foliage	5.5	6.9	8.1	8.9
Root estimate				

CAI (m³/ha/yr)

Net production (t/ha/yr)

	a	b	c
Stem wood		$4.32 \rbrace + 0.37^d$	
Stem bark	$\rbrace 7.9$	$0.53 \rbrace + 1.11^e$	$\rbrace 4.5$
Branches		$0.71 + 1.04^d$	
Fruits etc.	$\rbrace 4.0^d$		$\rbrace 3.0^d$
Foliage		$0.15 + 2.13^d + 0.43^f$	
Root estimate			

Seventy-five trees were sampled. Stand biomass values for the above plot of 809 m² (measured in 1955, '62, '72 and '77 in columns left to right) were derived from regressions on D. There were over 200 standing dead trees weighing about 10 t/ha, plus about 3.7 t/ha of understorey biomass. There was also 2.1, 2.8, 3.4 and 3.8 t/ha of dead branches in columns left to right. Nutrient contents were determined.
a. Woody increment and litterfall in this column refer to ages 44 to 51 years.
b. Increments and litterfall at a mean age of 51 years; taken from Attiwell (1981).
c. Woody increment and litterfall at ages 61 to 66.
d. Litterfall.
e. Mortality.
f. Consumption.

Feller, M.C. (1980). Biomass and nutrient distribution in two eucalypt forest eco-systems. *Aust. J. Ecol.* <u>5</u>, 309-333.

37°38'S 145°35'E   (alt. given below)   Australia, 60 km NE of Melbourne.

Krasnozems to podzols pH 3.9-5.0	*Eucalyptus regnans* (96%)[a] with *Acacia* spp.	*Eucalyptus obliqua* and *Eucalyptus dives* } (99%)[a]
	560 m	180 m
Age (years)	39	39
Trees/ha	550[b]	830[b]
Tree height (m)	40-45	20-25
Basal area (m²/ha)	52.1	65.8
Leaf area index		
Stem volume (m³/ha)		

Dry biomass (t/ha)

Stem wood	} 574.4	263.0
Stem bark		81.3
Branches	36.7	24.1
Fruits etc.		
Foliage	3.0	3.9
Root estimate	63.2	45.4

CAI (m³/ha/yr)

Net production (t/ha/yr)

Stem wood		
Stem bark		
Branches		
Fruits etc.		
Foliage		
Root estimate		

Five or six trees of each *Eucalyptus* species and 5 *Acacias* were sampled during the autumn and winter and roots were excavated in three 1 m² pits per site. Stand values for the above 400 m² plots were derived from regressions on D and D²H. There was 27.0 and 19.1 t/ha of standing dead wood in columns left and right, respectively. Nutrient contents were determined.
a. Percentage of total basal area.
b. There were an additional 100 and 30 trees/ha (in the left and right columns, res-pectively) of non-*Eucalyptus* or *Acacia* species, which were not included in the biomass estimates.

Ashton, D.H. (1976). Phosphorus in forest ecosystems at Beenak, Victoria. *J. Ecol.*
<u>64</u>, 171-186.

---

ca. 38°S 146°E   500-550 m   Australia, 65 km E of Melbourne, Beenak.

	Eucalyptus regnans with understorey shrubs. Wet sclerophyll forest on south-facing slope; brown, red-brown loams.	*Eucalyptus sieberii* with understorey shrubs. Dry sclerophyll forest on north-facing slope; red podzolic soils.
Age (years)	27	27
Trees/ha		
Tree height (m)	up to 60	up to 45
Basal area (m² /ha)		
Leaf area index	$6.0 + 3.0^a$	$5.2 + 5.8^a$
Stem volume (m³ /ha)		

Dry biomass (t/ha)

Stem wood	$\left.\begin{array}{r}713.1\\[4pt]\\[4pt]71.2\end{array}\right\} + 36.5^a$	$\left.\begin{array}{r}804.1\\[4pt]\\[4pt]67.0\end{array}\right\} + 40.5^a$
Stem bark		
Branches		
Fruits etc.		
Foliage	$8.1 + 2.2^a$	$12.1 + 4.8^a$
Root estimate		

CAI (m³ /ha/yr)

Net production (t/ha/yr)

Stem wood		
Stem bark		
Branches		
Fruits etc.		
Foliage		
Root estimate		

Five trees of *E. regnans* and 6 of *E. sieberii* were sampled in the autumn, and roots
were excavated from 20 soil blocks. Stand values were derived from regressions on
D. Phosphorus contents were determined.
a. Understorey shrubs.

Turner, J. (1980). Nitrogen and phosphorus distributions in naturally regenerated *Eucalyptus* spp. and planted Douglas-fir. *Aust. For. Res.* 10, 289-294.

35°06'S   141°01'E   680 m   Australia, New South Wales, near Tumut.

Red, permeable
soils.

Eucalyptus radiata,
Eucalyptus dalrympleana,
Eucalyptus spp. with *Acacia* spp.
and other understorey shrubs

Age (years)	ca. 45
Trees/ha	1590
Tree height (m)	
Basal area (m²/ha)	
Leaf area index	
Stem volume (m³/ha)	

Dry biomass (t/ha)		
Stem wood	70.0	
Stem bark	22.5	
Branches	6.8	$+ 29.6^{a}$
Fruits etc.		
Foliage	4.7	
Root estimate		

CAI (m³/ha/yr)	
Net production (t/ha/yr) Stem wood	
Stem bark	
Branches	
Fruits etc.	
Foliage	
Root estimate	

Five trees were sampled, and stand values for six circular plots, each 20 m in diameter, were derived from regressions on D. Nitrogen and phosphorus contents were determined.

a. Understorey shrubs.

Rogers, R.W. and Westman, W.E. (1981). Growth rhythms and productivity of a coastal subtropical *Eucalyptus* forest. *Aust. J. Ecol.* 6, 85-98.

Westman, W.E. and Rogers, R.W. (1977a). Biomass and structure of a subtropical Eucalypt forest, North Stradbroke Island. *Aust. J. Bot.* 25, 171-191.

Westman, W.E. and Rogers, R.W. (1977b). Nutrient stocks in a subtropical Eucalypt forest, North Stradbroke Island. *Aust. J. Ecol.* 2, 447-460.

Rogers, R.W. and Westman, W.E. (1977). Seasonal nutrient dynamics of litter in a subtropical Eucalypt forest, North Stradbroke Island. *Aust. J. Bot.* 25, 47-58.

27°30'S 155°30'E   100 m   Australia, Queensland, North Stradbroke Island.

Infertile podzols, with a sandy substratum.	*Eucalyptus signata* (55%)[a] *Eucalyptus umbra* (18%)[a] with *Tristania conferta*, *Banksia aemula et al.*, with understorey shrubs.

Age (years)	
Trees/ha	660
Tree height (m)	12-15
Basal area (m²/ha)	
Leaf area index	2.46[b]
Stem volume (m³/ha)	

Dry biomass (t/ha)

Stem wood	51.9 ⎫
Stem bark	6.6 ⎬ + 2.6[c]
Branches	38.9 ⎭
Fruits etc.	0.2
Foliage	2.5[d]
Root estimate	61.9[e] + 10.0[c]

CAI (m³/ha/yr)

Net production (t/ha/yr)

Stem wood	7.23 ⎫
Stem bark	0.81 ⎬ + 3.06[f]
Branches	5.58 ⎭
Fruits etc.	0.17
Foliage	0.04 + 2.62[f] + 0.02[g]
Root estimate	8.26

Thirty-one trees were sampled and roots were excavated. Stand biomass values for a 0.25 ha plot were derived from regressions on D. Increments were estimated by re-measurement after 2 years and by measuring tagged shoots. There was 16.5 t/ha of standing dead wood. Nutrient contents were determined.

a. Percentage of the total tree biomass.
b. The authors' 2-sided LAI value has been halved.
c. Understorey shrubs.
d. Leaves of the three major species only.
e. Including 21.7 t/ha of root crowns.
f. Mortality and litterfall, measured over 2 years.
g. Consumption.

Keay, J. and Turton, A.G. (1970). Distribution of biomass and major nutrients in a
maritime pine plantation. *Aust. For.* <u>34</u>, 39-48.

31°50'S 115°40'E   ca. 50 m   Australia, near Perth, Gnangara Plantation.

Plantation.
Infertile soils,
low in phosphate

Pinus pinaster[a]

	50 kg/ha P and 55 kg/ha S applied at age 11	No fertilizers applied at age 11
Age (years)	14	14
Trees/ha	1013	1013
Tree height (m)		
Basal area (m²/ha)		
Leaf area index	6.4[b]	5.2[b]
Stem volume (m³/ha)	183	153

Dry biomass (t/ha)

Stem wood	59.8	50.7
Stem bark	11.3	9.6
Branches	25.4	19.0
Fruits etc.		
Foliage	24.4	19.8
Root estimate		

CAI (m³/ha/yr)

Net production (t/ha/yr)

Stem wood	
Stem bark	
Branches	
Fruits etc.	
Foliage	
Root estimate	

Four trees were sampled from each treatment in summer. Stand values for one 390 m²
plot per treatment were derived from regressions on D. Nutrient contents were
determined.
a. Both treatments received 139 k/ha superphosphate at planting, were pruned to
   2.1 m height at age 10, and were thinned from 2500 to 1013 trees/ha at age 12.
b. All-sided LAI values were 17.8 and 14.6 in columns left and right, respectively.

Turton, A.G. and Keay, J. (1970). Changes in dry weight and nutrient distribution in maritime pine after fertilization. *Aust. For.* <u>34</u>, 84-96.

Keay, J., Turton, A.G. and Campbell, N.A. (1970). Fertilizer response of maritime pine on a lateritic soil. *Aust. For.* <u>33</u>, 248-258.

32°28'S   115°50'E   100 m   Australia, SE of Perth.

Plantation.	*Pinus pinaster*	
Deep yellow, loamy, ironstone, gravelly lateritic podzols.	Pruned to 2.1 m height at age 11	
	206 kg/ha N, 90 kg/ha P at age 13	No fertilizers applied at age 13

Age (years)		16.5	16.5
Trees/ha		1800	1800
Tree height (m)			
Basal area (m²/ha)			
Leaf area index			
Stem volume (m³/ha)		35.2	28.6
Dry biomass (t/ha)	Stem wood	} 20.7	} 12.5
	Stem bark		
	Branches	18.4[a]	11.4[a]
	Fruits etc.		
	Foliage	12.5	4.5
	Root estimate		
CAI (m³/ha/yr)			
Net production (t/ha/yr)	Stem wood		
	Stem bark		
	Branches		
	Fruits etc.		
	Foliage		
	Root estimate		

Six trees were sampled from each treatment in January. Stand values for three 0.081 ha plots per treatment were derived from regressions on D. Nutrient contents were determined.
a. Including stem biomass above the base of the crowns.

Forrest, W.G. and Ovington, J.D. (1970). Organic matter changes in an age series of *Pinus radiata* plantations. *J. appl. Ecol.* <u>7</u>, 177–186.

Forrest, W.G. (1973). Biological and economic production in radiata pine plantations. *J. appl. Ecol.* <u>10</u>, 259–267.

32°20'S   148°14'E   200–500 m   Australia, New South Wales, Tumut.

Plantations.

Red sandy loams derived *in situ* from granite

Pinus radiata

Pruned to 3 m height at age 8

Age (years)	3	5	7	9	12
Trees/ha	1483	1492	1458	1470	1560
Tree height (m)	1.4	3.1	7.9	12.1	15.6
Basal area (m²/ha)		2.0	16.0	25.0	32.9
Leaf area index					
Stem volume (m³/ha)					

Dry biomass (t/ha)

	3	5	7	9	12
Stem wood	0.3	2.1	21.1	48.2	80.7
Stem bark	0.1	0.3	2.7	5.6	8.8
Branches	0.2	1.2	14.9	9.9	18.7
Fruits etc.	0.0	0.0	0.4	0.5	0.7
Foliage	0.5	2.1	11.6	8.8	9.5
Root estimate			ca.9.0[a]	ca.12.9[a]	

CAI (m³/ha/yr)

Net production (t/ha/yr)

	3	5	7	9	12
Stem wood		0.9	9.5	13.6	10.8
Stem bark		0.1	1.2	1.5	1.1
Branches		0.5[b]	2.5[b]	2.5[b]	2.5[b]
Fruits etc.		0.1	0.5	0.7	0.8
Foliage		ca.0.7[c]	ca.3.4[c]	ca.5.1[c]	ca.4.6[c]
Root estimate					

Nine trees were sampled per stand. Biomass values for plots of 0.101 or 0.081 ha in each stand were derived from regressions on H and basal area times H. Production values given above are increments between ages 3–5, 5–7, 7–9 and 9–12 in columns left to right.
a. Roots over 5 mm diameter, assumed to comprise 15% of the total biomass.
b. Excluding woody litterfall, and averaging branch increment during years 5 to 12.
c. Half foliage biomass.

Turner, J. (1980). Nitrogen and phosphorus distributions in naturally regenerated *Eucalyptus* spp. and planted Douglas-fir. *Aust. For. Res.* <u>10</u>, 289-294.

35°06'S 141°01'E   680 m   Australia, New South Wales, near Tumut.

Plantation.

Red, permeable soils.

Pseudotsuga menziesii

		Unfertilized
Age (years)		50
Trees/ha		1110
Tree height (m)		
Basal area (m²/ha)		
Leaf area index		
Stem volume (m³/ha)		
Dry biomass (t/ha)	Stem wood	319.1
	Stem bark	37.9
	Branches	30.5
	Fruits etc.	
	Foliage	16.7
	Root estimate	
CAI (m³/ha/yr)		
Net production (t/ha/yr)	Stem wood	
	Stem bark	
	Branches	
	Fruits etc.	
	Foliage	
	Root estimate	

Five trees were sampled, and stand values for six circular plots 20 m in diameter were derived from regressions on D. Nitrogen and phosphorus contents were determined.

Kestemont, P. (1975). "Biomasse, Nécromasse et Productivité Aériennes Ligneuses de quelques Peuplements Forestiers en Belgique." Thesis. Faculty of Sciences, Free University of Brussels.

Belgium	50°01'N 5°05'E 265 m Mache valley Daverdisse *Alnus glutinosa* with *Carpinus* sp. *Crateagus* sp. *et al.* Infertile gley	50°28'N 4°18'E 200-300 m Hainant Chappelle-lez-Herlaimont *Betula pendula* (88%)[a] *Salix* spp., *Quercus robur*, *Castanea sativa et al.* Humus rich gley pH 5.5
Age (years)	13-18	14
Trees/ha	ca.30000[b]	4920
Tree height (m)	9	10
Basal area (m²/ha)	23.8-39.6	20.2
Leaf area index		
Stem volume (m³/ha)		
Dry biomass (t/ha)		
Stem wood	} 59.3	} 73.5
Stem bark		
Branches		
Fruits etc.		0.5
Foliage	2.8	3.2
Root estimate	4.3	21.3
CAI (m³/ha/yr)		
Net production (t/ha/yr)		
Stem wood	} 5.87[d]	} 9.4 + 3.0[c]
Stem bark		
Branches		
Fruits etc.		0.5
Foliage	2.79[e]	3.2[e]
Root estimate		2.0

Many trees were sampled and roots were excavated. Stand biomass values were derived from regressions on D. There was 7.3 t/ha of standing dead wood at Chappelle-lez-Herlaimont.

a. Percentage of total woody biomass.
b. There were 5719 *A. glutinosa* per hectare.
c. Woody litterfall.
d. Excluding woody litterfall.
e. Leaf biomass; leaf litterfall was 2.5 and 2.8 t/ha/yr in columns left and right, respectively.

Kestemont, P. (1975). "Biomasse, Nécromasse et Productivité Aériennes Ligneuses de quelques Peuplements Forestiers en Belgique." Thesis. Faculty of Sciences, Free University of Brussels.

Duvigneaud, P. and Kestemont, P. (1977). "Productivité Biologique en Belgique." Publ. Ministère de l'Education Nationale et de la Culture Française et par het Ministerie van Nationale Opvoeding en Nederlandse Cultuur.

50°02'N 5°14'E   350-400 m   Belgium, Mirwart.

Fagus sylvatica
with *Quercus robur*, *Quercus petraea*,
Carpinus betulus, *Acer campestre et al.*

Plantations.

Brown loams
pH 3.8-3.9

	Woodland with herbaceous ground vegetation (Hêtraie herbeuse)	Woodland without herbaceous ground vegetation (Hêtraie nue)
Age (years)	144	ca.130
Trees/ha	160	190 (163 *Fagus*)
Tree height (m)	30.8	27-30
Basal area (m²/ha)	31.0	28.9
Leaf area index	6.4	6.5

Stem volume (m³/ha)

Dry biomass (t/ha)

Stem wood	} 318.5^a	} 213.6^a
Stem bark		
Branches	50.8^a	122.4^a
Fruits etc.	1.3	0.5
Foliage	3.0	2.9
Root estimate	74.0	68.0

CAI (m³/ha/yr)

Net production (t/ha/yr)

Stem wood	} 4.14 } d	} 7.12 + 0.77^b
Stem bark		
Branches	1.83 + 1.24^b }	
Fruits etc.	1.31	0.54
Foliage	3.00^c	2.92^c
Root estimate	1.37	1.82

Many trees were sampled and roots were excavated in each plantation. Stand biomass values were derived from regressions on D. There was 1.8 t/ha of standing dead wood in the left column.
a. 'Bois fort' regarded here as stems, 'bois menu' as branches.
b. Woody litterfall.
c. Leaf biomass; leaf litterfall was 2.08 t/ha/yr in the left column.
d. Alternatively, Duvigneaud and Kestemont (1977) gave values of 6.80 and 0.93 t/ha/yr for total above-ground wood increment and woody litterfall, respectively.

Duvigneaud, P., Kestemont, P. and Ambroes, P. (1971). Productivité primaire des forêts tempérées d'essences feuillues caducifolices en Europe occidentale. In: "Productivity of Forest Ecosystems" (P. Duvigneaud, ed.) pp. 259-270. UNESCO, Paris.

Duvigneaud, P. and Froment, A. (1969). Recherches sur l'écosystème forêt. Serie E. Forêts de haute Belgique. Contribution No.5. Eléments biogènes de l'édaphotope et phytocénose forestière. *Bull. Inst. Sci. nat. Belg.* 45, 1-48.

Duvigneaud, P. and Kestemont, P. (1977). See p.24.

Belgium, the Famenne.	50°11'N 5°01'E ca.250 m Ferage *Quercus petraea* with coppiced understorey of *Corylus avellana,* *Carpinus betulus, et al.* Mull soil, pH 5.4	50°07'N 5°15'E ca.100 m Wavreille *Quercus robur* with coppiced understorey of *Corylus avellana,* *Carpinus betulus, et al.* Mull soil, pH 6.2
Age (years)	117 10^a	120 20^a
Trees/ha	163 + 1500^a	111 + 12000^a
Tree height (m)	24 4^a	27 7^a
Basal area (m²/ha)		
Leaf area index	5.7	
Stem volume (m³/ha)	300	304

Dry biomass (t/ha)	Stem wood	161.7 ⎫ 18.5 ⎬ + 18.1^a 58.3 ⎭	210.0 ⎫ ⎬ + 29.3^a 88.2 ⎭
	Stem bark		
	Branches		
	Fruits etc.	1.2	1.0
	Foliage	3.5	4.0
	Root estimate	53.0	51.2

CAI (m³/ha/yr)

Net production (t/ha/yr)	Stem wood	2.31 ⎫ 0.16 ⎬ + 2.06^{ab} 5.39^b ⎭	2.22 ⎫ ⎬ + 2.20^{ab} 5.12^b ⎭
	Stem bark		
	Branches		
	Fruits etc.	1.24	1.02
	Foliage	3.50^c	3.96^c
	Root estimate	1.68	0.80

Many trees were sampled and roots were excavated. Stand biomass values were derived from regressions on D. Stumps and big root biomasses were estimated as 12% and 30%, respectively, of the above-ground biomass of the trees plus understorey.

a. Understorey shrubs; values given for LAI, foliage, roots and fruits etc. (but not stem volumes) include the understorey.

b. Including woody litterfall, estimated as 2.0 and 2.3 t/ha/yr in the 1971 paper, in columns left and right, respectively, but total wood death estimated as 8.5 t/ha/yr at Wavreille in the 1977 report.

c. Leaf biomass; leaf litterfall was about 3.1 and 3.5 t/ha/yr in columns left and right, respectively.

Duvigneaud, P., Kestemont, P. and Ambroes, P. (1971). Productivité primaire des
  forêts tempérées d'essences feuillues caducifolices en Europe occidentale. In:
  "Productivity of Forest Ecosystems" (P. Duvigneaud, ed.) pp. 259-270. UNESCO,
  Paris.
Duvigneaud, P. and Froment, A. (1969). Recherches sur l'écosystème forêt. Serie E.
  Forêts de haute Belgique. Contribution No.5. Eléments biogènes de l'édaphotope
  et phytocénose forestière. *Bull. Inst. Sci. nat. Belg.* 45, 1-48.
Duvigneaud, P. and Kestemont, P. (1977). See p.24.

Belgium, the Famenne.	50°09'N 5°06'E$_d$ ca.50 m Villers		50°03'N 4°59'E ca.100 m Vonêche	
	Quercus robur, Quercus petraea with coppiced understorey of *Corylus avellana,* *Carpinus betulus et al.* 'Moder' and pseudo-gleys, pH 5.2		*Q. petraea, Q. robur,* *Betula pubescens* with coppiced understorey of *Corylus avellana,* *Carpinus betulus, et al.* Acid podzols and gleys, pH 3.5	
Age (years)	90	20^a	135	
Trees/ha	185	8780^a	422	1170^a
Tree height (m)	20		22	
Basal area (m²/ha)				
Leaf area index	4.6			
Stem volume (m³/ha)	148		188	
Dry biomass (t/ha)				
Stem wood	93.2 ⎫		120.9 ⎫	
Stem bark	⎬ + 26.7^a		⎬ + 2.0^a	
Branches	37.0 ⎭		75.8 ⎭	
Fruits etc.	0.8		0.8	
Foliage	3.2		3.2	
Root estimate	31.1		36.2	
CAI (m³/ha/yr)				
Net production (t/ha/yr)				
Stem wood	1.33 ⎫		0.67 ⎫	
Stem bark	⎬ + 2.31^{ab}		⎬ + 0.13^{ab}	
Branches	3.40^b ⎭		4.17^b ⎭	
Fruits etc.	0.79		0.83	
Foliage	3.19^c		3.16^c	
Root estimate	1.03		0.57	

Many trees were sampled and roots were excavated. Stand biomas values were derived
from regressions on D. Stumps and big roots biomasses were estimated as 12% and 30%
respectively, of the above-ground biomass of the trees plus understorey.

a. Understorey coppices; values given for LAI, foliage, roots and fruits etc. (but
  not stem volume) include the understorey.
b. Including woody litterfall, estimated as about 1.8 t/ha/yr in both stands in the
  1971 paper.
c. Leaf biomass; leaf litterfall was about 2.8 t/ha/yr in both stands.
d. Averages given here of two sites at Villers.

Kestemont, P. (1971). Productivité primaire des taillis simples et concept de nécromasse. In: "Productivity of Forest Ecosystems" (P. Duvigneaud, ed.) pp. 271-279. UNESCO, Paris.

Kestemont, P. (1975). "Biomasse, Nécromasse et Productivité Aériennes Ligneuses de quelques Peuplements Forestiers en Belgique." Thesis. Faculty of Sciences, Free University of Brussels.

Belgium	49°53'N 4°55'E 250 m Semois region, Orchimont	50°02'N 5°14'E 350-400 m Mirwart
	Quercus robur (75%)[a] *Betula pendula* (15%)[a] *Sorbus aucuparia* (3%)[a]	*Q. robur* (86%)[a] *Quercus petraea,* *Fagus sylvatica et al.*
		Plantation
	Brown mull-gley, pH 3.9-4.5	Brown soils, pH 3.8-3.9
Age (years)	20-25	66
Trees/ha	6050[b]	958
Tree height (m)	8	22
Basal area (m²/ha)		32.1
Leaf area index		
Stem volume (m³/ha)		
Dry biomass (t/ha)		
Stem wood	} 71.4 + 5.3[c]	} 130.5[d]
Stem bark		
Branches		36.9[d]
Fruits etc.	0.8	0.5
Foliage	3.1	3.6
Root estimate	19.2	41.8
CAI (m³/ha/yr)		
Net production (t/ha/yr)		
Stem wood	} 6.64 + 0.75[e]	} 3.58
Stem bark		
Branches		3.58 + 1.07[e]
Fruits etc.	0.77	0.49
Foliage	3.07[f]	3.60[f]
Root estimate	1.28	1.70

Many trees were sampled, and roots were excavated, in each stand. Stand biomass values were derived from regressions on D. There was 2.9 t/ha of standing dead wood at Mirwart.

a. Percentage of the total biomass.
b. There were 7000 stems/ha.
c. Understorey coppice, mainly of *Corylus avellana*; all other biomass values and production values in this column include the understorey.
d. 'Bois fort' regarded here as stems, 'bois menu' as branches.
e. Woody litterfall.
f. Leaf biomass; leaf litterfall was 2.7 and 3.0 t/ha/yr in columns left and right, respectively.

Froment, A. and 8 others (1971). La chênaie mélangée calcicole de Virelles-
Blaimont, en haute Belgique. In: "Productivity of Forest Ecosystems" (P.
Duvigneaud, ed.) pp.635-665. UNESCO, Paris.

Duvigneaud, P., Danaeyer-de-Smet, S., Ambroes, P., and Timperman, J. (1969). Aperçu
préliminaire sur les biomasses, la productivité et le cycle des éléments biogènes.
Bull. Soc. bot. Belg. 102, 317-323.

Ambroes, P. (1969). La biomasse aèrienne de la strate arborescente. *Bull. Soc. bot.
Belg.* 102, 325-338.

50°04'N   4°21'E   245 m   Belgium, Blaimont region, Virelles.

Humus over
calcareous rock.
pH 6.4-7.4

Quercus robur (35%),[a] *Carpinus betulus* (32%)[a]

Fagus sylvatica (25%)[a] and *Acer campestre* (3%)[a]

with understorey coppice of *Corylus avellana et al.*

Age (years)	35-75	
Trees/ha	1486[b]	
Tree height (m)	11-23	0.5-3.0[c]
Basal area (m² /ha)		
Leaf area index	6.8	
Stem volume (m³ /ha)	129	

Dry biomass (t/ha)

Stem wood	64.5	
Stem bark	8.8	
Branches	35.9	+ 2.5[c]
Fruits etc.	0.5	
Foliage	3.5	
Root estimate	34.6 + 0.7[c]	

CAI (m³ /ha/yr)

Net production (t/ha/yr)

Stem wood	2.64	
Stem bark	0.34	+ 0.22[c]
Branches	3.14 + 1.81[d]	
Fruits etc.	0.26	
Foliage	3.46[e]	
Root estimate	2.00 + 0.33[c]	

Forty-eight trees were sampled in September-October and roots were excavated. Stand
biomass values were derived from regressions on D. Nutrient contents were determined.

a. Percentage of total above-ground tree biomass.
b. Including 260 dominants.
c. Understorey; LAI and foliage increment values include the understorey; tree age
   and stem volume data refer to the overstorey only.
d. Woody and miscellaneous litterfall.
e. Leaf biomass; leaf litterfall was 3.17 t/ha/yr.

Kestemont, P. (1975). "Biomasse, Nécromasse et Productivité Aériennes Ligneuses de quelques Peuplements Forestiers en Belgique." Thesis. Faculty of Sciences, Free University of Brussels.

Duvigneaud, P. and Kestemont, P. (1977). "Productivité Biologique en Belgique." Publ. Ministère de l'Education Nationale et de la Culture Française et par het Ministerie van Nationale Opvoeding en Nederlandse Cultuur.

50°02-03'N   5°16'E   340-400 m   Belgium, Mirwart.

Plantations.

Acid brown soils, pH 3.9	*Picea abies*	*Pseudotsuga menziesii* with a few *Fagus sylvatica*
Age (years)	55	70
Trees/ha	1061	217
Tree height (m)	19	36.5
Basal area (m²/ha)	41.5	58.4
Leaf area index		
Stem volume (m³/ha)		

Dry biomass (t/ha)

	Picea abies	*Pseudotsuga menziesii*
Stem wood	168.4 (or 166.4)[a]	334.0
Stem bark	1.6	33.0
Branches	16.6	29.0
Fruits etc.	0.1	0.1
Foliage	16.1	7.7
Root estimate	31.0 (or 70.0)[a]	67.0

CAI (m³/ha/yr)

Net production (t/ha/yr)

	Picea abies	*Pseudotsuga menziesii*
Stem wood	6.63	14.0
Stem bark	0.30	1.0
Branches	3.61 + 0.41[b] (or 4.08)[a]	5.0 + 0.2[b]
Fruits etc.	0.09 (or 0.32)[a]	0.1
Foliage	2.27[c] (or 2.46)[a]	2.0[c] (or 2.6)[a]
Root estimate	1.77 (or 4.08)[a]	3.4 (or 3.9)[a]

Many trees were sampled, and roots were excavated, in each plantation. Stand biomass values were derived from regressions on D. There was 7.8 and 10.3 t/ha of standing dead wood in the left and right columns, respectively.
a. Updated values, given by Duvigneaud and Kestemont (1977).
b. Woody litterfall.
c. New foliage; foliage litterfall measured over one year was 2.78 and 0.95 t/ha/yr in the left and right columns, respectively.

Klinge, H. (1978). Litter production in tropical ecosystems. *Malay. Nat. J.* <u>30</u>, 415-422.

Klinge, H. (1977). Fine litter production and nutrient return to the soil in three natural forest stands of eastern Amazonia. *Geo. Eco. Trop.* <u>1</u>, 159-167.

Klinge, H., Rodrigues, W.A., Brunig, E. and Fittkau, E.S. (19$\overline{75}$). Biomass and structure in a central Amazonian rain forest. In: "Tropical Ecological Systems. Trends in Terrestrial and Aquatic Research" (F.B. Golley and E. Medina, eds) pp.115-122. Springer-Verlag, New York, Heidelberg and Berlin.

---

ca.3°06'S 60°00'W   100 m   Brazil, Manaus, Walter Engler Forest Reserve.

Loamy yellow latosol, unflooded

About 470 species of trees with a stem diameter of at least 25 cm.

Tropical rainforest

'Terra firma'

Age (years)	Mature
Trees/ha	400
Tree height (m)	25
Basal area (m²/ha)	ca.30.7
Leaf area index	
Stem volume (m³/ha)	385

Dry biomass (t/ha)		
Stem wood	} 299
Stem bark	
Branches	98
Fruits etc.	
Foliage	9.3
Root estimate	130

CAI (m³/ha/yr)

Net production (t/ha/yr)	
Stem wood	
Stem bark	
Branches	$1.4^a + 7.6^a$
Fruits etc.	0.4^a
Foliage	5.6^a
Root estimate	

The fresh weight of all vegetation was determined in 43 plots, each 50 m², and dry biomass values were estimated from fresh/dry weight ratios.

a. Litterfall only, measured over 2 years; woody litterfall is divided into small (1.4 t/ha/yr) plus estimated large woody litterfall (7.6 t/ha/yr).

Garelkov, D. (1973). Biological productivity of some beech forest types in
  Bulgaria. In: "IUFRO Biomass Studies", pp. 307-314. College of Life Sciences and
  Agriculture, University of Maine, Orono, USA.

42-43°N   23-25°E   1400-1600 m   Bulgaria, W. Balkan Mountains.

Plantations.

Brown forest soil   *Fagus sylvatica*

Age (years)	ca.100	ca.100	ca.100
Trees/ha	2580	2000	1200
Tree height (m)	14.5	17.2	23.7
Basal area (m²/ha)	38.5	41.8	48.3
Leaf area index			
Stem volume (m³/ha)	273	352	460
Dry biomass (t/ha) Stem wood	} 169.6	} 280.0	} 364.7
Stem bark			
Branches	24.2	31.6	49.1
Fruits etc.			
Foliage	3.8	2.9	4.7
Root estimate	54.7	37.5	49.7
CAI (m³/ha/yr)	2.3	3.6	4.7
Net production (t/ha/yr) Stem wood	} 1.7	} 2.8	} 3.6
Stem bark			
Branches	} (6.3)[a]	} (5.4)[a]	} (9.3)[a]
Fruits etc.			
Foliage	3.8	2.9	4.7
Root estimate			

Eighteen trees were sampled and roots were excavated. Stand biomass values were
derived using regression methods.
a. Tentative estimates of total net production including branches and roots, but
   excluding woody litterfall and any mortality.

Rozanov, B.G. and Rozanova, I.M. (1964). Biological circulation of nutrient elements of bamboo in the tropical forests of Burma. *Bot. Zbl.* <u>49</u>, 348–357. [Quoted
by Bazilevich, N.I. and Rodin, L.E. (1966) in *For. Abstr.* <u>27</u>, 357–368, and (1967)
In: "Production and Mineral Cycling in Terrestrial Vegetation." (Table 46).
Oliver and Boyd, Edinburgh and London.]

16–20°N 94–97°E   --   Burma.

Yellow-brown
latosols

	Oxytenanthera albociliata Wet-zone bamboo	*Dendrocalamus strictus* Dry-zone bamboo
Age (years)		
Trees/ha		
Tree height (m)		
Basal area (m²/ha)		
Leaf area index		
Stem volume (m³/ha)		

Dry biomass (t/ha)		
Stem wood		
Stem bark	} 168.5	} 41.0
Branches		
Fruits etc.		
Foliage	10.6	7.2
Root estimate		

CAI (m³/ha/yr)

Net production (t/ha/yr)		
Stem wood		
Stem bark	} $7.0 + 2.0^{a}$	} $2.7 + 0.8^{a}$
Branches		
Fruits etc.		
Foliage	10.5^{a}	7.2^{a}
Root estimate		

Nutrient contents were determined.
a. Litterfall.

Post, L.J. (1970). Dry matter production of mountain maple and balsam fir in north western New Brunswick. *Ecology* <u>51</u>, 548-550.

---

ca.47°30'N 68°20'W   300 m   Canada, New Brunswick, Green River.

Silty clay loams
derived from
soft shale   *Acer spicatum*

Stands which had been clear-felled in different years.

Age (years)	8	11	13	16	18	21	23	26
Trees/ha	33300	22800	18300	18400	8500	6700	6700	6800
Tree height (m)	1.9	2.7	3.2	3.4	4.7	5.5	5.3	5.4
Basal area (m²/ha)								
Leaf area index								
Stem volume (m³/ha)								

Dry biomass (t/ha)

Stem wood	7.6	9.0	12.5	13.7	11.8	17.4	16.3	20.9
Stem bark	2.0	2.2	2.9	3.1	2.6	3.7	3.5	4.5
Branches	2.4	3.9	6.8	7.5	7.2	11.2	10.4	13.7
Fruits etc.								
Foliage	0.9	0.9	1.0	1.1	0.9	1.2	1.1	1.4
Root estimate								

CAI (m³/ha/yr)

Net production (t/ha/yr)

Stem wood								
Stem bark								
Branches								
Fruits etc.								
Foliage								
Root estimate								

Forty-four trees were sampled, and stand values for three plots of 0.004 to 0.010 ha in each stand were derived from regressions on D^2H. Stands measured at ages 8, 13, 18 and 23 were measured again after 3 years giving the values for ages 11, 16, 21 and 26. Mean annual increments between ages 4 and 26 were about 1.1 t/ha/yr of foliage, and 0.81, 0.15 and 0.51 t/ha/yr of stem wood, stem bark and branches, respectively, excluding mortality and woody litterfall.

MacLean, D.A. and Wein, R.W. (1977). Nutrient accumulation for postfire jack pine and hardwood succession patterns in New Brunswick. *Can. J. For. Res.* 7, 562-578.

MacLean, D.A. and Wein, R.W. (1978). Litter production and forest floor nutrient dynamics in pine and hardwood stands of New Brunswick, Canada. *Holarctic Ecol.* 1, 1-15.

47°30'N 65°15'W   15-170 m   Canada, New Brunswick, Gloucester and Northumberland Counties.

Glacial till   *Acer rubrum, Betula papyrifera, Prunus pensylvanica*
                              and *Populus tremuloides*

Age (years)	7	7	7	10	13	17
Trees/ha	2960	7120	12590	9160	7200	10040
Tree height (m)						
Basal area (m²/ha)	0.1	0.3	1.5	1.9	2.5	2.1
Leaf area index						
Stem volume (m³/ha)						

Dry biomass (t/ha)

Stem wood } Stem bark	} 0.4	} 0.6	} 3.0	} 3.0	} 3.3	} 2.5
Branches } Fruits etc. } Foliage } Root estimate	} 0.4	} 0.9	} 4.9	} 4.9	} 5.8	} 4.4

CAI (m³/ha/yr)

Net production (t/ha/yr)

Stem wood						
Stem bark						
Branches			0.2[a]			0.2[a]
Fruits etc.						
Foliage			1.9[a]			1.0[a]
Root estimate						

Over 200 trees were sampled in July-August. The biomasses of ten 25 m² plots per stand were derived from regressions on D. Nutrient contents were determined.
a. Litterfall only.

Continued from p.34.

Same as p.34.

Age (years)	18	20	25	29	37
Trees/ha	4490	16960	5720	7680	4920
Tree height (m)					
Basal area (m²/ha)	2.0	7.0	6.1	12.2	10.2
Leaf area index					
Stem volume (m³/ha)					
Dry biomass (t/ha) Stem wood / Stem bark	}2.6	}8.8	}6.2	}12.8	}9.0
Branches / Fruits etc. / Foliage / Root estimate	}4.8	}19.8	}16.4	}27.4	}21.1
CAI (m³/ha/yr)					
Net production (t/ha/yr) Stem wood					
Stem bark					
Branches		0.1[a]		0.2[a]	
Fruits etc.					
Foliage		1.5[a]		1.4[a]	
Root estimate					

See p.34

James, T.D.W. and Smith, D.W. (1977). Short-term effects of surface fire on the biomass and nutrient standing crop of *Populus tremuloides* in southern Ontario. *Can. J. For. Res.*7, 666-679.
Bray, J.R. and Dudkiewicz, L.A. (1963). The composition, biomass and productivity of two *Populus* forests. *Bull. Torrey bot. Club* 90, 298-308.

Canada	43°55'N 80°30'W 420-460 m Ontario, West Luther.		ca.45°14'N 78°55'W 200-400 m Dorset.
	Populus tremuloides		*Populus grandidentata* (67%)[a] *P. tremuloides* (26%)[a] *et al.*
	Poorly drained organic soil		Sandy uplands
	Lightly burned	Unburned	(Bray and Dudkiewicz 1963)
Age (years)	30	30	ca.40
Trees/ha	110[b]	144[b]	1036
Tree height (m)	9	9	
Basal area (m²/ha)			18.4
Leaf area index	0.37	0.37	
Stem volume (m³/ha)			
Dry biomass (t/ha)			
Stem wood	} 5.0	} 6.1	} 47.2
Stem bark			
Branches	1.6	2.4	9.1
Fruits etc.			
Foliage	0.3	0.4	1.6
Root estimate			
CAI (m³/ha/yr)			
Net production (t/ha/yr)			
Stem wood			} >1.3[c]
Stem bark			
Branches			>0.2[c]
Fruits etc.			
Foliage			1.6
Root estimate			

At West Luther 5 ramets were sampled in each of two burned and two unburned 2250 m² plots in July-August. Stand values were derived from regressions on D. Nutrient contents were determined.
At Dorset three *P. grandidentata* were sampled, and 6 *P. tremuloides* were sampled elsewhere, all during July-September. Stand biomass values were derived by assigning biomass in proportion to the 'effective canopy area' per tree of 4 trees at each of 40 random points.
a. Percentage of the total basal area.
b. Numbers of ramets.
c. New twigs, plus old wood biomass divided by its age, and excluding woody litter-fall and mortality.

Peterson, E.B., Chan, Y.H. and Cragg, J.B. (1970). Aboveground standing crop, leaf area, and calorific value in an aspen cline near Calgary, Alberta. *Can. J. Bot.* **48**, 1459-1469.

ca.51°N 115°W   1430 m   Canada, Alberta, 72 km W of Calgary.

On 40° west-facing slope   *Populus tremuloides*

	Top of slope	Mid-slope	Mid-slope	Bottom of slope
Age (years)	66–89	66–89	66–89	66–89
Trees/ha	3800	3800	1200	1000
Tree height (m)	6.7	8.7	11.2	15.1
Basal area (m²/ha)	16.0[a]	26.2[a]	13.2[a]	49.2[a]
Leaf area index	1.5	1.8	0.8	3.1
Stem volume (m³/ha)				
Dry biomass (t/ha)				
Stem wood	} 31.6	} 51.5	} 31.3	} 153.7
Stem bark				
Branches	5.7	6.2	2.7	19.1
Fruits etc.				
Foliage	1.4	1.7	0.7	2.9
Root estimate				
CAI (m³/ha/yr)				
Net production (t/ha/yr)				
Stem wood				
Stem bark				
Branches				
Fruits etc.				
Foliage				
Root estimate				

Forty-nine trees were sampled in August. Stand values, for four 50 m² plots in each of the four stands, were derived from regressions on D²H and stem diameter at the base of the crowns. There was 2.6, 3.5, 1.9 and 11.3 t/ha of dead branches in columns left to right.

a. Estimated from the authors' data on stem diameter frequency distributions.

Pollard, D.F.W. (1972). Above-ground dry matter production in three stands of trembling aspen. *Can. J. For. Res.* 2, 27-33.

46°00'N   77°26'W   121-300 m   Canada, Ontario, near Chalk River.

Sandy soils   *Populus tremuloides et al.*

Age (years)	5	15	52
Trees/ha	31200	9400	494
Tree height (m)	2.0-7.5	4.0-13.6	20-26
Basal area (m²/ha)			
Leaf area index	2.4	2.9	1.6
Stem volume (m³/ha)			

Dry biomass (t/ha)

Stem wood			
Stem bark	}21.5	}51.2	}91.8
Branches			
Fruits etc.			
Foliage	2.6	2.6	1.5
Root estimate			

CAI (m³/ha/yr)

Net production (t/ha/yr)

Stem wood			
Stem bark	}7.2^a	}6.2^a	}0.8^a
Branches			
Fruits etc.			
Foliage	2.6	2.6	1.5
Root estimate			

Sixteen to 26 trees (or stems) were sampled per age. Stand biomass values for plots of 100 m² (ages 5 and 15) or 400 m² (age 52) were derived from regressions on D. *a*. Excluding any woody litterfall and mortality.

Kimmins, J.P. and Krumlik, G.J. (1973). Comparisons of the biomass distribution and tree form of old virgin forests at medium and high elevations in the mountains of south coastal British Columbia, Canada. In: "IUFRO Biomass Studies", pp. 315-335. College of Life Sciences and Agriculture, University of Maine, Orono, USA.

Weetman, G.F. and Webber, B. (1972). The influence of wood harvesting on the nutrient status of two spruce stands. *Can. J. For. Res.* **2**, 351-369.

Canada	49°41'N 122°55'W 1500 m British Columbia, Squamish.	46°07'N 74°36'W -- Quebec, near St Jovite.
	Abies amabilis (69%)[a] *Tsuga mertensiana* (31%)[a]	*Abies balsamea* *Picea rubens*
	Leached humic podzols pH 3.7-4.6	Deep well-drained ferro-humic podzols (Weetman and Webber 1972)
Age (years)	300-350	Mixed
Trees/ha	221	
Tree height (m)	28-31	
Basal area (m²/ha)		
Leaf area index		
Stem volume (m³/ha)		174[c]

		Squamish	St Jovite
Dry biomass (t/ha)	Stem wood	370.0	} 92.4
	Stem bark	90.5	
	Branches	53.3	16.0
	Fruits etc.		
	Foliage	28.4[b]	23.6
	Root estimate		
CAI (m³/ha/yr)			6[c]
Net production (t/ha/yr)	Stem wood		
	Stem bark		
	Branches		
	Fruits etc.		
	Foliage		
	Root estimate		

Twenty trees were sampled at Squamish, and stand values for a 1.16 ha plot were derived from regressions on D and D²H. Stand values at St Jovite were estimated by using published regressions of biomass on D and H; nutrient contents were determined.
a. Percentage of the total tree number.
b. Including twigs.
c. Merchantable volume.

Baskerville, G.L. (1965). Dry matter production in immature balsam fir stands. *Forest Sci. Monogr.* 9.

Baskerville, G.L. (1966). Dry matter production in immature balsam fir stands: roots, lesser vegetation, and total stand. *Forest Sci.* 12, 49-53.

47°51'N 68°20'W   500 m   Canada, New Brunswick, Green River Watershed.

Podsolized, slightly stoney silt loam

Abies balsamea (89-98%),[a]

Picea mariana and *Picea glauca* (0-10%),[a]

Betula papyrifera and *Betula alleghaniensis* (1-5%)[a]

Stands selected with different stem densities.

Age (years)		51	42	42	43	42	42
Trees/ha		1840	2642	4168	4797	7646	12491
Tree height (m)		11.0	9.8	9.4	9.8	9.8	8.3
Basal area (m²/ha)		34.1	35.0	44.1	51.3	57.3	66.6
Leaf area index							
Stem volume (m³/ha)		220	209	262	277	319	360
	Stem wood	61.6	59.2	68.4	81.3	92.3	103.2
	Stem bark	8.9	9.9	10.0	11.7	13.8	14.5
Dry biomass (t/ha)	Branches	18.9	17.2	17.2	16.8	16.5	16.5
	Fruits etc.	0.6	0.7	0.6	0.7	0.6	0.5
	Foliage	17.4	16.3	17.1	18.3	18.7	19.7
	Root estimate	35.7	35.3	41.9	44.2	47.1	51.9
CAI (m³/ha/yr)		10.2	9.7	10.5	11.7	13.3	14.2
	Stem wood	2.93[b]	3.22[b]	3.33[b]	3.80[b]	4.39[b]	4.91[b]
Net production (t/ha/yr)	Stem bark	0.34[b]	0.32[b]	0.32[b]	0.34[b]	0.41[b]	0.41[b]
	Branches	1.40[c]	1.31[c]	1.37[c]	1.44[c]	1.67[c]	1.82[c]
	Fruits etc.	0.11	0.16	0.11	0.14	0.14	0.09
	Foliage	4.70	4.39	4.52	4.91	5.09	5.40
	Root estimate						

Sixteen to 18 *Abies*, 1 to 4 *Picea* and 4 *Betula* were sampled per stand, and totals of 89 *Abies*, 2 *Picea* and 7 *Betula* root systems were excavated. Unrecoverable fine roots were assumed to weigh 5.7 t/ha. Stand biomass values were derived for 5 plots per stand (each plot containing 24 trees) from regressions on D and D²H. Foliage increment was taken as 26% of the crown weight of *Abies*, and 35% of *Picea*, averaged over 4 years.

a. Percentage of total basal area (*B. alleghaniensis* syn. *B. lutea*).
b. Including estimates of mortality.
c. Branch biomass divided by branch age, excluding woody litterfall.

Weetman, G.F. and Harland, R. (1964). Foliage and wood production in unthinned black spruce in northern Quebec. *Forest Sci.* <u>10</u>, 80-88.
Weetman, G.F. and Webber, B. (1972). The influence of wood harvesting on the nutrient status of two spruce stands. *Can. J. For. Res.* <u>2</u>, 351-369.
Rencz, A.N. and Auclair, A.N.D. (1980). Dimension analysis of various components of black spruce in subarctic lichen woodland. *Can. J. For. Res.* <u>10</u>, 491-497.
Rencz, A.N. and Auclair, A.N.D. (1978). Biomass distribution in a subarctic *Picea mariana - Cladonia alpestris* woodland. *Can. J. For. Res.* <u>8</u>, 168-176.

Canada, Quebec.	49°20'N 68°10'W -- N of Baie Comeau *Picea mariana* Deep, well-drained ferro-humic podzols (Weetman and Harland 1964)	54°55' N 66°55'W over 500 m 22 km NE of Schefferville *P. mariana* with a few *Picea glauca, Larix laricina* and understorey shrubs Poorly drained podzols (Rencz and Auclair 1980, 1978)
Age (years)	65	Over 80
Trees/ha		525
Tree height (m)	13.3^a	7.5 (2 to 16)
Basal area (m²/ha)	41.8	
Leaf area index	4.3^b	
Stem volume (m³/ha)	223	
Dry biomass (t/ha) Stem wood	76.9	$\left.\begin{array}{l} 8.5 \ (\text{or } 8.8)^d \\ \ \\ 3.4 \ (\text{or } 4.6)^d \end{array}\right\} + 4.4^e$
Stem bark	11.6	
Branches	10.3	
Fruits etc.		0.7 (or 0.5)^d
Foliage	8.3	3.8 (or 3.7)^d + 0.7^e
Root estimate		9.6^f (or 7.8)^{df} + 3.3^e
CAI (m³/ha/yr) Stem wood		
Net production (t/ha/yr) Stem bark		
Branches		
Fruits etc.		
Foliage	1.5^c	0.53^g
Root estimate		

Weetman and Harland (1964) sampled 20 trees, estimated stem volumes from yield tables, and derived biomass values for three 0.08 ha plots from regressions on D. Nutrient contents were determined (Weetman and Webber 1972). Rencz and Auclair (1980, 1978) sampled 15 trees in June-September, excavated roots and derived values for a 0.20 ha plot from regressions on D. There was 2.2 t/ha of dead branches and 9.7 t/ha of lichens and mosses.
a. Dominant tree height. *b.* All-sided LAI 9.9. *c.* Litterfall, measured over 4 years.
d. Alternative values (in brackets) using quadratic rather than logarithmic regression equations. *e.* Understorey shrubs.
f. Including root crowns, weighing 3.8 t/ha (or 1.8 t/ha using the alternative equations).   *g.* Biomass of new foliage.

Moore, T.R. and Verspoor, E. (1973). Aboveground biomass of black spruce stands in subarctic Quebec. *Can. J. For. Res.* 3, 596–598.

56°30'N 69°15'W   200–500 m   Canada, Quebec, Cambrian Lake.

Picea mariana

	Picea-lichen woodlands			*Picea*-moss forests		
Age (years)						
Trees/ha	1270	3080	2510	4840	5095	5350
Tree height (m)	3.5^a	3.5^a	4.5^a	5.5^a	7.5^a	6.5^a
Basal area (m²/ha)	2.4	11.8	16.1	19.5	36.6	20.1
Leaf area index						
Stem volume (m³/ha)						

Dry biomass (t/ha)	Stem wood						
	Stem bark						
	Branches	9.6	21.1	29.3	82.3	163.4	78.4
	Fruits etc.						
	Foliage						
	Root estimate						

	CAI (m³/ha/yr)						
Net production (t/ha/yr)	Stem wood						
	Stem bark						
	Branches						
	Fruits etc.						
	Foliage						
	Root estimate						

Twelve trees were sampled from the *Picea*-lichen woodlands and 10 trees were sampled from the *Picea*-moss forests. Stand values were derived from regressions on D for plots of 314 m² in the *Picea*-lichen woodland and 79 m² in the *Picea*-moss forest. *a*. Approximate mean heights; the tallest trees were about twice these values.

Gordon, A.G. (1981). In: "Dynamic Properties of Forest Ecosystems" (D.E. Reichle, ed.) pp. 576-579. Cambridge University Press, Cambridge, London, New York and Melbourne.

45°14-32'N   78°16-49'W   465-503 m   Canada, Ontario.

Podzols
pH 4.1 to 4.5

Picea rubens, with *Picea mariana, Picea glauca, Abies balsamea, Tsuga canadensis et al.*

	Poorly drained			
Age (years)	84	130	212	246
Trees/ha	3311	9058	3065	3879
Tree height (m)	14.9	20.7	25.6	18.3
Basal area (m² /ha)	32.3	45.7	59.4	46.8
Leaf area index	11.6	7.1	16.9	15.0

Stem volume (m³ /ha)

Dry biomass (t/ha)

Stem wood	76.0	125.7	344.7	189.2
Stem bark	8.3	12.8	31.6	17.6
Branches	21.1	20.5	64.5	36.4
Fruits etc.	0.2	0.2	0.1	0.1
Foliage	15.9	10.0	21.2	20.5
Root estimate	28.0	47.5	104.5	62.0

CAI (m³ /ha/yr)

Net production (t/ha/yr)

Stem wood	$1.92 \left.\right\} + 0.12^a$	$1.02 \left.\right\} + 0.15^a$	$5.26 \left.\right\} + 0.06^a$	$3.68 \left.\right\} + 0.10^a$
Stem bark	$0.20 \left.\right\} + 0.11^b$	$0.12 \left.\right\} + 0.11^b$	$0.52 \left.\right\} + 0.29^b$	$0.41 \left.\right\} + 0.17^b$
Branches	$0.30 \quad + 0.23^a$	$0.35 \quad + 0.40^a$	$1.01 \quad + 0.27^a$	$0.71 \quad + 0.68^a$
Fruits etc.	0.25^a	0.05^a	0.11^a	0.38^a
Foliage	$0.19 \quad + 1.20^a$	$0.27 \quad + 1.42^a$	$0.35 \quad + 2.26^a$	$0.36 \quad + 2.19^a$
Root estimate	0.72	0.65	1.56	1.06

There was 5.8, 8.3, 18.2 and 8.5 t/ha of standing dead wood in columns left to right.
a. Litterfall, excluding frass litterfall estimated to be 0.05 to 0.19 t/ha/yr.
b. Mortality.

Hegyi, F. (1972). Dry matter production in jack pine stands in northern Ontario. *For. Chron.* 48, 193–197.
Morrison, I.K. (1973). Distribution of elements in aerial components of several natural jack pine stands in northern Ontario. *Can. J. For. Res.* 3, 170–179.
Morrison, I.K. (1974). Dry matter and element content of roots of several natural stands of *Pinus banksiana* Lamb. in northern Ontario. *Can. J. For. Res.* 4, 61–64.
Foster, N.W. (1974). Annual macro-element transfer from *Pinus banksiana* Lamb. forest to soil. *Can. J. For. Res.* 4, 470–476.

46°25'N 83°23'N   ca.200 m   Canada, Ontario, Mississagi River, Wellstown.

Glaciofluvial coarse stoney medium sand, of low base status, over weak podzol

Pinus banksiana

Site class I (trees 19 m tall at age 50)

Age (years)	20	20	30	30	65	65
Trees/ha	2224	1347	3002	1137	568	346
Tree height (m)	8.2	7.9	13.1	13.1	22.9	20.4
Basal area (m²/ha)	4.8	2.8	28.9	16.6	23.7	12.5
Leaf area index						
Stem volume (m³/ha)						

Dry biomass (t/ha): Stem wood, Stem bark, Branches, Fruits etc., Foliage

| | 13.2 | 7.5 | 99.2 | 63.4 | 105.8 | 54.0 |

Root estimate: 15.9 11.4

CAI (m³/ha/yr)

Net production (t/ha/yr): Stem wood, Stem bark, Branches, Fruits etc., Foliage, Root estimate

ca.3.73[a]

Up to 6 trees were sampled per site (77 in all). Roots of 8 trees were excavated. Stand values for one 0.08 ha plot per stand were derived from regressions on D. Nutrient contents were determined.
a. Total litterfall measured over 2 years (from Foster 1974, for 30-year-old stand with 3285 trees/ha).

Continued from p.44.

Same as p.44.

Site class II (trees 16 m tall at age 50)

Age (years)	20	20	30	30	65	65
Trees/ha	1940	605	3842	1742	519	358
Tree height (m)	6.4	6.4	11.0	11.6	18.9	18.3
Basal area (m² /ha)	5.9	2.4	27.6	17.6	17.8	10.5
Leaf area index						
Stem volume (m³ /ha)						

Dry biomass (t/ha)

	Stem wood						
	Stem bark						
	Branches	16.0	6.9	88.5	61.0	76.0	43.6
	Fruits etc.						
	Foliage						
	Root estimate			13.2	9.9		

CAI (m³ /ha/yr)

Net production (t/ha/yr)

	Stem wood	
	Stem bark	
	Branches	
	Fruits etc.	ca.3.73[a]
	Foliage	
	Root estimate	

See p.44

MacLean, D.A. and Wein, R.W. (1976). Biomass of jack pine and mixed hardwood stands in northeastern New Brunswick. *Can. J. For. Res.* <u>6</u>, 441-447.

MacLean, D.A. and Wein, R.W. (1978). Litter production and forest floor nutrient dynamics in pine and hardwood stands of New Brunswick, Canada. *Holarctic Ecol.* <u>1</u>, 1-15.

47°30'N 65°15'W   15-170 m   Canada, New Brunswick, Gloucester and Northumberland Counties

Glacial till   *Pinus banksiana*

Age (years)	13	16	29	29	31	37
Trees/ha	3040	2320	3040	6560	2200	2520
Tree height (m)						
Basal area (m²/ha)	0.2	0.3	13.5	24.7	13.0	20.4
Leaf area index						
Stem volume (m³/ha)						

Dry biomass (t/ha)

Stem wood	} 0.3	} 0.6	} 9.8	} 19.0	} 8.9	} 13.5
Stem bark						
Branches						
Fruits etc.	} 0.4	} 1.2	} 32.6	} 59.4	} 31.3	} 51.9
Foliage						
Root estimate				10-16	10-16	10-16

CAI (m³/ha/yr)

Net production (t/ha/yr)

Stem wood						
Stem bark						
Branches	0.1[a]	0.1[a]				
Fruits etc.						
Foliage	0.3[a]	1.5[a]				
Root estimate						

Over 200 trees were sampled in July-August. The biomass of ten 25 m² plots per stand were derived from regressions on D. Nutrient contents were determined.
a. Litterfall only.

Continued from p.46.

Same as p.46.

	37	38	40	44	49	57
Age (years)	37	38	40	44	49	57
Trees/ha	4840	2600	6000	3440	3480	2440
Tree height (m)						
Basal area (m²/ha)	26.7	11.5	28.2	28.4	17.8	24.8
Leaf area index						
Stem volume (m³/ha)						

Dry biomass (t/ha)

	37	38	40	44	49	57
Stem wood } Stem bark	19.0	8.7	21.7	18.3	13.3	15.9
Branches } Fruits etc. } Foliage	64.2	27.1	67.9	65.4	43.1	60.2
Root estimate						

CAI (m³/ha/yr)

Net production (t/ha/yr)

	37	38	40	44	49	57
Stem wood						
Stem bark						
Branches						0.2^a	0.4^a
Fruits etc.						
Foliage						1.4^a	1.4^a
Root estimate						

See p.46.

Doucet, R., Berglund, J.V. and Farnsworth, C.E. (1976). Dry matter production in 40-year-old *Pinus banksiana* stands in Quebec. *Can. J. For. Res.* 6, 357–367.

47°05'N 73°30'W   420 m   Canada, Quebec, NW of Trois Rivières, Mattawin River.

Deep fluvio-glacial   *Pinus banksiana*, with some *Picea mariana et al.*
sandy podzols

'Good quality' sites

Age (years)	44	44	44	44	44	44
Trees/ha	3163	3163	2026	2026	1235	1186
Tree height (m)	15.6	14.6	15.4	15.7	15.0	15.3
Basal area (m²/ha)	26.3	26.7	25.5	25.9	17.3	20.1
Leaf area index						
Stem volume (m³/ha)						

Dry biomass (t/ha)

Stem wood	74.6	78.6	70.7	74.0	44.4	53.7
Stem bark	7.7	8.0	7.7	7.8	5.0	5.9
Branches	9.1	8.5	10.4	10.1	7.1	7.7
Fruits etc.	0.3	0.3	0.3	0.3	0.2	0.2
Foliage	4.8	4.5	5.1	4.8	3.4	3.7
Root estimate						

CAI (m³/ha/yr)

Net production (t/ha/yr)

Stem wood	2.42^a	2.53^a	2.33^a	2.37^a	1.48^a	1.72^a
Stem bark	0.16^a	0.15^a	0.16^a	0.16^a	0.11^a	0.12^a
Branches	0.64^a	0.63^a	0.67^a	0.67^a	0.42^a	0.47^a
Fruits etc.	0.03	0.02	0.02	0.02	0.08	0.05
Foliage	1.36^b	1.36^b	1.43^b	1.45^b	0.92^b	1.08^b
Root estimate						

Six trees were sampled in August–September in each stand, plus 6 *P. mariana* and 2 trees of each of the broadleaved species. Stand biomass values for two 0.02 ha plots per stand were derived from regressions on D and H.
a. Excluding woody litterfall and any mortality.
b. New foliage biomass.

Continued from p.48.

Same as p.48.

'Medium quality' sites

Age (years)	44	44	44	44	44	44
Trees/ha	5140	3954	3311	2718	1631	1433
Tree height (m)	13.2	13.6	13.0	13.8	13.2	13.4
Basal area (m²/ha)	26.8	27.0	21.7	23.4	16.2	15.0
Leaf area index						
Stem volume (m³/ha)						
Dry biomass (t/ha)						
Stem wood	62.1	65.4	49.3	57.7	39.2	36.5
Stem bark	7.8	8.0	6.1	6.7	4.5	4.2
Branches	12.3	11.7	12.3	11.4	10.5	10.3
Fruits etc.	0.9	1.0	0.8	1.2	0.9	0.9
Foliage	7.8	7.3	7.1	6.9	5.8	5.7
Root estimate						
CAI (m³/ha/yr)						
Net production (t/ha/yr)						
Stem wood	2.71^a	2.77^a	2.10^a	2.38^a	1.58^a	1.48^a
Stem bark	0.21^a	0.20^a	0.16^a	0.16^a	0.11^a	0.10^a
Branches	0.80^a	0.76^a	0.78^a	0.73^a	0.67^a	0.65^a
Fruits etc.	0.03	0.03	0.03	0.04	0.04	0.03
Foliage	1.71^b	1.74^b	1.69^b	1.87^b	1.64^b	1.58^b
Root estimate						

See p.48.

Johnstone, W.D. (1972). Total standing crop and tree component distributions in
   three stands of 100-year-old lodgepole pine. In: "Forest Biomass Studies",
   pp. 81-89. College of Life Sciences and Agriculture, University of Maine, Orono,
   USA.

51°06'N 115°04'W   1400 m   Canada, Alberta, Kananakis Forest Research Station.

Well-drained, calcareous, grey podzol	*Pinus contorta*, with few *Picea glauca var. albertiana*		
		a	
Age (years)	100	100	100
Trees/ha	2521	717	12257
Tree height (m)	16.7	20.3	5.7
Basal area (m²/ha)	52.3	34.9	35.9
Leaf area index			
Stem volume (m³/ha)	445	357	182
Dry biomass (t/ha) Stem wood	} 213.3	} 153.1	} 68.2
Stem bark			
Branches	13.9	23.5	14.0
Fruits etc.			
Foliage	12.5	14.6	7.4
Root estimate	42.2[b]	35.1[b]	20.5[b]
CAI (m³/ha/yr)			
Net production (t/ha/yr) Stem wood			
Stem bark			
Branches			
Fruits etc.			
Foliage			
Root estimate			

Over 300 trees were sampled and 72 root systems were excavated. Stand values for
plots of 0.04 ha (two left columns) or 0.08 ha (right column) were derived from
regressions on D²H.
a. 31% of the basal area of this stand was removed as thinnings at age 70.
b. Including stumps to 30 cm above ground level.

Webber, B.D. (1977). Biomass and nutrient distribution patterns in a young *Pseudotsuga menziesii* ecosystem. *Can. J. For. Res.* 7, 326-334.

Webber, B.D. (1973). "Plant Biomass and Nutrient Distribution in a young *Pseudotsuga menziesii* Forest Ecosystem." Unpublished Ph.D. thesis, Oregon State University, and Canadian For. Service, Victoria, B.C, Internal Report BC-41.

---

ca.48°30'N 123°20'W   300 m   Canada, Vancouver Island, near Victoria.

Orthic, dystric brunisol over glacial till, pH 5.2-6.0	*Pseudotsuga menziesii* (97%),[a] *Thuja plicata* and *Tsuga heterophylla*, with understorey shrubs

---

Age (years)		15-20
Trees/ha		3000[b]
Tree height (m)		2-12
Basal area (m²/ha)		
Leaf area index		
Stem volume (m³/ha)		
	Stem wood	42.9
	Stem bark	7.0
Dry biomass (t/ha)	Branches	12.7
	Fruits etc.	
	Foliage	9.6 (or 10.9)[d]
	Root estimate	

$\Big\}$ +4.0[c]

CAI (m³/ha/yr)		
	Stem wood	
	Stem bark	
Net production (t/ha/yr)	Branches	
	Fruits etc.	
	Foliage	2.09[e] (or 2.78)[de]
	Root estimate	

Over 50 trees were sampled in late summer, including 7 large *P. menziesii*. Stand values for a 0.04 ha plot were derived from regressions on D²H. Nutrient contents were determined.

a. Percentage of the total tree biomass.
b. Stems over 5 cm diameter; there were 13000 smaller trees per hectare.
c. Understorey shrubs.
d. Alternative values taken from Webber (1973).
e. New foliage biomass.

52 CANADA *Tsuga*

Kimmins, J.P. and Krumlik, G.J. (1973). Comparison of the biomass distribution and tree form of old virgin forests at medium and high elevations in the mountains of south coastal British Columbia, Canada. In: "IUFRO Biomass Studies", pp. 317-335. College of Life Sciences and Agriculture, University of Maine, Orono, USA.

Krumlik, G.J. and Kimmins, J.P. (1973). Studies of biomass distribution and tree form in old virgin forests in the mountains of south coastal British Columbia. In: "IUFRO Biomass Studies", pp. 361-374. College of Life Sciences and Agriculture, University of Maine, Orono, USA.

49°13'N 122°36'W   720 m   Canada, British Columbia, Haney.

Leached humic podzols pH 3.7-4.6

Tsuga heterophylla (59%)[a]
Thuja plicata (28%)[a]
Chamaecyparis nookatensis

Age (years)	115-400
Trees/ha	144
Tree height (m)	21
Basal area (m²/ha)	
Leaf area index	
Stem volume (m³/ha)	

Dry biomass (t/ha)

Stem wood	27.1
Stem bark	5.7
Branches	8.7
Fruits etc.	
Foliage	3.0[b]
Root estimate	

CAI (m³/ha/yr)

Net production (t/ha/yr)

Stem wood	
Stem bark	
Branches	
Fruits etc.	
Foliage	
Root estimate	

Thirteen trees were sampled and stand values for a 0.79 ha plot were derived from regressions on D and D²H.
a. Percentage of the total tree number.
b. Including twigs.

Veblen, T.T., Schlegel, F.M. and B. Escobar, R. (1980). Dry matter production of two species of bamboo (*Chusquea culeou* and *C. tenuiflora*) in south-central Chile. *J. Ecol.* <u>68</u>, 397-404.

39°33'S 72°03' W   700 m   Chile, San Pablo.

Deep soils
derived from   *Chusquea culeou* (bamboo)
volcanic ash
                        beneath a few overstorey trees

Age (years)	20-30	20-30
Trees/ha	ca.200,000	ca.170,000
Tree height (m)	5-6	5-6
Basal area (m²/ha)		
Leaf area index		
Stem volume (m³/ha)		

Dry biomass (t/ha)

Stem wood		
Stem bark	} 130.5^a	} 127.4^a
Branches		
Fruits etc.		
Foliage	26.1^b	24.0^b
Root estimate		

CAI (m³/ha/yr)

Net production (t/ha/yr)

Stem wood		
Stem bark	} 5.57^c	} 6.95^c
Branches		
Fruits etc.		
Foliage	$0.12 + 4.27^d$	$0.14 + 4.27^d$
Root estimate		

About 100 culms were sampled in the autumn and stand biomass values for the above plots of 25 m² and 50 m² (left and right columns, respectively) were derived from regressions on culm diameter at 1 m height.
a. Culms.
b. Leaves plus leaf sheaths.
c. New growth, excluding mortality.
d. Litterfall measured over one year.

A 20-30 year-old stand of *Chusquea tenuiflora* beneath a dense overstorey had a total biomass of only 12.9 t/ha.

Fölster, H., De Las Salas, G. and Khanna, P. (1976). A tropical evergreen forest site with perched water table, Magdalena Valley, Colombia. Biomass and bioelement inventory of primary and secondary vegetation. *Oecol. Plant.* 11, 297-320.

6°50'N 73°55'W   100 m   Colombia, Magdalen Valley, 50 km S of Barrancas Bermeja.

Infertile, acid,
surface water   *Bellucia grossularioides, Miconia minutiflora et al.*
gleys

Regrowth after clear-felling

Age (years)	2	5	16
Trees/ha		3700	1222[a]
Tree height (m)	3	12	to 20
Basal area (m²/ha)			20.0[a]
Leaf area index			
Stem volume (m³/ha)			

Dry biomass (t/ha)

Stem wood	} 0.0	} 48.9	} 176.8
Stem bark			
Branches	9.8	$11.2 + 3.9^c$	$16.0 + 3.0^c$
Fruits etc.			
Foliage	$5.2 + 0.4^b$	$3.3 + 0.6^c$	$6.3 + 0.2^b + 0.6^c$
Root estimate			

CAI (m³/ha/yr)

Net production (t/ha/yr)

Stem wood
Stem bark
Branches
Fruits etc.
Foliage
Root estimate

Stand values for the 2 and 5-year-old stands were estimated by harvesting all trees within three plots of 16 m² and 64 m², respectively. Twenty-three trees were sampled from the 16-year-old stand, and values for 3 plots of 64 m² were derived from regressions on D²H.
a. Trees over 10 cm D.
b. Palms.
c. Woody undergrowth.

Fölster, H., De Las Salas, G. and Khanna, P. (1976). A tropical evergreen forest site with perched water table, Magdalena Valley, Colombia. Biomass and bioelement inventory of primary and secondary vegetation. *Oecol. Plant.* <u>11</u>, 297-320.

6°50'N 73°55'W   100 m   Colombia, Magdalen Valley, 50 km S of Barrancas Bermeja.

Infertile acid, surface water gleys	Evergreen seasonal tropical rainforests	
	Jessenia polycarpa (palm) *Pseudolmedia rigida et al.*	*Clathotropis brachipetalum,* *Couma macrocarpa et al.*
	Terrace forest	Slope forest

		Terrace forest	Slope forest
	Age (years)	Mature	Mature
	Trees/ha	720^b	600^b
	Tree height (m)	6-14, 20-30^a	to over 30
	Basal area (m²/ha)	21.6^b	32.0^b
	Leaf area index		
	Stem volume (m³/ha)		
Dry biomass (t/ha)	Stem wood	} $120.0 + 10.6^c$	} $248.0 + 1.7^c$
	Stem bark		
	Branches	$34.8 + 2.8^d$	$66.8 + 0.3^d$
	Fruits etc.		
	Foliage	$4.2 + 5.8^c + 0.4^d$	$7.0 + 1.4^c + 0.6^d$
	Root estimate		
	CAI (m³/ha/yr)		
Net production (t/ha/yr)	Stem wood		
	Stem bark		
	Branches		
	Fruits etc.		
	Foliage		
	Root estimate		

Forty-three trees were sampled and stand values for the above two 2500 m² plots were derived from regressions on D²H, using separate regressions for palms. There was 3.4 t/ha of dead trees in the left column. Nutrient contents were determined.
a. Lower and upper storeys.
b. Trees over 10 cm D.
c. Palms.
d. Woody undergrowth.

Kubicek, F. and 6 others (1977). "Primary Productivity of an Oak-Hornbeam Ecosystem." Slovenská Akademia Vied, Bratislava.
Kubicek, F. (1974). Leaf number, leaf area index and leaf production of hornbeam (*Carpinus betulus* L.). *Biológia, Bratisl.* 29, 39-49.
Kubicek, F. (1972). Leaf litter in the oak-hornbeam forest. *Biológia, Bratisl.* 27, 775-783.
Biskupsky, V. (1981). In: "Dynamic Properties of Forest Ecosystems" (D.E. Reichle, ed.), p. 580. Cambridge University Press, Cambridge, London, New York, Melbourne.

48°11'N 17°54'E   209 m   Czechoslovakia, Nitra, Bab.

Well-drained loess soil pH 5.8-6.1	*Carpinus betulus, Acer campestre, Quercus cerris, Quercus petraea et al.*	
	(Kubicek 1972, 1974, 1977)	(Biskupsky 1981)
Age (years)		50-70
Trees/ha		733
Tree height (m)		19
Basal area (m²/ha)		25.6
Leaf area index	3.8^a	5.2
Stem volume (m³/ha)		
Dry biomass (t/ha)		
Stem wood		100.4
Stem bark		14.3
Branches	165	42.8
Fruits etc.		1.7
Foliage		3.4
		$+ 8.2^d$
Root estimate	79	75.1
CAI (m³/ha/yr)		
Net production (t/ha/yr)		
Stem wood		2.88^e
Stem bark	$5.5 + 0.9^b + 1.9^b$	0.41^e
Branches		4.01^e
Fruits etc.	0.9^b	
Foliage	3.5^c	3.4
		$+ 0.14^{de}$
Root estimate		

a. Trees only; including shrubs the LAI was 4.4
b. Litterfall of twigs (0.9 t/ha/yr), miscellaneous matter (1.9 t/ha/yr) and seeds etc. (0.9 t/ha/yr).
c. Mean leaf litterfall measured over 5 years was 2.9 t/ha/yr, of which 49% was *Quercus* spp.
d. Understorey shrubs.
e. Excluding woody litterfall.

Vyskot, M. (1972). Aerial biomass of silver fir (*Abies alba* Mill.). *Acta Universitatis agriculturae (BRNO) Series C.* 41, 243-294.

Vyskot, M. (1973). Root biomass of silver fir (*Abies alba* Mill.). *ibid.* 42, 215-61.

Vyskot, M. (1976a). Biomass production of the tree layer in a floodplain forest near Lednice. In: "Oslo Biomass Studies" pp. 177-202 and pp. 205-209. College of Life Sciences and Agriculture, University of Maine, Orono, USA.

Vyskot, M. (1976b). "Tree Storey Biomass in Lowland Forests in Southern Moravia." Rozpravy Ceskosl. Akad. Ved, Rada Matemat. a Prirodnich Ved. Rocnik 86 - Sesit 10.

Czechoslovakia	49°19'N 16°40'E 460 m Olomucany Forest District	48°48'N 16°26' E 162 m S. Moravia, Lednice.
	Abies alba Shelterwood. Brown forest sandy clays	*Quercus robur* (19%)[a] *Fraxinus excelsior* (19%)[a] *Tilia* sp., *Acer* sp. *et al.* River floodplain sands and loams
Age (years)	51	96
Trees/ha	1667	854
Tree height (m)	ca.17	25-30
Basal area (m²/ha)		
Leaf area index	3.4[b]	4.6
Stem volume (m³/ha)	216	465
Dry biomass (t/ha)		
Stem wood	} 103.8	218.4
Stem bark		29.8
Branches	18.7	62.7
Fruits etc.		
Foliage	6.8	3.5
Root estimate	14.7	46.0[c]
CAI (m³/ha/yr)		21.3
Net production (t/ha/yr)		
Stem wood		} 11.19[d]
Stem bark		
Branches		3.46[d]
Fruits etc.		
Foliage		3.23
Root estimate		

In both studies stand biomass values were derived by multiplying the means of 5 sampled dominant trees, 5 co-dominants and 5 subdominants by the numbers of trees in each of these classes. Roots of all 15 sample trees were excavated.

a. Percentage of the total stem volume per hectare.
b. All-sided LAI was 7.8.
c. Including 25.3 t/ha of stumps.
d. Excluding woody litterfall and mortality.

Möller, C.M. (1945). Untersuchungen über Laubmenge, Stoffverlust und Stoffproduktion des Waldes. *Forst. ForsVaes. Danm.* <u>17</u>, 1-287.

Möller, C.M., Müller, D. and Nielsen, J. (1954b). Graphic presentation of dry matter production of European beech. *Forst. ForsVaes. Danm.* <u>21</u>, 327-335.

Möller, C.M., Müller, D. and Nielsen, J. (1954a). Loss of branches in European beech. *Forst. ForsVaes. Danm.* <u>21</u>, 253-271.

56°00'N 12°20'E   200 m   Denmark, Nödelbo.

Red alluvial soils.   *Fagus sylvatica*

Age (years)	47	54	58	118	150	200
Trees/ha	1433	956	1266	271	300	154
Tree height (m)	14.6	16.1	14.5	24.8	22.1	26.0
Basal area (m²/ha)	18.8	20.5	18.2	30.0	29.6	27.9
Leaf area index	4.1	4.3	5.0	5.4	6.6	5.4
Stem volume (m³/ha)	172	204	165	429	378	415
Dry biomass (t/ha)						
Stem wood	}129	}153	}124	}322	}284	}311
Stem bark						
Branches						
Fruits etc.						
Foliage	2.1	2.2	2.5	2.6	2.9	2.6
Root estimate						
CAI (m³/ha/yr)	11.1	12.7	8.2	10.1	6.1	5.9
Net production (t/ha/yr)						
Stem wood	}8.3+1.2[a]	}9.5+1.2[a]	}6.2+1.2[a]	}7.6[b]	}4.6[b]	}4.4[b]
Stem bark						
Branches						
Fruits etc.						
Foliage	2.1	2.2	2.5	2.6	2.9	2.6
Root estimate						

Stand biomass values were derived by multiplying mean tree values by the numbers of trees per hectare. Stem wood specific gravity was assumed to be 0.75 g/c.c. Branch values were not estimated.
a. Woody litterfall was estimated to be 1.2 t/ha/yr in stands 25 to 85 years old.
b. Excluding woody litterfall.

[Möller (1945) reported similar estimates of stem and leaf biomass and production for over 20 other stands of *F. sylvatica* in Denmark.]

Holm, E. and Jensen, V. (1981). In: "Dynamic Properties of Forest Ecosystems"
   (D.E. Reichle, ed.), p. 581. Cambridge University Press, Cambridge, London, New
   York and Melbourne.

56°18'N 10°29'E   11-28 m   Denmark, Hestehaven.

Grey brown   *Fagus sylvatica*
podzol,
pH 4.5-5.8

Age (years)	85-90	
Trees/ha	370	
Tree height (m)	28.6	
Basal area (m²/ha)	28.5	
Leaf area index	5.0	
Stem volume (m³/ha)		

Dry biomass (t/ha)

Stem wood	163.0	
Stem bark	7.4	
Branches	43.2	$+ 5.4^a$
Fruits etc.		
Foliage	2.1	
Root estimate	43.2	

CAI (m³/ha/yr)

Net production (t/ha/yr)

Stem wood	4.91
Stem bark	0.26
Branches	$4.45 + 0.90^b$
Fruits etc.	0.23^b
Foliage	2.69^b
Root estimate	

---

a. Understorey shrubs.
b. Litterfall.

Boysen Jensen, P. (1932). "Die Stoffproduktion der Pflanzen." Verlag von Gustav Fischer, Jena, Germany.

55°26'N 11°34'E   25 m   Denmark, Soro, Lille Bogeskov.

Plantations

Fraxinus excelsior

Age (years)	12	14
Trees/ha	5300	ca.4000
Tree height (m)		
Basal area (m² /ha)		
Leaf area index	5.4	5.4
Stem volume (m³ /ha)	54	72

Dry biomass (t/ha)

Stem wood	} 25	} 33
Stem bark		
Branches		
Fruits etc.		
Foliage	2.7	2.7
Root estimate		

Net production (t/ha/yr)

CAI (m³ /ha/yr)		8.9
Stem wood		} 4.1
Stem bark		
Branches		0.6[a]
Fruits etc.		
Foliage		2.7[a]
Root estimate		

Stand biomass values were obtained by multiplying mean tree values by the number of trees per hectare. Increments were estimated between ages 12 and 14.
a. Litterfall only; the branch increment was not estimated.

Mälkönen, E. (1977). "Annual Primary Production and Nutrient Cycle in a Birch Stand." Communicationes Instituti Forestalis Fenniae (Helsinki) No.91.

61°37'N 24°09'E   160 m   Finland, Orivesi.

Fine sandy   *Betula pubescens* (84%),[a] *Betula verrucosa* (16%)[a]
moraine

Age (years)		40
Trees/ha		1012
Tree height (m)		20.6[b]
Basal area (m²/ha)		18.0
Leaf area index		
Stem volume (m³/ha)		155
Dry biomass (t/ha)	Stem wood	67.1
	Stem bark	11.4
	Branches	8.6
	Fruits etc.	
	Foliage	2.9
	Root estimate	24.1
CAI (m³/ha/yr)		6.1
Net production (t/ha/yr)	Stem wood	2.74
	Stem bark	0.30
	Branches	0.79 + 0.20[c] ⎫
	Fruits etc.	0.06[c] ⎬ + 0.21[c]
	Foliage	2.90[d] ⎭
	Root estimate	1.85

Twenty trees were sampled in August. Six root systems were excavated and fine roots were core sampled. Stand biomass values for a 0.16 ha plot were derived from regressions on D²H. There was 0.25 t/ha of dead branches. The root biomass value includes stumps. Roots were assumed to grow at the same relative rates as aboveground parts. Nutrient contents were determined.
a. Percentage of the tree number (*Betula verrucosa* syn. *B. pendula*).
b. Dominant tree height.
c. Litterfall measured over one year; 0.21 t/ha/yr was miscellaneous litterfall.
d. Biomass of new leaves; leaf litterfall was 1.88 t/ha/yr.

Havas, P. (1981). In: "Dynamic Properties of Forest Ecosystems" (D.E. Reichle, ed.), p. 582. Cambridge University Press, Cambridge, London, New York, Melbourne.

66°22'N 29°00'E   270 m   Finland, Oulu.

Poorly drained   *Picea abies*
podzol,
pH 4.7

Age (years)		260
Trees/ha		550
Tree height (m)		16.2
Basal area (m²/ha)		22.2
Leaf area index		4.9
Stem volume (m³/ha)		

Dry biomass (t/ha)		
	Stem wood	56.6
	Stem bark	10.7
	Branches	17.1 } $+ 0.13^a$
	Fruits etc.	0.2
	Foliage	6.6
	Root estimate	37.5

CAI (m³/ha/yr)

Net production (t/ha/yr)		
	Stem wood	
	Stem bark	} 0.10^b
	Branches	
	Fruits etc.	0.20^b
	Foliage	1.25^b
	Root estimate	

There was 4.2 t/ha of standing dead wood.
a. Understorey shrubs.
b. Litterfall only.

Mälkönen, E. (1974). "Annual Primary Production and Nutrient Cycle in some Scots Pine Stands." Communicationes Instituti Forestalis Fenniae (Helsinki) No.84.

60°31' to 61°40' N   23°51' to 24°19' E   125-140 m   Finland, Tammela.

Podzols with 2 cm humus layer. Fine or coarse sands with stones.

Pinus sylvestris

Age (years)	28	47	45
Trees/ha	2911	845	1420
Tree height (m)	6.4	12.0	15.4
Basal area (m²/ha)	7.7	12.1	19.9
Leaf area index			
Stem volume (m³/ha)	30.2	75.5	148.8
Dry biomass (t/ha) — Stem wood	9.6	27.1	55.6
Stem bark	1.9	3.3	5.3
Branches	3.3	6.8	7.4
Fruits etc.			
Foliage	2.3	3.5	4.4
Root estimate	7.0	11.0	19.3
CAI (m³/ha/yr)	2.3	5.0	5.9
Net production (t/ha/yr) — Stem wood	0.78	1.79	2.35
Stem bark	0.09	0.10	0.13
Branches	0.68 + 0.07[a]	0.77 + 0.19[a]	0.87 + 0.32[a]
Fruits etc.			
Foliage	0.90[b]	1.38[b]	1.74[b]
Root estimate	0.96	1.07	1.29

Twenty trees were sampled in each stand. Root systems were excavated and fine roots were core sampled. Stand biomass values for the above plots of about 0.1 ha were derived from regressions on D or D²H. There was 0.8, 1.2 and 3.2 t/ha of dead branches in columns left to right. Root biomass values include the stumps. Roots were assumed to grow at the same relative rates as above-ground parts. Nutrient contents were determined.

a. Woody litterfall, measured over 2 years.

b. Biomass of new foliage plus the increase in weight of old foliage; foliage litterfall was 0.49, 0.80 and 1.11 t/ha/yr in columns left to right.

Paavilainen, E. (1980). "Effect of Fertilization on Plant Biomass and Nutrient
Cycle on a Drained Dwarf Shrub Pine Swamp." Communicationes Instituti Forestalis
Fenniae (Helsinki) No.98.

62°04'N 24°34'E   5-50 m   Finland, Vilppula, Jaakkoinso

Pinus sylvestris (dwarf trees)

Drained peatlands, pH 3.1-3.6	Fertilizer treatments given 13 years previously			Fertilizer treatments given 13 years before, plus NPK 3 years previously		
	Nil	PK	NPK	Nil	PK	NPK
Age (years)	Mature	Mature	Mature	Mature	Mature	Mature
Trees/ha	688	800	784	592	757	784
Tree height (m)						
Basal area (m²/ha)	16.6	13.1	16.1	13.7	14.0	14.3
Leaf area index						
Stem volume (m³/ha)	116	79	115	99	92	95
Dry biomass (t/ha)						
Stem wood	51.8	36.7	50.3	41.0	41.1	41.4
Stem bark	3.3	2.6	3.3	3.0	3.1	3.1
Branches	10.4	7.4	9.8	8.1	8.6	7.9
Fruits etc.	1.2	0.3	0.9	1.1	0.8	0.9
Foliage	4.1	3.5	4.1	4.0	4.5	4.3
Root estimate	27.8	22.6	27.9	25.3	25.5	27.6
CAI (m³/ha/yr)						
Stem wood	1.81	1.63	1.64	1.98	2.33	2.02
Stem bark	0.29	0.23	0.25	0.27	0.31	0.32
Net production (t/ha/yr)						
Branches	$0.88\big\}$	$0.62\big\}$	$0.83\big\}$	$0.61\big\}$	$0.65\big\}$	$0.60\big\}$
Fruits etc.	$0.58\big\}^{+1.06^a}$	$0.15\big\}^{+0.75^a}$	$0.46\big\}^{+0.63^a}$	$0.56\big\}^{+0.68^a}$	$0.42\big\}^{+0.42^a}$	$0.44\big\}^{+0.79^a}$
Foliage	1.13^b	0.95^b	1.07^b	1.18^b	1.41^b	1.20^b
Root estimate	1.73	1.52	1.64	1.94	2.15	2.10

Three trees were sampled per plot in April-May, including the fine roots. Stand
biomass values for 625-700 m² plots were derived from regressions on D^2H and using
published regressions for stumps and thick roots. Roots were assumed to grow at the
same relative rates as above-ground parts. There was 4.4, 2.7, 4.1, 2.8, 2.5 and
2.6 t/ha of dead branches in columns left to right. Nutrient contents were determined.
a. Non-foliage litterfall (i.e. total litterfall, measured over 4 years, minus new
   foliage biomass).
b. New foliage biomass.

Auclair, D. and Métayer, S. (1980). Méthodologie de l'évaluation de la biomasse aérienne sur pied et de la production en biomasse des taillis. *Acta Oecologica/Oecol. Applic.* <u>1</u>, 357-377.

ca.47°50'N 1°50'E   50-200 m   France, near Orléans.

Coppices

	Carpinus betulus	*Betula pubescens*
Age (years)	35	25
Trees/ha	5604[a]	6849[a]
Tree height (m)	11.8	11.4
Basal area (m²/ha)	23.9	22.8
Leaf area index		
Stem volume (m³/ha)		

		Carpinus betulus	*Betula pubescens*
Dry biomass (t/ha)	Stem wood	} 109	} 80
	Stem bark		
	Branches		
	Fruits etc.		
	Foliage	3.5	2.8
	Root estimate		

CAI (m³/ha/yr)

		Carpinus betulus	*Betula pubescens*
Net production (t/ha/yr)	Stem wood	} 5.5[b]	} 5.6[b]
	Stem bark		
	Branches		
	Fruits etc.		
	Foliage	3.5	2.8
	Root estimate		

163 stems were sampled and stand biomass values for the above plots of about 0.1 ha were derived from regressions on stem circumference.
a. Number of stems per hectare, with several stems per root system.
b. Excluding woody litterfall and mortality.

Lemée, G. (1978). La hêtraie naturelle de Fontainebleau. In: "Problèmes d'Ecologie: Structure et Fonctionnement des Ecosystèmes Terrestres" (M. Lamotte and F. Bourlière, eds), pp.75-127. Masson, Paris, New York, Barcelona and Milan.

48°24'N 2°42'E   135 m   France, Fontainebleau, 50 km SE of Paris.

Mixtures of brown leached soils and podzols, pH 3.4-4.8		*Fagus sylvatica* Means of large woodland areas	
	7 plots of 2500 m²	'Tillaie'	'Gros-Fouteau'
Age (years)	150-270	150-270	150-270
Trees/ha	ca.350	ca.350	ca.350
Tree height (m)	24-40	24-40	24-40
Basal area (m²/ha)	ca.27.5	ca.27.5	ca.27.5
Leaf area index	ca. 6.6	ca. 6.6	ca. 6.6

Stem volume (m³/ha)

Dry biomass (t/ha)

	7 plots of 2500 m²	'Tillaie'	'Gros-Fouteau'
Stem wood	} 232	} 185	} 195 } + 4.8[a]
Stem bark			
Branches	58	46	48
Fruits etc.	0.5		
Foliage	3.5[b]	3.0[b]	3.2[b] + 0.7[a]
Root estimate	49	39	41

CAI (m³/ha/yr)

Net production (t/ha/yr)

Stem wood	}
Stem bark	} $4.76^c + 1.06^d$
Branches	}
Fruits etc.	0.50^d
Foliage	3.45^b
Root estimate	0.80

Twenty trees were sampled and many windthrown trees were measured. Stand biomass values were derived from regressions on D. Root biomass was assumed to be 17% of the above-ground biomass. Nutrient contents were determined.
a. Understorey shrubs, mostly *Ilex* spp.
b. Leaf litterfall multiplied by 1.1 to account for losses by decay and consumption.
c. Including estimated mortality.
d. Litterfall.

Lossaint, P. and Rapp, M. (1978). La forêt Méditerranéenne de chênes verts (*Quercus ilex* L.). In: "Problèmes d'Ecologie: Structure et Fonctionnement des Ecosystèmes Terrestres" (M. Lamotte and F. Bourlière, eds) pp. 129-185. Masson, Paris, New York, Barcelona and Milan.

Auclair, D. and Métayer, S. (1980). Méthodologie de l'évaluation de la biomasse aérienne sur pied et de la production en biomasse des taillis. *Acta Oecologica/ Oecol. Applic.* 1, 357-377.

France	43°36'N 3°53'E 185 m 14 km NE of Montpellier, Roquet	ca.47°50'N 1°50'E 50-200 m near Orléans
	Quercus ilex	*Quercus robur*
	Infertile, red-yellow soils and rendzinas pH 6.7-7.8	Coppice (Auclair and Métayer 1980)
Age (years)	ca.150	40
Trees/ha	1400	3064[a]
Tree height (m)	10-12	8.9
Basal area (m²/ha)	38.8	20.4
Leaf area index	4.4	
Stem volume (m³/ha)		

Dry biomass (t/ha)

Stem wood		
Stem bark	$\left.\right\}$ 235	
Branches	27[c] $\left.\right\}$ + 1.9[b]	$\left.\right\}$ 67
Fruits etc.	1	
Foliage	7.0 + 0.3[b]	2.2
Root estimate	40-50	

CAI (m³/ha/yr)

Net production (t/ha/yr)

Stem wood		
Stem bark	$\left.\right\}$ 2.2 + 0.8[d]	$\left.\right\}$ 2.7[f]
Branches		
Fruits etc.	0.8	
Foliage	4.5[e]	2.2
Root estimate		

At Roquet an area with 216 trees was clear-felled and the biomass of all trees was determined. Roots were excavated to 40 cm depth in several 25 cm² soil blocks. Nutrient contents were determined. At Orléans many stems were sampled and stand biomass values for a plot of about 0.1 ha were derived from regressions on stem circumference.

a. Number of stems/ha.
b. Understorey shrubs.
c. Including 2 t/ha of sprouts (suckers).
d. Woody litterfall.
e. New foliage biomass; leaf litterfall was 2.5 t/ha/yr.
f. Excluding woody litterfall and mortality.

Mounet, J.P. (1978). "Production de quelques Ecosystèmes à Chêne Pubescent: Evaluation de la Biomasse Aérienne des Chênes Pubescents." Ph.D. thesis. Université de
Droit d'Economie et des Sciences, Aix-Marseille.

44°31'N 5°05'E   (alt. given below)   France, Saint Maurice, near Dieulefit.

Calcareous   *Quercus pubescens* (syn. *Quercus lanuginosa*)
soils

	570 m	660 m	670 m	730 m	860 m	880 m
Age (years)						
Trees/ha	4800	6600	6300	6200	4550	1240
Tree height (m)	3.4-6.5	2.2-3.8	1.6-5.8	2.4-4.8	4.3-8.2	5.6-10.7
Basal area (m²/ha)	21.7	17.6	23.3	22.2	28.0	35.4
Leaf area index						
Stem volume (m³/ha)						

Dry biomass (t/ha)

	Stem wood	Stem bark	Branches	Fruits etc.	Foliage	Root estimate

570 m	660 m	670 m	730 m	860 m	880 m
66.8	46.3	60.6	63.6	76.7	158.1

CAI (m³/ha/yr)

Net production (t/ha/yr)

	Stem wood	Stem bark	Branches	Fruits etc.	Foliage	Root estimate

---

Sixty-seven trees were sampled, and stand biomass values were obtained by multiplying mean tree values by the numbers of trees per hectare separately for different diameter classes.

Ranger, J. (1978). Recherches sur les biomasses comparées de deux plantations de Pin laricio de Corse avec ou sans fertilisation. *Ann. Sci. For.* <u>35</u>, 93–115.

46°45'N 0°20'E ca.100 m France, Vienne, 20 km N of Poitiers, Moulières.		
Plantations	*Pinus nigra* ssp. *laricio*, *Quercus robur*, *Picea abies* (80%)[a]	(98%)[a]
	Received fertilizers before planting	Unfertilized
Age (years)	15	15
Trees/ha	3780[b]	3443
Tree height (m)	7.9	5.7
Basal area (m²/ha)	36.6	19.2
Leaf area index		
Stem volume (m³/ha)		
Dry biomass (t/ha) Stem wood	64.0[bc]	31.4[c]
Stem bark	13.0[bc]	7.7[c]
Branches	25.7[b]	8.1
Fruits etc.		
Foliage	13.8	5.6
Root estimate	8.6	2.8
CAI (m³/ha/yr)		
Net production (t/ha/yr) Stem wood		
Stem bark		
Branches		
Fruits etc.		
Foliage		
Root estimate		

Ten trees were sampled per treatment in October, and roots were excavated. Stand values for the above two 0.045 ha plots were derived from regressions on stem circumference.
a. Percentage of the total basal area accounted for by *P. nigra*.
b. Including 668 *P. nigra* felled at age 13.
c. Including the stumps.

Mounet, J.P. (1978). "Production de quelques Ecosystèmes à Chêne Pubescent: Evaluation de la Biomasse Aérienne des Chênes Pubescents." Ph.D. thesis. Université de Droit d'Economie et des Sciences, Aix-Marseille.

---

ca.44°30'N   5°30'E   400–700 m   France, ca. 50 km S of Grenoble.

Pinus sylvestris

Age (years)						
Trees/ha	2250	3600	1600	640	2750	1100
Tree height (m)	3.5–7.1	2.5–7.3	2.0–3.8	1.6–5.1	4.0–7.6	5–15
Basal area (m²/ha)	27.6	28.1	8.5	7.6	26.2	33.1
Leaf area index						
Stem volume (m³/ha)						

Dry biomass (t/ha)							
	Stem wood						
	Stem bark						
	Branches	77.1	37.3	23.7	18.9	122.0	108.4
	Fruits etc.						
	Foliage						
	Root estimate						

CAI (m³/ha/yr)

Net production (t/ha/yr)	
	Stem wood
	Stem bark
	Branches
	Fruits etc.
	Foliage
	Root estimate

Fifty-four trees were sampled and stand values were obtained by multiplying mean tree values by the numbers of trees per hectare separately for different diameter classes.

Cabanettes, A. (1979). "Croissance, Biomasse et Productivité de *Pinus pinea* L. en Petite Camargue." Ph.D. thesis. Académie de Montpellier, Université des Sciences et Techniques du Languedoc, France.

Cabanettes, A. and Rapp, M. (1978). Biomasse, minéralomass et productivité d'un écosystème à Pins pignon (*Pinus pinea* L.) du littoral méditerranéen. I Biomasse. *Oecol. Plant.* 13, 271-286.

Rapp, M. and Cabanettes, A. (1980). (As above) II Composition chimique et minéralomasse. *Oecol. Plant.* 15, 151-164.

France	43°40'N 4°15'E 50 m 35 km E of Montpellier, Saint Laurent d'Aigouze *Pinus pinea* Unmanaged stand Sandy soil, pH 7.8-8.4	ca.48°45'N 6°30'E -- 15 km NE of Nancy *Pseudotsuga menziesii* Unthinned plantation of Yelm, Washington provenance (Oswald and Pardé 1981, personal communication)
Age (years)	up to 35	24
Trees/ha	800	1500
Tree height (m)	10.4	18.8
Basal area (m²/ha)	33.9	44.8
Leaf area index		
Stem volume (m³/ha)		379
Dry biomass (t/ha) — Stem wood		
Dry biomass (t/ha) — Stem bark	144.1[a]	158
Dry biomass (t/ha) — Branches		21
Dry biomass (t/ha) — Fruits etc.		
Dry biomass (t/ha) — Foliage	12.7	14
Dry biomass (t/ha) — Root estimate	22.0	
CAI (m³/ha/yr)	6.9	
Net production (t/ha/yr) — Stem wood		
Net production (t/ha/yr) — Stem bark	$5.25 + 0.85$[b]	
Net production (t/ha/yr) — Branches		
Net production (t/ha/yr) — Fruits etc.	0.84[b]	
Net production (t/ha/yr) — Foliage	$0.45 + 6.13$[b] $+ 0.81$[c]	
Net production (t/ha/yr) — Root estimate	1.1	

Nineteen *P. pinea* were sampled in this and previous studies and one root system and one soil monolith were excavated. Stand biomass values were derived from regressions on tree circumference. Roots were assumed to have the same relative growth rates as above-ground parts. Nutrient contents were determined.

Three trees of *P. menziesii* were sampled within each of 2 plots, and stand values were obtained by multiplying mean tree values by the number of trees per hectare. There was 19 t/ha of dead branches.

a. Including 97.5 t/ha of 'bois fort' and 17.0 t/ha of bark.

b. Litterfall, measured over 4 years.

c. Pre-fall losses.

Ellenberg, H. (ed.) (1971). "Integrated Experimental Ecology. Methods and Results of Ecosystem Research in the German Solling Project." (Chapters B, D and E) Chapman and Hall Ltd, London; Springer-Verlag, Berlin, Heidelberg and New York.

Ellenberg, H. (1981). In: "Dynamic Properties of Forest Ecosystems" (D.E. Reichle, ed.) pp. 666-668. Cambridge University Press, Cambridge, London, New York and Melbourne.

51°45-49'N   9°34-36'E   430-500 m   Germany, Federal Republic, 55 km NW of Göttingen, Solling plateau.

Brown earths
pH 3.5-3.7   *Fagus sylvatica*

Age (years)	59	80	122
Trees/ha	3620	1190	243
Tree height (m)	15.1[a]	20.3[a]	26.5[a]
Basal area (m²/ha)	30.4	25.2	28.3
Leaf area index	6.5[b]	6.7[b]	5.9[b]
Stem volume (m³/ha)	159[c]	219[c]	348[c]

Dry biomass (t/ha)				
Stem wood	102.9	121.2	222.9
Stem bark	7.2	8.4	15.5
Branches	41.5	25.9	32.5
Fruits etc.	0.3	0.3	0.4
Foliage	3.2	3.3	3.1
Root estimate	24.0[d]	22.1[d]	30.0[d]

CAI (m³/ha/yr)

Net production (t/ha/yr)				
Stem wood	7.17 ⎫	5.52 ⎫	6.07 ⎫
Stem bark	0.50 ⎬ + 0.40[e]	0.38 ⎬ + 0.35[e]	0.42 ⎬ + 0.53[e]
Branches	0.99 ⎭	0.45 ⎭	0.78 ⎭
Fruits etc.	0.25[e]	0.28[e]	0.77[e]
Foliage	3.16[e] + 0.17[f]	3.27[e] + 0.18[f]	2.98[e] + 0.19[f]
Root estimate	1.26[d]	0.63[d]	0.66[d]

Twenty-seven trees were sampled in winter. Stand biomass values were derived from regressions on D²H. The 122-year-old trees were measured in a 1.0 ha plot; the other plots were 0.1 ha. There was 0.7 t/ha of standing dead wood in the 122-year-old stand.

a. Stand heights
b. Estimated in April-May over 2-3 years.
c. Volumes of stems over 7 cm D measured two years previously (from Ellenberg 1971).
d. Roots over 5 mm diameter only.
e. Litterfall, measured over 2-3 years (woody litterfall from Ellenberg 1971), excluding any mortality.
f. Consumption.

Ellenberg, H. (1981). In: "Dynamic Properties of Forest Ecosystems" (D.E. Reichle, ed.) pp. 669-671. Cambridge University Press, Cambridge, London, New York, and Melbourne.

Droste zu Hülshoff, B. von (1970). Struktur, Biomasse und Zuwachs eines älteren Fichtenbestandes. *Forstwiss. ZentBl.* <u>89</u>, 162-171.

Germany, Federal Republic. (alt. given below)	51°44-49'N 9°34-35'E 55 km NW of Göttingen, Solling plateau			48°04'N 11°59'E near München Ebersberger Forest
Plantations Brown forest soils pH 3.2-4.6	*Picea abies* (Ellenberg 1981)			(Droste zu Hülshoff 1970)
	390 m	505 m	440 m	ca.600 m
Age (years)	34	87	115	76
Trees/ha	1490	595	300	800
Tree height (m)	17.5[a]	24.9[a]	31.3[a]	28
Basal area (m²/ha)	35.5	44.8	37.7	57.3
Leaf area index				9.4[e]
Stem volume (m³/ha)				728
Dry biomass (t/ha) — Stem wood	96.7	182.5	180.1	} 268.0
Stem bark	8.4	15.9	15.7	
Branches	18.7	28.2	24.6	38.3
Fruits etc.				
Foliage	18.9	17.9	12.7	15.9
Root estimate	34.6[b]	71.7[b]	74.9[b]	
CAI (m³/ha/yr)				16
Net production (t/ha/yr) — Stem wood	4.50[c]	4.93[c]	3.68[c]	} 5.88[c]
Stem bark	0.39[c]	0.43[c]	0.32[c]	
Branches	0.63[c]	0.60[c]	0.39[c]	3.29[c]
Fruits etc.				
Foliage	2.92[d]	3.39[d]	3.08[d]	6.34
Root estimate	1.59[b]		0.85[b]	

At Solling many trees were sampled and stand biomass values were derived from regressions on D and D²H. The plots were 0.10, 1.00 and 0.25 ha in columns left to right. There was 0.3 t/ha of standing dead wood in the 87-year-old stand.
At Ebersberger Forest five trees in each of several size classes were sampled and stand biomass values for a 0.12 ha plot were obtained by multiplying by the number of trees per hectare in each size class. Branch and foliage increments were estimated by dividing their biomasses by their mean ages.
a. Stand heights.   *b.* Roots over 5 mm diameter only.
c. Excluding woody litterfall and any mortality.
d. Foliage litterfall.
e. All-sided LAI was 21.6.

Greenland, D.J. and Kowal, J.M.L. (1960). Nutrient content of the moist tropical forest of Ghana. *Pl. Soil* <u>12</u>, 154-174.

Nye, P.H. (1961). Organic matter and nutrient cycles under moist tropical forest. *Pl. Soil* <u>13</u>, 333-346.

John, D.M. (1973). Accumulation and decay of litter and net production of forest in tropical West Africa. *Oikos* <u>24</u>, 430-435.

6°08'N 0°51'W   150 m   Ghana, Kade.

Reddish yellow latosols   *Diospyros* spp., *Strombosia glaucescens et al.* (35 spp.)

Semi-deciduous secondary tropical rainforest

Age (years)	30-50
Trees/ha	5305[a]
Tree height (m)	15, 20-40, 50-60[b]
Basal area (m²/ha)	33.3
Leaf area index	
Stem volume (m³/ha)	

Dry biomass (t/ha)		
	Stem wood	
	Stem bark	
	Branches	232.8[c]
	Fruits etc.	
	Foliage	
	Root estimate	54.1

CAI (m³/ha/yr)

Net production (t/ha/yr)		
	Stem wood	
	Stem bark	
	Branches	3.5[d] (or 1.9)[e]
	Fruits etc.	0.4[d] (or 0.4)[e]
	Foliage	7.0[d] (or 7.4)[e]
	Root estimate	

All trees were felled within two areas of about 0.24 ha and all were weighed apart from a few large trees whose volumes and densities were estimated. Stumps were excavated on half of each plot, and roots were excavated in 14 areas of 1.35 m². Nutrient contents were determined.

a. Trees over 7.6 cm D.

b. Lower storey, mid-storey, and emergents.

c. Including 173.2 t/ha of living wood, 14.5 t/ha lianes, 25.5 t/ha leaves and twigs and 19.6 t/ha above-ground stumps.

d. Litterfall only.

e. Litterfall measured by John (1973), where 1.9 is the sum of small wood and miscellaneous litter.

Elkington, T.T. and Jones, B.M.G. (1974). Biomass and primary productivity of birch (*Betula pubescens* S. Lat.) in south-west Greenland. *J. Ecol.* <u>62</u>, 821-830.

61°06'N   45°58'W   120-135 m   Greenland, Nordre Sermilik Fjord, Eqaluit.

Weathered granitic sand. 22° N-facing sheltered slope   *Betula pubescens (B. pubescens* × *B. glandulosa)*

Age (years)	79^a
Trees/ha	2150^b
Tree height (m)	2.5 (1.1-3.6)
Basal area (m²/ha)	
Leaf area index	1.57
Stem volume (m³/ha)	

Dry biomass (t/ha)		
	Stem wood	$\left.\right\}11.5^b$
	Stem bark	
	Branches	35.8
	Fruits etc.	
	Foliage	1.2
	Root estimate	6.4

CAI (m³/ha/yr)		
Net production (t/ha/yr)	Stem wood	$\left.\right\}0.14$
	Stem bark	
	Branches	0.64^c
	Fruits etc.	
	Foliage	1.20
	Root estimate	0.09

Fifteen trees were sampled in August and roots were excavated. Stand biomass values for a 330 m² plot were derived from regressions on stem and branch circumferences.

a. Mean age of sampled horizontal stems (caudices).

b. Caudices, from which vertical branches arose; there were 7700 such branches per hectare.

c. Excluding woody litterfall.

Kunkel-Westphal, I. and Kunkel, P. (1979). Litterfall in a Guatemalan primary forest, with details of leaf-shedding by some common tree species. *J. Ecol.* <u>67</u>, 665-686.

15°30'N 90°27'W (alt. given below)   Guatemala, Sierra de Chamá, NW of Cobán.

Acid, sandy   *Calaphyllum* sp., *Quercus* spp., *Swartzia* sp. *et al.* (147 species)
loam over
clay loam   Lower montane rainforest

	900 m	1000–1015 m
Age (years)	over 70	over 70
Trees/ha		
Tree height (m)	4, 15-20, 30-40^a	4, 15-20, 30-40^a
Basal area (m²/ha)	46	40
Leaf area index		
Stem volume (m³/ha)		

		900 m	1000–1015 m
Dry biomass (t/ha)	Stem wood		
	Stem bark		
	Branches	457-499	324-353
	Fruits etc.		
	Foliage		
	Root estimate		

		900 m	1000–1015 m
CAI (m³/ha/yr)			
Net production (t/ha/yr)	Stem wood		
	Stem bark		
	Branches	1.46^b	2.10^b
	Fruits etc.	0.11^b	0.04^b
	Foliage	6.74^b + 1.27^c	7.30^b + 1.42^c
	Root estimate		

Stand biomass values of the above 0.1 ha plots were derived from tree basal areas and heights using regressions published by Edwards and Grubb (1977).
a. Lower storey, mid-storey and emergents.
b. Litterfall only, measured over 2 years.
c. Loss from decomposition.

Jakucs, P. (1981). In: "Dynamic Properties of Forest Ecosystems" (D.E. Reichle, ed.) p. 586. Cambridge University Press, Cambridge, London, New York, Melbourne.

47°54'N 20°28'E   250-280 m   Hungary, Sikfokut.

Brown forest soil, pH 4.0-6.2

Quercus petraea and *Quercus cerris* with understorey shrubs

Age (years)	65-68
Trees/ha	
Tree height (m)	17.4
Basal area (m²/ha)	15.1
Leaf area index	8.1
Stem volume (m³/ha)	

Dry biomass (t/ha)

Stem wood	$\left.\begin{array}{c} \\ \end{array}\right\}140.4$ $\left.\begin{array}{c} \\ \end{array}\right\} + 3.8^{a}$
Stem bark	
Branches	59.1
Fruits etc.	
Foliage	$3.4 + 0.3^{a}$
Root estimate	35.6

CAI (m³/ha/yr)

Net production (t/ha/yr)

Stem wood	$\left.\begin{array}{c} \\ \\ \end{array}\right\} 3.00 + 0.46^{b}$
Stem bark	
Branches	
Fruits etc.	
Foliage	3.71^{b}
Root estimate	

---

a. Understorey shrubs.
b. Litterfall.

Singh, R.P. and Sharma, V.K. (1976). Biomass estimation in five different aged
  plantations of *Eucalyptus tereticornis* Smith in western Uttar Pradesh. In: "Oslo
  Biomass Studies" pp. 145-161. College of Life Sciences and Agriculture, University of Maine, Orono, USA.

---

ca.29°N 77°W   200-400 m   India, Uttar Pradesh.

Plantations.
Canal-side   *Eucalyptus tereticornis*
plantings.

Age (years)	5	6	7	8	9
Trees/ha	1670	1110	700	1360	840
Tree height (m)					
Basal area (m²/ha)	18^a	15^a	38^a	50^a	42^a
Leaf area index					
Stem volume (m³/ha)					

		5	6	7	8	9
Dry biomass (t/ha)	Stem wood	}53.7	}41.3	}54.9	}101.0	}139.2
	Stem bark					
	Branches	10.1	6.3	11.4	28.1	30.9
	Fruits etc.					
	Foliage	6.7	3.4	4.6	16.1	8.0
	Root estimate	10.6	9.7	10.9	22.2	18.6

CAI (m³/ha/yr)

		5	6	7	8	9
Net production (t/ha/yr)	Stem wood					
	Stem bark					
	Branches					
	Fruits etc.					
	Foliage					
	Root estimate					

Nine trees were sampled in each plantation in November-December, and roots were
excavated. Stand biomass values for ten 100 m² plots in each plantation were
derived from regressions on D^2H, H, and stem diameter measured at ground level and
at the base of the crowns.
a. Estimated from the authors' data on numbers of trees in three stem girth classes.

Foruqi, Q. (1981). In: Dynamic Properties of Forest Ecosystems" (D.E. Reichle, ed.) pp. 589-591. Cambridge University Press, Cambridge, London, New York, Melbourne.

---

ca.27°N 83°53'E   81 m   India, Uttar Pradesh, Gorakhpur Forest.

Plantations.   *Shorea robusta*

Age (years)	5	8	14	26	30	40
Trees/ha	3150	2568	1660	1620	1496	1134
Tree height (m)	4.7	7.4	11.1	15.5	17.2	21.2
Basal area (m²/ha)	7.6	18.7	20.4	46.3	58.6	66.6
Leaf area index	2.1	5.8	6.3	6.9	11.4	13.4
Stem volume (m³/ha)						

Dry biomass (t/ha)

	5	8	14	26	30	40
Stem wood / Stem bark	}11.6	}45.2	}73.1	}190.9	}292.3	}482.2
Branches	1.4	7.4	8.7	17.6	30.1	42.2
Fruits etc.						
Foliage	2.5	6.8	6.2	8.1	13.5	15.8
Root estimate	4.4	15.9	16.9	55.1	63.3	100.9

CAI (m³/ha/yr)

Net production (t/ha/yr)

	5	8	14	26	30	40
Stem wood						
Stem bark						
Branches						
Fruits etc.						
Foliage	2.3[a]	5.9[a]	5.8[a]	7.9[a]	8.9[a]	9.9[a]
Root estimate						

Three trees of average size were sampled in each stand. Stand biomass values for the above 0.4 ha plots were obtained by multiplying mean tree values by the numbers of trees per hectare. Nutrient contents were determined.
a. Leaf litterfall.

Ramam, S.S. (1981). In: "Dynamic Properties of Forest Ecosystems" (D.E. Reichle, ed.) pp. 595-597. Cambridge University Press, Cambridge, London, New York, Melbourne.

---

ca.27°N 83°30'E   81 m   India, Uttar Pradesh, Gorakhpur Forest.

Plantations.   *Shorea robusta*

Age (years)	10	16	22	28	35	38
Trees/ha	2229	1461	1687	1594	1741	742
Tree height (m)	8.6	12.3	14.2	17.1	17.6	20.9
Basal area (m²/ha)	15.2	14.0	19.9	31.2	41.6	31.6
Leaf area index	6.4	6.2	5.8	11.7	14.0	8.5
Stem volume (m³/ha)						

Dry biomass (t/ha)

	10	16	22	28	35	38
Stem wood / Stem bark	} 31.7	} 37.9	} 96.9	} 145.5	} 226.9	} 207.0
Branches	3.2	4.5	3.7	16.4	14.5	33.0
Fruits etc.						
Foliage	3.6	3.5	3.7	5.5	7.0	6.4
Root estimate	12.2	13.7	23.6	38.8	62.2	47.5

CAI (m³/ha/yr)

Net production (t/ha/yr)

	10	16	22	28	35	38
Stem wood						
Stem bark						
Branches						
Fruits etc.						
Foliage	2.90[a]	2.60[a]	2.80[a]	3.80[a]	5.10[a]	4.90[a]
Root estimate						

Three trees of average size were sampled in each stand. Stand biomass values for the above 0.4 ha plots were obtained by multiplying mean tree values by the numbers of trees per hectare. Nutrient contents were determined.
a. Leaf litterfall.

Singh, R.P. (1975). Biomass, nutrient and productivity structure of a stand of dry deciduous forest of Varanasi. *Trop. Ecol.* <u>16</u>, 104-109.

Misra, R. (1972). A comparative study of net primary productivity of dry deciduous forest and grassland of Varanasi, India. In: "Tropical Ecology, with an Emphasis on Organic Productivity." (P.M. Golley and F.B. Golley, eds) pp. 279-293. Institute of Ecology, University of Georgia, Athens, USA.

Sharma, V.K., Singh, K.P. and Bandhu, D. (1981). In: "Dynamic Properties of Forest Ecosystems" (D.E. Reichle, ed.) pp. 587-588. Cambridge University Press.

---

ca.25°20'N 83°00'E   300-350 m   India, Varanasi, Chakia Forest.

Alluvial sandy
to clay loams.
pH 5.7-6.8

Shorea robusta, with *Buchanania lanzan*,
Anogeissus latifolia et al.

Tropical dry deciduous forest

Age (years)	38	10 to 120	60
Trees/ha	729	1019	664
Tree height (m)	3.0-12.7	2.0-15.0	5.0-18.0
Basal area (m²/ha)	11.4	12.8	30.6
Leaf area index			6
Stem volume (m³/ha)			

Dry biomass (t/ha)

Stem wood	} 20.5	} 29.1	} 133.9 (or 141.1)[c]
Stem bark			
Branches	5.8	11.6	54.3 (or 58.2)[c]
Fruits etc.			0.1
Foliage	2.4	3.6	5.8 (or 6.2)[c]
Root estimate	6.7	9.5	32.7 (or 34.3)[c]

CAI (m³/ha/yr)

Net production (t/ha/yr)

Stem wood		} 1.01[b]	} 5.76
Stem bark			
Branches	0.58[a]	0.40[b]	2.33 }+ 1.12[a]
Fruits etc.			
Foliage	1.51[a]	0.33 + 3.03[a]	5.66[a]
Root estimate			0.29

Regressions methods were used to estimate stand biomass values. Nutrient contents were determined.

a. Litterfall.

b. Excluding woody litterfall and any mortality.

c. Alternative values from Misra (1972); unbracketed values in this column are from Sharma *et al.* (1981).

Singh, K.P. and Misra, R (eds) (1979). "Structure and Functioning of Natural, Modified and Silvicultural Ecosystems of Eastern Uttar Pradesh." Tech. Report to UNESCO, MAB. Banaras Hindu University, Varanasi - 221005, India.

Singh, A.K., Pandey, V.N. and Misra, K.N. (1980). Stand composition and phytomass distribution of a tropical deciduous teak (*Tectona grandis*) plantation of India. *J. Jap. For. Soc.* <u>62</u>, 128-137.

India, Uttar Pradesh, Varanasi Forest.

Plantations. Impoverished, reddish-brown, leached, sandy loams.	24°52-58'N 83°3-12'E 140-380 m		25°03'N 83°13'E ca.50 m
		Tectona grandis	
	Fenced stand	Unfenced stand	(Singh *et al.* 1980)
Age (years)	15	15	15
Trees/ha	467	467	305
Tree height (m)	3.5-10.6	3.5-10.6	7.8
Basal area (m²/ha)			3.6
Leaf area index	1.11[a]	0.96[a]	
Stem volume (m³/ha)			
Dry biomass (t/ha) Stem wood	} 8.71	} 3.92	} 7.66
Stem bark			
Branches	3.43	1.44	2.30
Fruits etc.			
Foliage	2.10	0.95	1.71
Root estimate	3.29	1.52	2.85
CAI (m³/ha/yr)			
Net production (t/ha/yr) Stem wood	} 1.20 }b	} 0.45 }b	
Stem bark			
Branches	0.61	0.19	
Fruits etc.			
Foliage	2.10	0.95	
Root estimate	0.46	0.15	

Fifteen trees were sampled in each study in December-February and roots were excavated. Stand biomass values for five 500 m² plots per stand were derived by Singh and Misra (1979) from regressions on D and by Singh *et al.* (1980) by multiplying mean tree values in each of 5 girth classes by the numbers of trees per class.
a. Maximum seasonal value attained in October.
b. Including woody litterfall, totalling 0.31 and 0.45 t/ha/yr in columns left and right, respectively; corresponding leaf litterfall was 1.57 and 1.12 t/ha/yr.

Foruqi, Q. (1981). In: "Dynamic Properties of Forest Ecosystems" (D.E. Reichle, ed.) pp. 592-594. Cambridge University Press, Cambridge, London, New York, Melbourne.

---

ca.27°N 83°53'E   81 m   India, Uttar Pradesh, Gorakhpur Forest.

Plantations.   *Tectona grandis*

Age (years)	5	8	14	26	30	40
Trees/ha	2068	1943	1022	791	682	545
Tree height (m)	9.2	9.9	13.2	17.1	17.8	19.2
Basal area (m²/ha)	15.3	25.3	32.6	40.6	54.9	97.9
Leaf area index	8.6	14.6	10.6	11.5	17.4	17.2
Stem volume (m³/ha)						
Dry biomass (t/ha)						
Stem wood / Stem bark	}33.0	}63.1	}119.9	}200.0	}237.8	}451.7
Branches / Fruits etc.	6.6	21.0	26.7	52.2	43.8	133.7
Foliage	10.0	16.8	12.3	13.3	20.3	20.0
Root estimate	11.5	27.6	31.2	64.7	75.0	137.9
CAI (m³/ha/yr)						
Net production (t/ha/yr)						
Stem wood						
Stem bark						
Branches						
Fruits etc.						
Foliage	6.0[a]	10.9[a]	10.8[a]	11.2[a]	14.1[a]	14.3[a]
Root estimate						

Three trees of average size were sampled in each stand. Stand biomass values for the above 0.4 ha plots were obtained by multiplying mean tree values by the numbers of trees per hectare. Nutrient contents were determined.
a. Leaf litterfall.

Singh, K.P. and Misra, R. (eds) (1979). "Structure and Functioning of Natural,
  Modified and Silvicultural Ecosystems of Eastern Uttar Pradesh." Tech. Report to
  UNESCO, MAB. Banaras Hindu University, Varanasi-221005, India.

24°52-58'N 83°3-12'E   140-380 m   India, Uttar Pradesh, Varanasi Forest.

	Anogeissus latifolia, Diospyros melanoxylon, Buchanania lanzan, Pterocarpus marsupium et al., with understorey shrubs	
Reddish-brown, leached, sandy loams.	Tropical dry deciduous forest	
	Fenced stand (24%, 10%, 13%, 10%)^a	Unfenced stand (15%, 13%, 12%, 14%)^a
Age (years)		
Trees/ha	1174 + 132^b	936 + 554^b
Tree height (m)	5-6, 7-9, 16-17^c	5-6, 7-9, 16-17^c
Basal area (m² /ha)	18.0 + 0.5^b	15.1 + 2.2^b
Leaf area index	3.50^d+ 0.26^{bd}	2.93^d+ 0.72^{bd}
Dry biomass (t/ha) Stem volume (m³ /ha)		
Stem wood	} 32.4 + 0.8^b	} 29.3 + 2.3^b
Stem bark		
Branches	39.2 + 0.6^b	32.5 + 1.7^b
Fruits etc.		
Foliage	4.7 + 0.2^b	4.1 + 0.6^b
Root estimate	20.7 + 0.6^b	16.9 + 1.7^b
Net production (t/ha/yr) CAI (m³ /ha/yr)		
Stem wood	} 1.77 + 0.07^b } e	} 1.16 + 0.10^b } e
Stem bark		
Branches	2.63 + 0.05^b }	2.26 + 0.10^b }
Fruits etc.		
Foliage	4.75 + 0.23^b	4.13 + 0.64^b
Root estimate	3.40 + 0.14^b	2.81 + 0.27^b

A total of 211 trees were sampled in December-February, and thick roots were excavated. Fine roots were excavated in 5 soil monoliths. Stand biomass values for six
500 m² plots per stand were derived from regressions on D. Nutrient contents were
determined.
a. Percentage of the total basal area of the tree layer occupied by *A. latifolia,*
  D. melanoxylon, B. lanzan and *P. marsupium*, respectively, written left to right
  within the brackets.   *b.* Understorey shrubs.
c. Lower, middle and upper storeys.   *d.* Maximum values, attained in October.
e. Including woody litterfall, totalling (for both storeys) 1.44 and 1.18 t/ha/yr
  in columns left and right, respectively; corresponding total leaf litterfall was
  4.25 and 3.87 t/ha/yr.

Carey, M.L. and O'Brien, D. (1979). Biomass, nutrient content and distribution in a stand of Sitka spruce. *Irish For.* 1, 25-35.

Carey, M.L. and Farrell, E.P. (1978). Production, accumulation and nutrient content of Sitka spruce litterfall. *Irish For.* 35, 35-44.

53°00'N 6°30'W   (alt. given below)   Ireland, County Wicklow.

Picea sitchensis

Plantations.

	Peaty gley pH 3.8-5.1	Thinned 2 years previously Brown podzols pH 3.8-5.1	
	Glenmalure 350 m	Glenealy 200 m	Ballinglen 300 m
Age (years)	33	39	47
Trees/ha	3760	1216	583
Tree height (m)	15.9[a]	15.7[a]	24.6[a]
Basal area (m²/ha)	74.7	35.5	38.8
Leaf area index			
Stem volume (m³/ha)			

		Glenmalure		
Dry biomass (t/ha)	Stem wood	201.5		
	Stem bark	17.7		
	Branches	34.5		
	Fruits etc.			
	Foliage	14.6		
	Root estimate	56.5		

CAI (m³/ha/yr)

Net production (t/ha/yr)	Stem wood			
	Stem bark			
	Branches			
	Fruits etc.	8.86[b]	4.04[b]	3.85[b]
	Foliage			
	Root estimate			

Eight trees were sampled from Glenmalure and roots were excavated. Stand biomass values for three 0.1 ha plots were derived from regressions on basal area per tree. Nutrient contents were determined.
a. Top heights.
b. Total litterfall measured over one year.

Visona, L., Naviglio, L., Simonetto, L., Azzollini, I. and Giovannardi, R. (1975). Researches on beech forest. I Structure and biomass of the beechwood in the Mount Terminillo IBP Station, Monit Reatini, Lazio. *Annal. Bot.* <u>34</u>, 143-170.

42°30'N 13°00'E   1709 m   Italy, Lazio, Mount Terminillo.

Brown mull-type forest soil, pH 5.5	*Fagus sylvatica* Coppiced stand
Age (years)	60
Trees/ha	3590
Tree height (m)	10.3
Basal area (m² /ha)	41.3
Leaf area index	
Stem volume (m³ /ha)	

		Dry biomass (t/ha)
Dry biomass (t/ha)	Stem wood	} 108
	Stem bark	
	Branches	36
	Fruits etc.	
	Foliage	2-4
	Root estimate	

CAI (m³ /ha/yr)

Net production (t/ha/yr)	Stem wood
	Stem bark
	Branches
	Fruits etc.
	Foliage
	Root estimate

One hundred and three stems were sampled in the autumn, and stand values for a 0.35 ha plot were derived from regressions on D²H.

Cantiani, M. (1974). Prime indagini sulla biomassa dell'abete bianco. III Tavola di produttivita della biomassa arborea. Ricerche Sperimentali di Dendrometria e di Auxometria. Bull.5. Instit. Assestamento Forestale, Univ. Firenze, Italy.

Hellrigl, B. (1974). Relazioni e tavola della biomassa arborea. I Tavola di produttivita della biomassa arborea. Ricerche Sperimentali di Dendrometria e di Auxometria. Bull.5. Instit. Assestamento Forestale, Univ. Firenze, Italy.

43°44'N   11°34'E   500-1200 m   Italy, Vallombrosa region, near Florence.

Plantations.
Average quality   *Abies alba*
sites.

Subject to a moderate thinning regime.

Age (years)	10	15	20	25	30	35
Trees/ha	25000	17500	2548	2180	1902	1679
Tree height (m)			9.2	11.6	13.9	15.8
Basal area (m²/ha)			24.1	32.3	37.9	43.6
Leaf area index						
Stem volume (m³/ha)	39	105	173	249	335	416

Dry biomass (t/ha)		10	15	20	25	30	35
	Stem wood	}14.7	}39.2	}64.7	}92.8	}124.8	}154.9
	Stem bark						
	Branches	4.3	11.5	15.0	17.9	20.0	21.6
	Fruits etc.						
	Foliage	0.8	2.1	6.6	9.6	10.0	13.3
	Root estimate						

CAI (m³/ha/yr)

Net production (t/ha/yr)		10	15	20	25	30	35
	Stem wood						
	Stem bark						
	Branches		7.04[a]	7.40[a]	7.76[a]	8.14[a]	8.52[a]
	Fruits etc.						
	Foliage						
	Root estimate						

One hundred and twenty-three trees were sampled varying in age from 27 to 105 years Stand biomass values were derived from regressions on D and H applied to yield tables of the numbers of trees per hectare in different diameter and height classes. *a.* Including thinnings, but excluding all litterfall.

Continued from p.87.

Same as p.87.

Age (years)	40	45	50	55	60	65
Trees/ha	1500	1320	1189	1060	940	821
Tree height (m)	17.7	19.3	20.9	22.2	23.5	24.6
Basal area (m²/ha)	47.9	52.3	55.6	58.6	60.6	61.9
Leaf area index						
Stem volume (m³/ha)	500	580	655	723	786	844

		40	45	50	55	60	65
Dry biomass (t/ha)	Stem wood	}186.2	}216.1	}243.8	}269.1	}292.6	}314.2
	Stem bark						
	Branches	23.8	25.7	28.2	30.3	32.5	34.3
	Fruits etc.						
	Foliage	14.3	15.5	15.8	16.4	16.7	16.8
	Root estimate						

		40	45	50	55	60	65
CAI (m³/ha/yr)							
Net production (t/ha/yr)	Stem wood						
	Stem bark						
	Branches	}8.70[a]	}8.66[a]	}8.40[a]	}8.12[a]	}7.94[a]	}7.62[a]
	Fruits etc.						
	Foliage						
	Root estimate						

See p.87.

Continued from p.88.

Same as p.87.

Age (years)	70	75	80	85	90	95
Trees/ha	759	694	633	594	549	512
Tree height (m)	25.6	26.5	27.2	27.9	28.5	29.0
Basal area (m²/ha)	64.2	65.5	67.0	68.3	69.6	70.5
Leaf area index						
Stem volume (m³/ha)	896	943	986	1025	1060	1095

Dry biomass (t/ha)		70	75	80	85	90	95
	Stem wood	} 333.4	} 351.0	} 366.9	} 381.3	} 394.4	} 406.1
	Stem bark						
	Branches	36.0	37.8	39.4	40.9	42.2	43.4
	Fruits etc.						
	Foliage	17.4	17.5	17.5	17.6	17.7	17.8
	Root estimate						

CAI (m³/ha/yr)

Net production (t/ha/yr)		70	75	80	85	90	95
	Stem wood						
	Stem bark						
	Branches	} 7.40^a	} 7.16^a	} 6.92^a	} 6.74^a	} 6.56^a	} 6.38^a
	Fruits etc.						
	Foliage						
	Root estimate						

See p.87.

Müller, D. and Nielsen, J. (1965). Production brute, pertes par respiration et production nette dans la forêt ombrophile tropicale. *Forst. ForsVaes. Danm.* 29, 60–160.

5°20'N 4°10'E   50 m   Ivory Coast, near Abidjan, Languededru.

Poor, sandy
soil.
pH 3.8–4.4

Strombosia pustulata, Conopharyngia durissima, Funtumia latifolia and 42 other species.

Lowland tropical rainforest

Age (years)	
Trees/ha	836
Tree height (m)	12.7^a (5 to 50)
Basal area (m²/ha)	31.2
Leaf area index	3.2
Stem volume (m³/ha)	421^b

Dry biomass (t/ha)

Stem wood	} 178^c
Stem bark	
Branches	62^c
Fruits etc.	
Foliage	2.5
Root estimate	

CAI (m³/ha/yr)	13.1^b

Net production (t/ha/yr)

Stem wood	} $7.5 + 1.9^d$
Stem bark	
Branches	
Fruits etc.	
Foliage	2.1^e
Root estimate	

A total of 52 trees was sampled. Stand values were derived in various ways from stem volumes and diameters measured in areas of 0.09 ha and 0.16 ha on three occasions over 5 years. Only trees over 3 cm diameter and 1.3 m height were included.
a. Weighted mean.
b. Including the branches.
c. Assuming that 26% of the above-ground wood was branches.
d. Woody litterfall, excluding mortality.
e. Assuming shade leaves lived for 2 years, and sun leaves lived for one year.

Bernhardt-Reversat, F., Huttel, C. and Lemée, G. (1978). La forêt sempervirente de basse Côte d'Ivoire. In: "Problèmes d'Ecologie: Structure et Fonctionnement des Ecosystèmes Terrestres" (M. Lamotte and F. Boulière, eds) pp. 313-345. Masson, Paris, New York, Barcelona and Milan.

Lemée, G, Bernhardt-Reversat, F. and Huttel, C. (1975). Recherches sur l'écosystème de la forêt subéquatoriale de basse Côte d'Ivoire. Parts I-VII. *Terre Vie* <u>29</u>, 169-264.

5°23-42'N 4°2-6'W   20-70 m   Ivory Coast, near Abidjan.

Poor, sandy soils. pH 4.1-5.1

Dacryodes klaineana, Strombosia glaucescens, Allanblackia sp., *Coula* sp., *Diospyros* sp. *et al.* (over 120 spp.)

Lowland tropical rainforest

	Le Banco		Yapo
	On a plateau	In a valley	On a plateau
Age (years)	Mature		Mature
Trees/ha	265[a]		427[a]
Tree height (m)	5-50		5-50
Basal area (m²/ha)	30		31
Leaf area index	8-10		8-10
Stem volume (m³/ha)	560		500
Dry biomass (t/ha)			
Stem wood	} 360		} 330
Stem bark			
Branches	105 + 24[b]		95
Fruits etc.			
Foliage	9		8
Root estimate	49		
CAI (m³/ha/yr)			
Net production (t/ha/yr)			
Stem wood	} 4.60 + 2.58[c]	} 3.05 + 1.09[c]	} 4.65 + 1.45[c]
Stem bark			
Branches			
Fruits etc.	1.10[c]	0.66[c]	1.05[c]
Foliage	8.19[c]	7.43[c]	7.12[c]
Root estimate	0.7	0.5	0.7

A total of 2614 trees representing 120 species were sampled, and some root systems were excavated. Stand values for the above 0.25 ha plots were derived using regression methods. Increments were derived by remeasurement of 250 trees over 5 to 7 years. Roots were assumed to have the same relative growth rates as above-ground woody parts. Nutrient contents were determined.

a. Trees over 40 cm circumference.
b. Lianes.
c. Litterfall measured over 2-3 years, including estimated pre-fall decay (20% of leaf litterfall).

Tanner, E.V.J. (1980). Studies on the biomass and productivity in a series of montane rainforests in Jamaica. *J. Ecol.* <u>68</u>, 573–588.

Tanner, E.V.J. (1980). Litterfall in montane rainforests of Jamaica and its relation to climate. *J. Ecol.* <u>68</u>, 833–848.

Tanner, E.V.J. (1977). Four montane rainforests of Jamaica: a quantitative characterization of the floristics, the soils and the foliar mineral levels, and a discussion of the interrelations. *J. Ecol.* <u>65</u>, 883–918.

8°05'N 176°30'W   (alt. given below)   Jamaica, The Blue Mountains.

Tropical evergreen montane rainforests

	Mor ridge 1615 m	Mull ridge 1615 m	Well-developed mull ridge 1530 m	Wet slope 1570 m	Gap forest 1590 m
Age (years)					
Trees/ha	6200[a]		6400[a]		
Tree height (m)	5–7	8–13	ca.10–15	8–13	12–16
Basal area (m²/ha)	65	65	65	46	48
Leaf area index	5.0[b]		5.7[b]		
Stem volume (m³/ha)					

Dry biomass (t/ha)

Stem wood					
Stem bark	} 218		} 327		
Branches		} 312		} 230	} 238
Fruits etc.					
Foliage	8.3[c]		6.7[c]		
Root estimate	54				

CAI (m³/ha/yr)

Net production (t/ha/yr)

Stem wood					
Stem bark	} 0.5 + 1.5[e]	} 0.2+2.0[d] +0.2[e]	} 0.2+2.0[d] +0.2[e]	} 2.0+2.7[d] +1.2[e]	} 3.5+0.6[d] +0.9[e]
Branches					
Fruits etc.					
Foliage	4.9[e]	5.5[e]	5.5[e]	4.4[e]	5.5[e]
Root estimate					

Thirty-five trees were sampled in the 100 m² Mor Ridge plot, plus all small vegetation and roots in one 17.5 m² pit. All 64 trees and other vegetation were harvested in the 100 m² well-developed Mull Ridge plot. Biomass values for the other plots were derived from regressions on basal area per tree. Wood increments were estimated from girth increments on 60 trees. Nutrient contents were determined.
a. Trees over 10 cm D.
b. Including epiphytes and ferns, but excluding herbs.
c. Including 0.3–0.4 t/ha of leaves on saplings, tree ferns and climbers.
d. Mortality (not estimated in the Mor Ridge plot).
e. Litterfall, measured over 1–2 years.

Fujimori, T. and Yamamoto, K. (1967). Productivity of *Acacia dealbata* stands. A report of 4 years old stands in Okayama Prefecture. *J. Jap. For. Soc.* <u>49</u>, 143-149.

---

ca. 35°N 134°E   --   Japan, Okayama Prefecture.

Seeded stands.   *Acacia dealbata*
Poor site.

Age (years)	4	4	4
Trees/ha	4000	4000	2000
Tree height (m)	6.7	5.8	5.3
Basal area (m²/ha)	17.2	13.1	5.8
Leaf area index			
Stem volume (m³/ha)	49.1	34.3	13.8

Dry biomass (t/ha)

Stem wood	} 24.2	} 18.0	} 7.0
Stem bark			
Branches	7.4	7.9	3.6
Fruits etc.	0.3	0.1	0.2
Foliage	4.4	3.9	2.6
Root estimate	16.2[a]	12.9[a]	5.8[a]

CAI (m³/ha/yr)

Net production (t/ha/yr)

Stem wood		
Stem bark	} 18.2[b]	} 9.8[b]
Branches		
Fruits etc.	0.3	0.1
Foliage	2.2[c]	1.9[c]
Root estimate	9.0[a]	5.1[a]

Sixteen trees were sampled from each of the two stands with 4000 trees/ha. Stand biomass values for the above plots of 122 to 286 m² were derived from regressions on D²H and by proportional basal area allocation; values given here are the means of the two estimates.
a. Assuming top/root ratios to be 2.3.
b. Excluding woody litterfall and any mortality.
c. Assumed to be half the foliage biomass.

Ando, T. and Takeuchi, I. (1973). Growth and production structure of *Acacia mollissima* Wild. and *Acacia dealbata* Link. in Saijo experimental stand. *Bull. Govt Forest Exp. Stn Tokyo* 252, 149-159.

ca.33°30'N 132°30'E   --   Japan, Shikoku, Ehime Prefecture.

Plantations

	Acacia mollissima		*Acacia dealbata*	
Age (years)	5	5	5	5
Trees/ha	1155	2378	1052	2155
Tree height (m)	7.6	7.7	7.8	7.0
Basal area (m²/ha)	6.8	11.2	7.8	8.4
Leaf area index				
Stem volume (m³/ha)	28.0	46.6	32.6	37.0
Dry biomass (t/ha) Stem wood	}15.6	}23.4	}13.4	}14.0
Stem bark				
Branches	6.3	6.4	7.2	5.7
Fruits etc.	0.0	0.0	0.1	0.2
Foliage	2.2	3.2	2.9	2.8
Root estimate	4.8	7.2	5.9	5.4
CAI (m³/ha/yr)	7.6	11.6	10.2	12.1
Net production (t/ha/yr) Stem wood	}4.2	}5.8	}4.3	}4.6
Stem bark				
Branches	2.5[a]	2.6[a]	3.5[a]	2.6[a]
Fruits etc.	0.0	0.0	0.1	0.2
Foliage	1.1[b]	1.6[b]	1.4[b]	1.4[b]
Root estimate	1.3	1.8	1.9	1.8

Eight trees were sampled per plot, and three root systems were excavated. Stand biomass values for the above 400 m² plots were estimated by proportional basal area allocation. Roots were assumed to grow at the same relative rate as above-ground woody parts.

a. Assumed to be equal to the stem increment within the crown; excluding any woody litterfall and mortality.

b. Assumed to be half the foliage biomass.

Tadaki, Y. (1968a). The primary productivity and the stand density control in
  Acacia mollissima stands. *Bull. Govt Forest Exp. Stn Tokyo* 216, 99-115.
Tadaki, Y. (1965a). Studies on production structure of forest. VIII Productivity
  of an *Acacia mollissima* stand in higher stand density. *J. Jap. For. Soc.* 47,
  384-391.
Tadaki, Y., Ogata, N. and Nagatomo, Y. (1963). Studies on production structure of
  forest. Some analyses on productivities of artificial stand of *Acacia mollissima*.
  J. Jap. For. Soc. 45, 293-301.

Japan	32°30'N 130°30'E 80 m			ca.33°30'N 130°30'E 60 m
	Kumamoto Prefecture			Fukuoka Prefecture
	Acacia mollissima			*A. mollissima*
	Fertilized plantations			Naturally seeded
Age (years)	3	5	7	4
Trees/ha	3450	5100	3150	14400
Tree height (m)	6.9	9.4	10.5	6.5
Basal area (m²/ha)	9.9	21.8	18.0	21.3
Leaf area index	7.9	8.8	6.9	9.9
Stem volume (m³/ha)	46.5	132.2	118.7	99.0
Dry biomass (t/ha)				
Stem wood	21.3	66.6	64.3	} 53.8
Stem bark	2.5	5.6	6.9	
Branches	7.0	11.0	12.7	7.8
Fruits etc.				
Foliage	8.1	9.0	7.0	10.1
Root estimate	4.6	12.5	12.6	9.3
CAI (m³/ha/yr)	23.9	34.6	29.0	37.0
Net production (t/ha/yr)				
Stem wood	} 13.5	} 20.8	} 19.4	} 20.0
Stem bark				
Branches	4.7[a]	5.5[a]	5.0[a]	4.7[a]
Fruits etc.				
Foliage	5.4[b]	4.5[b]	3.5[b]	5.1[b]
Root estimate	2.7	4.0	3.6	3.7

Eight to fifteen trees were sampled per stand. Stand biomass values for 2 or 3
plots of 100 m² at Kumamoto and one plot of only 50 m² at Fukuoka were derived by
proportional basal area allocation. Production values refer to increments in the
previous year. Values given here are from Tadaki (1968a) which updated those
published earlier. Roots were assumed to grow at the same relative rates as above-
ground woody parts.
a. Assumed to be equal to the stem increment within the crown; excluding any woody
  litterfall and mortality.
b. Assumed to be half the foliage biomass.

Furuno, T. and Uenishi, Y. (1977). Investigations on the productivity of Japanese fir (*Abies firma* Sieb. et Zucc.) and hemlock (*Tsuga sieboldii* Carr.) stand in Kyoto University Forest in Wakayama. IV On the growth of young Japanese cherry birch (*Betula grossa* Sieb. et Zucc.) stands regenerated on felling area. *Bull. Kyoto Univ. For.* 49, 41-52.

30°04'N   135°30'E   850-1000 m   Japan, Wakayama Prefecture, Kyoto University Forest.

Betula grossa (81%)[a]

and other broadleaved species

Age (years)	13
Trees/ha	20064
Tree height (m)	5.0
Basal area (m²/ha)	14.5
Leaf area index	
Stem volume (m³/ha)	52

Dry biomass (t/ha)		
	Stem wood	} 26.3
	Stem bark	
	Branches	6.5
	Fruits etc.	
	Foliage	3.2
	Root estimate	

CAI (m³/ha/yr)

Net production (t/ha/yr)		
	Stem wood	
	Stem bark	
	Branches	0.48^b
	Fruits etc.	0.07^b
	Foliage	$3.48^b + 0.03^b$
	Root estimate	

Thirty-six trees were sampled. Stand biomass values were derived from regressions on D²H; values above are the means of 11 plots varying in area from 4 to 72 m² (the authors reported individual plot values). Nutrient contents were determined.
a. Percentage of the total basal area.
b. Litterfall only; 0.07 was the miscellaneous litter fraction, 0.03 was frass litterfall.

Satoo, T. (1974a). Materials for the studies of growth in forest stands.
IX Primary production relations in a natural forest of *Betula maximowicziana* in
Hokkaido. *Bull. Tokyo Univ. For.* <u>66</u>, 109-117.

Satoo, T. (1970a). A synthesis of studies by the harvest method: primary production
relations in the temperate deciduous forests of Japan. In: "Analysis of Temperate
Forest Ecosystems" (D.E. Reichle, ed.) pp.55-72. Springer-Verlag, New York,
Heidelberg and Berlin.

43°13'N 142°27'E   260 m   Japan, Hokkaido, near Mount Asibetu.

Brown forest soils derived from volcanic ash.	*Betula maximowicziana* and other broadleaved species		
	$76\%^a$	$93\%^a$	$67\%^a$
Age (years)	47	47	47
Trees/ha	500	600	270
Tree height (m)	20.6	22.5	22.2
Basal area (m²/ha)	16.0	17.2	12.5
Leaf area index	4.1	5.2	3.7
Stem volume (m³/ha)	155	202	122
Dry biomass (t/ha)			
Stem wood	} 100.0	} 128.3	} 77.7
Stem bark			
Branches	14.8	12.1	11.1
Fruits etc.			
Foliage	2.2	2.6	1.8
Root estimate			
CAI (m³/ha/yr)			
Net production (t/ha/yr)			
Stem wood	} 2.90^b	} 3.65^b	} 2.83^b
Stem bark			
Branches	1.04^b	0.98^b	0.95^b
Fruits etc.			
Foliage	2.17	2.59	1.76
Root estimate			

Five trees were sampled per stand. Stand biomass values for each of the above 0.1
ha plots were derived by proportional basal area allocation.
a. Percentage of the total tree number that were *B. maximowicziana*.
b. Excluding woody litterfall and any mortality.

Tadaki, Y., Shidei, T., Sakasegawa, T. and Ogino, K. (1961). Studies on productive structure of forest. II Estimation of standing crop and some analyses on productivity of young birch stand (*Betula platyphylla*). *J. Jap. For. Soc.* <u>43</u>, 19-26.

Satoo, T. (1970a). A synthesis of studies by the harvest method: primary production relations in the temperate deciduous forests of Japan. In: "Analysis of Temperate Forest Ecosystems" (D.E. Reichle, ed.) pp. 55-72. Springer-Verlag, New York, Heidelberg and Berlin.

Japan	ca.43°00'N 144°00'E 70 m Hokkaido *Betula platyphylla* Natural regeneration (Tadaki *et al.* 1961)	ca.43°13'N 142°25'E 230-260 m Hokkaido, near Mount Asibetu *Betula ermanii* (Satoo 1970a)
Age (years)	10	22
Trees/ha	18954	
Tree height (m)	4.3	
Basal area (m²/ha)	10.6	
Leaf area index	2.9	5.6
Stem volume (m³/ha)		
Dry biomass (t/ha) Stem wood	} 14.0	} 50.8
Stem bark		
Branches	3.2	8.9
Fruits etc.		
Foliage	1.2	2.8
Root estimate		
CAI (m³/ha/yr)	7.5	
Net production (t/ha/yr) Stem wood	} 2.90[a]	} 5.28[a]
Stem bark		
Branches		
Fruits etc.		
Foliage	1.20	2.80
Root estimate		

Twenty-eight *B. platyphylla* trees were sampled and stand biomass values for 9 plots of about 25 m² each were derived from regressions on basal area per tree (the authors reported individual plot values). Fresh weights were converted to dry weights using a factor of 0.5 for woody parts and 0.3 for leaves. Stand values for *B. ermanii* were obtained by clear-felling a large plot; alternative stand biomass values estimated using 5 other methods were all within 5% of the values given above. *a*. Excluding woody litterfall and any mortality.

Kan, M., Saito, H. and Shidei, T. (1965). Studies of the productivity of evergreen broadleaved forests. *Bull. Kyoto Univ. For.* <u>37</u>, 55-75.

ca.34°30'N　136-137°E　300-400 m　Japan, Mie Prefecture.

Camellia japonica

Evergreen broadleaved single-storey forest

	(i)	(ii)	(iii)	(iv)
Age (years)	70	70	70	70
Trees/ha	4400	4600	9000	4300
Tree height (m)	10.1	10.0	8.4	10.2
Basal area (m²/ha)	33	27	25	33
Leaf area index	6.2	5.6	3.7	6.8
Stem volume (m³/ha)	184	145	122	176

Dry biomass (t/ha)

	(i)	(ii)	(iii)	(iv)
Stem wood	} 128	} 101	} 85	} 122
Stem bark				
Branches	49	43	24	47
Fruits etc.				
Foliage	7.5	7.1	5.3	7.6
Root estimate				

CAI (m³/ha/yr)	5.8			

Net production (t/ha/yr)

	(i)	(ii)	(iii)	(iv)
Stem wood	} 4.0^a	} 3.4^a		} 4.2^a
Stem bark				
Branches	0.2^a	0.2^a		0.2^a
Fruits etc.				
Foliage	7.5^b	7.1^b		7.6^b
Root estimate				

Stand biomass values in column (i) were obtained by harvesting and weighing all trees in a 180 m² plot. Stand biomass values in the other columns were derived from regressions on D²H, based on 40 trees sampled in August, for plots of 180 m² in column (ii) and 90 m² in columns (iii) and (iv).
a. Excluding woody litterfall and any mortality.
b. Assumed to be equal to the foliage biomass in August.

Miyata, I. and Shiomi, T. (1965). Ecological studies on the vegetation of Akiyoshi-dai limestone plateau. 1 Structure of the forest community of 'Chojaga-mori'. *Jap. J. Ecol.* <u>15</u>, 29-34.

---

ca.34°15'N 131°30'E   200-500 m   Japan, Yamaguchi Prefecture.

Calcareous soil

Camellia japonica (36%)[a], *Machilus thunbergii* (16%)[a] with 9 other evergreen and 15 deciduous broadleaved species.

A religious taboo forest on a plateau.

Age (years)	Mature	
Trees/ha		
Tree height (m)		
Basal area (m²/ha)	68	
Leaf area index	6.3	
Stem volume (m³/ha)		

Dry biomass (t/ha)	Stem wood	} 268.0
	Stem bark	
	Branches	
	Fruits etc.	
	Foliage	12.6
	Root estimate	

CAI (m³/ha/yr)

Net production (t/ha/yr)	Stem wood	
	Stem bark	
	Branches	
	Fruits etc.	
	Foliage	
	Root estimate	

Stand values for two 2 m wide transects across a 0.2 ha stand were derived using regressions published by Kitazawa *et al.* (1959) (see p.109). Only trees over 3 cm diameter were included.
a. Percentage of total tree number.

Saito, H., Shidei, T. and Kira, T. (1965). Dry matter production by *Camellia japonica* stands. *Jap. J. Ecol.* 15, 131-139.

ca.34°30'N 136°40'E 300-400 m   Japan, Mie Prefecture, Ise Shima Forest.

Fertile soils.	*Camellia japonica* Evergreen broadleaved forest				*C. japonica* *Quercus glauca* and other evergreen broadleaved species	
Age (years)	ca. 60	ca. 60	ca. 60	ca. 60	ca. 60	ca. 60
Trees/ha	4300	4600	9000	4300	14000	5600
Tree height (m)	9-11	9-11	9-11	9-11	9-11	9-11
Basal area (m²/ha)	32.7	27.1	24.8	32.8	59.6	51.5
Leaf area index	6.4	5.4	4.1	5.7	6.6	5.5
Stem volume (m³/ha)	173	143	114	170	250	231
Dry biomass (t/ha) Stem wood / Stem bark	130.0	107.0	86.0	128.0	191.0	167.0
Branches	47.3	37.1	23.0	43.5	50.2	53.8
Fruits etc.						
Foliage	7.3	6.1	5.0	7.4	7.5	6.3
Root estimate						
CAI (m³/ha/yr)	5.8	4.8	5.0	5.7	14.0	12.2
Net production (t/ha/yr) Stem wood / Stem bark / Branches	7.2^a	5.8^a	6.8^a	7.0^a	14.6^a	12.2^a
Fruits etc.						
Foliage	3.7^b	3.1^b	2.5^b	3.7^b	3.8^b	3.2^b
Root estimate						

Sixty-eight trees were sampled in all, and stand biomass values for plots of 100 to 400 m² were derived from regressions on D² and D²H.
a. Excluding woody litterfall and any mortality.
b. Assumed to be half the foliage biomass.

Kan, M., Saito, H. and Shidei, T. (1965). Studies of the productivity of evergreen broadleaved forests. *Bull. Kyoto Univ. For.* <u>37</u>, 55-75.

---

ca.32°30'N 130°40'E   --   Japan, Kumamoto Prefecture.

Castanopsis cuspidata

Evergreen broadleaved two-storeyed forest

Age (years)	40	40	40	40
Trees/ha	7800 + 4700	7200 + 2900	4450 + 2950	3000 + 4700
Tree height (m)	5.9 10.8	6.3 11.6	4.8 11.8	4.7 11.0
Basal area (m²/ha)	5.9 + 40.0	4.6 + 41.0	3.7 + 48.0	2.7 + 46.0
Leaf area index	2.1 + 8.6	1.6 + 8.5	1.3 + 8.5	0.9 + 7.6
Stem volume (m³/ha)	23 + 282	16 + 290	14 + 334	1 + 299

Dry biomass (t/ha)				
Stem wood	} 12.0 + 147.0	} 8.3 + 151.0	} 7.4 + 174.0	} 0.5 + 156.0
Stem bark				
Branches	3.4 + 29.0	2.5 + 35.0	2.2 + 44.0	1.6 + 30.0
Fruits etc.				
Foliage	2.4 + 10.0	1.9 + 9.9	1.5 + 9.8	1.1 + 8.8
Root estimate				

Net production (t/ha/yr)				
CAI (m³/ha/yr)			0.8 + 20.0	0.6 + 17.0
Stem wood			} 0.4^{a}+ 10.0^{a}	} 0.3^{a}+ 9.1^{a}
Stem bark				
Branches			3.2^{a}	2.1^{a}
Fruits etc.				
Foliage			0.6^{b}+ 11.0^{c}	0.5^{b}+ 9.9^{c}
Root estimate				

About 30 trees were sampled in October. Stand biomass values for plots of 100 to 400 m² were derived from regressions on D²H. Values above are given separately for the lower storey and the upper storey (left and right in each column), except for branch and foliage production.
a. Excluding woody litterfall and any mortality.
b. Foliage increment.
c. Foliage litterfall assumed to be equal to foliage biomass in October.

Kan, M., Saito, H. and Shidei, T. (1965). Studies of the productivity of evergreen broadleaved forests. *Bull. Kyoto Univ. For.* <u>37</u>, 55-75.

---

ca.32°30'N 130°40'E -- Japan, Kumamoto Prefecture.

Castanopsis cuspidata

Evergreen broadleaved single-storey forest

						a
Age (years)				12	12	
Trees/ha	2900	3100	3150	13800	20900	150000
Tree height (m)	11.8	10.8	10.8	6.2	6.1	0.9
Basal area (m²/ha)	44	27	27	18	25	17
Leaf area index	7.4	4.4	4.4	5.2	7.2	5.5
Stem volume (m³/ha)	294	175	177	63	90	56
Dry biomass (t/ha)						
Stem wood / Stem bark	}153.0	}91.0	}92.0	}33.0	}47.0	}29.0
Branches	53.0	16.0	17.0	8.4	11.0	9.1
Fruits etc.						
Foliage	8.6	5.1	5.2	6.0	8.4	6.4
Root estimate						
CAI (m³/ha/yr)				15	21	
Net production (t/ha/yr)						
Stem wood / Stem bark				}7.7[b]	}11.0[b]	
Branches				2.3[b]	3.2[b]	
Fruits etc.						
Foliage				$1.3^{c}+6.0^{d}$	$1.9^{c}+8.4^{d}$	
Root estimate						

About 30 trees were sampled in October. Stand biomass values for plots of 100 to 400 m² (and see below) were derived from regressions on D²H.
a. Values in this column refer to a dense thicket of 1.2 x 1.5 m which was completely harvested.
b. Excluding woody litterfall and any mortality.
c. Foliage increment.
d. Foliage litterfall assumed to be equal to foliage biomass in October.

Tadaki, Y., Ogata, N. and Takagi, T. (1962). Studies on the production structure of forest. III Estimation of standing crop and some analyses on productivity of young stands of *Castanopsis cuspidata*. *J. Jap. For. Soc.* 44, 350-359.
Tadaki, Y. (1965b). Studies on production structure of forests. VII The primary production of a young stand of *Castanopsis cuspidata*. *Jap. J. Ecol.* 15, 142-147.
Tadaki, Y. (1968b). Studies on the production structure of forests. XIV The third report on the primary production of a young stand of *Castanopsis cuspidata*. *J. Jap. For. Soc.* 50, 60-65.

32°50'N 130°40'E   80 m   Japan, Kumamoto Prefecture, near Kumamoto City.

Clayey soil with some humus.

Castanopsis cuspidata with a few *Quercus glauca*.

Coppiced evergreen broadleaved forest

	(Tadaki *et al.* 1962)		(Tadaki 1965b)	(Tadaki 1968b)
Age (years)	10	10	11	14
Trees/ha	58000	23592	42000	24667
Tree height (m)	4.1	4.4	5.2	5.5
Basal area (m²/ha)	23.1	14.5	22.3	29.6
Leaf area index	12.5	8.2	8.0	8.7
Stem volume (m³/ha)	74	49.3	84.6	123.0
Dry biomass (t/ha) Stem wood	} 36.3	} 22.9	} 42.6	} 58.7
Stem bark				
Branches	9.1	6.0	6.4	13.4
Fruits etc.				
Foliage	11.4	7.0	7.4	8.4
Root estimate			9.8	13.7
CAI (m³/ha/yr)	17.4	11.4		
Net production (t/ha/yr) Stem wood	} 9.1[a]	} 5.04[a]	} 11.8[b] + 0.9[c]	} 10.11 } +2.38[e]
Stem bark				
Branches				3.97
Fruits etc.				
Foliage			3.7[d]	3.75[f]
Root estimate			2.3[b]	2.45

Tadaki *et al.* (1962) harvested all trees within a 25 m² plot and derived stand values for 8 other 25 m² plots from regressions on basal area per tree. The far left column gives values for the harvested plot; the column with 23592 trees/ha gives the means of all 9 plots. Tadaki *et al.* (1962) reported individual plot values. At age 11 twenty-one trees were sampled, at age 14 all trees were sampled within a 15 m² plot, and in both instances stand values were derived from regressions on D^2 and D^2H.
a. Excluding woody litterfall.   b. Tadaki (1965b) estimated the increment of all woody parts (stems, branches and roots); here it is assumed that 17% of that increment was roots.
c. Branch litterfall.   d. Assumed to be half the foliage biomass.
e. Woody litterfall and mortality.   f. Foliage litterfall, measured over 3 years.

Kawanabe, S. (1977). A subtropical broadleaved forest at Yona, Okinawa. In: "Primary Productivity in Japanese Forests" (T. Shidei and T. Kira, eds) pp. 268-279. JIBP Synthesis vol. 16. University of Tokyo Press.

26°45'N   128°05'E   320 m   Japan, Okinawa Island.

Clay loam soils.

Castanopsis cuspidata var. *sieboldii* (54%)[a]

and over 20 other subtropical evergreen broadleaved species

Age (years)	30–85
Trees/ha	2897
Tree height (m)	9.4 (to 14)
Basal area (m²/ha)	47.9
Leaf area index	6.0
Stem volume (m³/ha)	

Dry biomass (t/ha)

Stem wood	} 136.7
Stem bark	
Branches	48.9
Fruits etc.	
Foliage	7.7
Root estimate	

CAI (m³/ha/yr)

Net production (t/ha/yr)

Stem wood	} 7.24[b]
Stem bark	
Branches	2.23 + 1.93[c]
Fruits etc.	
Foliage	3.85[d]
Root estimate	

Thirteen trees were sampled. Stand biomass values for a 400 m² plot were derived from regressions on D². Only trees at least 4.5 cm in diameter were included. Branches were assumed to grow at the same relative rate as the stems. Nutrient contents were determined.

a. Percentage of the total basal area.
b. Excluding any mortality.
c. Woody litterfall, assumed to be half the value of foliage litterfall.
d. Assumed to be half the foliage biomass (*cf.* Kan *et al.* 1965, see p.103, who assumed that foliage litterfall was equal to foliage biomass).

Satoo, T. (1968). Materials for the studies of growth in stands. VII Primary production and distribution of produced dry matter in a plantation of *Cinnamomum camphora*. *Bull. Tokyo Univ. For.* <u>64</u>, 241-275.

35°09'N   140°09'E   200 m   Japan, Chiba Prefecture.

Plantation.
Alluvial soil.   *Cinnamomum camphora*

Evergreen broadleaved forest

Age (years)	46
Trees/ha	1250
Tree height (m)	16.6
Basal area (m²/ha)	32.4
Leaf area index	4.3 (or 4.8)^a + 1.8^b
Stem volume (m³/ha)	324 (or 271)^a

Dry biomass (t/ha)		
Stem wood	$\left.\begin{array}{l} 157.0 \text{ (or } 150.8)^a \\ \\ 35.0 \text{ (or } 37.4)^a \end{array}\right\} + 2.3^b$	
Stem bark		
Branches		
Fruits etc.		
Foliage	4.1 (or 4.5)^a + 1.1^b	
Root estimate		

CAI (m³/ha/yr)	8.2 (or 8.2)^a

Net production (t/ha/yr)	
Stem wood	$\left.\begin{array}{l} 4.70^c \text{ (or } 4.74)^{ac} \\ \\ 4.86^c \text{ (or } 5.74)^{ac} \end{array}\right\} + 0.48^{bc}$
Stem bark	
Branches	
Fruits etc.	
Foliage	4.07^d (or 4.54)^{ad} + 1.14^{bd}
Root estimate	

Fifteen trees were sampled and stand biomass values for a 0.97 ha plot were derived from regressions on D.
a. Alternative values derived by proportional basal area allocation.
b. Understorey shrubs.
c. Excluding woody litterfall and any mortality.
d. Assumed to be equal to the foliage biomass.

Yasui, H. and Fujie, I. (1970). Studies on the productive structure of 'Shirakashi' (*Cyclobalanopsis myrsinaefolia* Oerst.) coppice forest managed by selection method. 7. On the growth in third circulation-period at the Shimoyamasa permanent plot. *Bull. Fac. Agr. Univ. Shimane* 4, 85-92.

ca.35°N 133°E   --   Japan, Shimane Prefecture, Hirosecho, Shimoyamasa.

Cyclobalanopsis myrsinaefolia

Evergreen oak coppice, thinned in 1965.

	1965	1966	1967	1968	1969
Age (years)					
Trees/ha					
Tree height (m)	8	8	8	9	9
Basal area (m²/ha)	14.7	17.3	20.0	24.3	28.9
Leaf area index					
Stem volume (m³/ha)	53	64	76	96	117
Dry biomass (t/ha) — Stem wood / Stem bark	} 37.0	} 44.7	} 53.7	} 68.1	} 83.5
Branches	11.5	13.9	16.4	20.2	24.6
Fruits etc.					
Foliage	6.1	7.4	8.7	10.7	13.1
Root estimate					
CAI (m³/ha/yr)		10.8	12.2	19.7	20.0
Net production (t/ha/yr) — Stem wood / Stem bark		} 7.7	} 9.0	} 14.4	} 15.4
Branches					
Fruits etc.					
Foliage					
Root estimate					

Sixteen stems were sampled and stand values were derived from regressions on D and H. The sample trees were 9 to 25 years old.

Yasui, H. and Fujie, I. (1971). Studies on the productive structure of 'Shirakashi' (*Cyclobalanopsis myrsinaefolia* Oerst.) coppice forest managed by selection method. 8. On the growth and the biomass at the Shirakashi sprout forest by clearing system. *Bull. Fac. Agr. Univ. Shimane* 5, 49-55.

---

ca.35°N   132°E   300 m   Japan, Shimane Prefecture, Hakuto cho, Takae.

Cyclobalanopsis myrsinaefolia

Evergreen oak coppices on 30-40° slopes

		8	15	25
Age (years)		8	15	25
Trees/ha		22286[a]	16417[a]	12108[a]
Tree height (m)		3.3-6.6[b]	4.0-8.9[b]	3.9-10.7[b]
Basal area (m²/ha)		17.0	25.2	40.0
Leaf area index		8.0	8.1	9.0
Stem volume (m³/ha)		56	111	202
Dry biomass (t/ha)	Stem wood	}53.0	}92.5	}131.5
	Stem bark			
	Branches	15.5	20.5	36.7
	Fruits etc.			
	Foliage	7.8	7.9	8.8
	Root estimate			
CAI (m³/ha/yr)				
Net production (t/ha/yr)	Stem wood			
	Stem bark			
	Branches			
	Fruits etc.			
	Foliage			
	Root estimate			

Eleven to 16 stems were sampled per stand and biomass values for the above 106-123 m² plots were derived from regressions on D.

a. Number of stems per hectare; there were 4245, 5159 and 3920 trees/ha in columns left to right.

b. Range of heights of sample trees.

Kimura, M. (1960). Primary production of the warm-temperate laurel forest in the southern part of Osumi peninsula, Kyushu, Japan. *Misc. Rep. Res. Inst. nat. Resour., Tokyo* 52/53, 36-47.

Kitazawa, Y., Kimura, M., Tezuka, Y., Kurasawa, H., Sakamoto, M. and Yoshino, M. (1959). Plant ecology of the southern part of Osumi peninsula. *Misc. Rep. Res. Inst. nat. Resour., Tokyo* 49, 19-36.

---

ca.31°15'N 131°00'E   500 m   Japan, Kyushu, Osumi peninsula.

Deep volcanic ash with well-developed brown forest soils.

Cyclobalanopsis acuta, Shiia sieboldii, Cyclobalanopsis stenophylla, Distylium racemosum et al.

Evergreen broadleaved forest

Age (years)	
Trees/ha	
Tree height (m)	$13-25^{a}$
Basal area (m²/ha)	
Leaf area index	8.8^{b}
Stem volume (m³/ha)	

Dry biomass (t/ha)

Stem wood	
Stem bark	3 + 36 + 273
Branches	
Fruits etc.	
Foliage	0.5 + 2.4 + 8.5
Root estimate	78^{bc}

CAI (m³/ha/yr)

Net production (t/ha/yr)

Stem wood	
Stem bark	$0.3^{d} + 1.4^{d} + 5.7^{d}$
Branches	
Fruits etc.	
Foliage	$0.5^{e} + 2.4^{e} + 8.5^{e}$
Root estimate	

Many trees were sampled and stand biomass values for plots totalling 0.25 ha were derived from regressions on D. Biomass and production values are given above for the lower plus middle plus upper storeys (written left to right).
a. Upper storey only.
b. All storeys.
c. Assumed to be 25% of the above-ground woody biomass value.
d. Excluding woody litterfall and any mortality.
e. Assumed to be equal to the foliage biomass.

Kira, T., Ono, Y. and Hosokawa, T. (eds) (1978). "Biological Production in a Warm-temperate Evergreen Oak Forest of Japan." JIBP Synthesis vol. 18. University of Tokyo Press.

32°10'N 130°28'E   400-637 m   Japan, Kagoshima Prefecture, Minimata.

Brown or
yellow-brown
forest soils.

Cyclobalanopsis spp. (especially *C. gilva*) and
Castanopsis spp. (especially *C. cuspidata*)

Evergreen broadleaved forest

Age (years)	ca. 50	ca. 50	ca. 50
Trees/ha	$(3138 + 4679)^a$	$(2580 + 4054)^a$	$(1684 + 5613)^a$
Tree height (m)	ca. 25	ca. 25	ca. 25
Basal area (m²/ha)			
Leaf area index	8.3	8.1	6.9
Stem volume (m³/ha)			

Dry biomass (t/ha)				
	Stem wood	} 289.3	} 316.8	} 289.1
	Stem bark			
	Branches	46.0	53.0	57.6
	Fruits etc.			
	Foliage	7.8	7.8	6.8
	Root estimate	83.8	92.4	86.7

CAI (m³/ha/yr)

Net production (t/ha/yr)				
	Stem wood	} $(3.74+0.74+0.23)^b$	} $(2.65+0.74+0.25)^b$	} $(3.26+0.74+0.23)^b$
	Stem bark			
	Branches	$(0.87+3.19+0.97)^b$	$(0.82+2.86+0.95)^b$	$(1.28+3.23+0.97)^b$
	Fruits etc.	0.58^c	0.58^c	0.58^c
	Foliage	$(3.25+0.23+0.08)^d$	$(4.20+0.29+0.08)^d$	$(3.86+0.27+0.08)^d$
	Root estimate	3.30	4.78	4.65

All trees were sampled within a 400 m² plot. Stand biomass values for the above three 0.16 ha plots were derived from regressions on D and H. Biomass values are means over 6 years, production values are means over 4 years. The biomass and increment values for thick roots were assumed to be 25% of those of above-ground woody parts.
a. Trees at least 4.5 cm D (3138, 2580 and 1684), plus smaller trees.
b. Increment plus woody litterfall plus pre-fall loss (written left to right); with pre-fall loss assigned to stems and branches in the same proportion as litterfall biomass.
c. Miscellaneous litterfall.
d. Foliage litterfall plus pre-fall loss plus consumption (written left to right).

Tadaki, Y., Hatiya, K. and Tochaiki, K. (1969). Studies on the production structure of forest. XV Primary productivity of *Fagus crenata* in plantation. *J. Jap. For. Soc.* <u>51</u>, 331–339.

ca.37°30'N 139°00'E   (alt. given below)   Japan, Niigata Prefecture.

Plantations.   *Fagus crenata*

		400 m	470 m	580 m
Age (years)		35	41	over 50
Trees/ha		5235	2186	2829
Tree height (m)		10.0	13.6	13.9
Basal area (m²/ha)		40.6	38.8	39.9
Leaf area index		7.7	7.6	7.8
Stem volume (m³/ha)		275	272	331
Dry biomass (t/ha)	Stem wood	}168.0	}166.5	}202.2
	Stem bark			
	Branches	31.4	43.5	39.5
	Fruits etc.			
	Foliage	4.8	4.7	4.9
	Root estimate	54.2	58.2	49.9
CAI (m³/ha/yr)		9.7	10.8	12.5
Net production (t/ha/yr)	Stem wood	}5.9	}6.6	}7.6
	Stem bark			
	Branches	$2.1 + 1.9^a$	$2.5 + 1.9^a$	$2.3 + 2.0^a$
	Fruits etc.			
	Foliage	4.8^b	4.7^b	4.9^b
	Root estimate	1.6	2.5	1.6

Six to eight trees were sampled per stand. Stand biomass values, for plots of 130, 352 and 438 m², for ages 35, 41 and 50, respectively, were derived from regressions on D²H. Increments were estimated for the previous year. Roots were assumed to grow at the same relative rates as above-ground woody parts.
a. Woody litterfall, assumed to be 40% of the value of foliage litterfall.
b. Foliage litterfall.

112   JAPAN   *Fagus*

Kakubari, Y. (1977). Beech forests in the Naeba Mountains. Distribution of primary productivity along the altitudinal gradient. In: "Primary Productivity in Japanese Forests" (T. Shidei and T. Kira, eds) pp. 201-212. JIBP Synthesis vol. 16. University of Tokyo Press.

36°51'N 138°41'E   (alt. given below)   Japan, Niigata Prefecture, Naeba Mountains.

Brown forest soils.

Fagus crenata with a few *Quercus mongolica*, *Magnolia obovata*, *Kalopanax septemlobus et al.*

	550 m (92%)[a]	700 m (60%)[a]	700 m (96%)[a]	900 m (38%)[a]
Age (years)	to 100	to 100	to 100	to 100
Trees/ha	375	327	235	680
Tree height (m)	18.8	15.6	24.6	11.2
Basal area (m²/ha)	24.9	24.5	32.7	22.7
Leaf area index	4.5	4.4	4.9	5.4

Stem volume (m³/ha)

Dry biomass (t/ha)

	550 m	700 m	700 m	900 m
Stem wood / Stem bark	}227.2	}250.6	}281.6	}243.9
Branches	45.9	54.0	60.0	99.8
Fruits etc.	0.1	0.2	0.1	0.1
Foliage	2.4	2.4	2.7	3.8
Root estimate	56.8	63.5	71.1	70.2

CAI (m³/ha/yr)

Net production (t/ha/yr)

	550 m	700 m	700 m	900 m
Stem wood / Stem bark	}1.78[b]	}1.86[b]	}2.87[b]	}1.59[b]
Branches	0.41+0.48[c]	0.47+0.48[c]	0.70+0.48[c]	0.78+0.96[c]
Fruits etc.	0.11[c]	0.11[c]	0.11[c]	0.12[c]
Foliage	3.30[d]	3.29[d]	3.31[d]	3.78[d]
Root estimate	0.46	0.49	0.74	0.49

Stand biomass values for plots of 800 to 2000 m² were derived from regressions on D²H. Root biomass was assumed to be 20% of the above-ground biomass. Only trees at least 4.5 cm D were included.
a. Percentage of total tree number that were *F. crenata*.
b. Including estimated mortality.
c. Litterfall.
d. Leaf increment; leaf litterfall was 2.37, 2.44, 2.76 and 3.97 t/ha/yr in columns left to right.

Continued from p.112.

Same as p.112.

	1100 m (29%)[a]	1300 m (45%)[a]	1500 m (68%)[a]	1500 m (78%)[a]
Age (years)	to 100	to 100	to 100	to 100
Trees/ha	1050	875	590	357
Tree height (m)	7.3	8.3	9.6	12.6
Basal area (m²/ha)	21.9	14.4	17.7	21.4
Leaf area index	5.2	3.3	3.2	3.3
Stem volume (m³/ha)				

Dry biomass (t/ha)

	1100 m	1300 m	1500 m	1500 m
Stem wood / Stem bark	}213.9	}161.0	}118.3	}133.9
Branches	86.1	42.2	31.0	35.6
Fruits etc.	0.1	0.0	0.2	0.2
Foliage	3.6	2.4	1.8	1.9
Root estimate	61.2	36.0	32.2	36.7

CAI (m³/ha/yr)

Net production (t/ha/yr)

	1100 m	1300 m	1500 m	1500 m
Stem wood / Stem bark	}1.10[b]	}1.17[b]	}0.78[b]	}0.69[b]
Branches	0.52+0.96[c]	0.33+0.13[c]	0.21+0.38[c]	0.19+0.38[c]
Fruits etc.	0.12[c]	0.00[c]	0.16[c]	0.16[c]
Foliage	3.78[d]	2.44[d]	2.48[d]	2.48[d]
Root estimate	0.34	0.26	0.22	0.19

Same as p.112, except:

d. Leaf increment; leaf litterfall was 3.62, 2.38, 1.85 and 1.90 t/ha/yr in columns left to right.

Maruyama, K. (1977). Beech forests in the Naeba Mountains. Comparison of forest structure, biomass and net productivity between the upper and lower parts of beech forest zone. In: "Primary Productivity in Japanese Forests." (T. Shidei and T. Kira, eds) pp. 186-201. JIBP Synthesis vol. 16. University of Tokyo Press.

Maruyama, K. (1971). Effect of altitude on dry matter production of primeval Japanese beech forest communities in Naeba Mountains. *Mem. Fac. Agric. Niigata Univ.* 9, 87-171.

ca.36°51'N   138°41'E   (alt. given below)   Japan, Niigata Prefecture, Naeba Mountains.

Brown forest soils.

Fagus crenata with a few *Quercus mongolica*, *Magnolia obovata*, *Kalopanax septemlobus et al.*

	650 m	650 m	700 m	700 m	700 m	700 m
Age (years)	to 100	to 100	to 100	to 100	to 100	to 100
Trees/ha	367	289	321	470	356	400
Tree height (m)	24.6	23.6	25.3	18.7	22.3	23.0
Basal area (m²/ha)	43.3	41.3	44.0	39.2	46.9	39.1
Leaf area index	5.8	5.5	5.6	4.6	6.2	5.3
Stem volume (m³/ha)	523	546	547	457	549	475

Dry biomass (t/ha)		650 m	650 m	700 m	700 m	700 m	700 m
	Stem wood	}275.1	}287.7	}299.7	}256.9	}301.1	}258.0
	Stem bark						
	Branches	56.1	58.6	58.9	54.4	60.9	50.9
	Fruits etc.						
	Foliage	3.0	2.9	3.0	2.5	3.2	2.7
	Root estimate	66.8	69.8	72.3	78.5	73.0	62.3

CAI (m³/ha/yr)

Net production (t/ha/yr)		650 m	650 m	700 m	700 m	700 m	700 m
	Stem wood	}5.41^a	}5.33^a	}3.99^a	}5.09^a	}5.48^a	}5.66^a
	Stem bark			2.81^a			
	Branches						
	Fruits etc.						
	Foliage	3.01^b	2.94^b	3.02^b	2.52^b	3.17^b	2.72^b
	Root estimate	1.68	1.65	1.96	1.52	1.73	1.68

A total of 59 trees were sampled in the autumn at various altitudes. Stand biomass values for the above plots of 448-1800 m² were derived from regressions on D^2H. Only trees at least 4.5 cm D were included. Root biomass was assumed to be 20% of the above-ground biomass, and root death was assumed to be 20% of above-ground litterfall.
a. Including estimated mortality, assumed to be 0.3% of the stem biomass; and including woody litterfall, assumed to be 3.8 kg/m³ of 'stand volume'.
b. Leaf biomass in the autumn.

Continued from p.114.

Same as p.114.

	900 m	1300 m	1500 m	1500 m	1500 m
Age (years)	to 100	to 100	to 100	to 100	to 100
Trees/ha	1016	959	639	422	820
Tree height (m)	8.7	11.3	11.5	13.2	10.6
Basal area (m²/ha)	40.4	38.5	37.5	33.5	34.7
Leaf area index	4.8	4.2	4.0	3.2	4.3
Stem volume (m³/ha)	328	289	301	243	342
Dry biomass (t/ha) Stem wood	} 181.0	} 192.3	} 161.6	} 119.4	} 167.7
Stem bark					
Branches	72.2	50.4	43.0	31.6	44.3
Fruits etc.					
Foliage	3.2	3.1	2.5	1.9	2.7
Root estimate	51.3	49.2	41.4	30.6	42.9
CAI (m³/ha/yr)					
Net production (t/ha/yr) Stem wood			} 1.64^{a}	} 3.26^{a}	} 3.56^{a}
Stem bark					
Branches			1.72^{a}		
Fruits etc.					
Foliage			2.47^{b}	1.94^{b}	2.72^{b}
Root estimate			1.13	1.04	1.26

See p.114.

Kawahara, T., Tadaki, Y, Takeuchi, I, Sato, A., Higuchi, K. and Kamo, K. (1979). Productivity and cycling of organic matter in natural *Fagus crenata* and two planted *Chamaecyparis obtusa* forests. *Jap. J. Ecol.* <u>29</u>, 387-395.

Ogino, K. (1977). A beech forest at Ashiu - biomass, its increment and net production. In: "Primary Production of Japanese Forests" (T. Shidei and T. Kira, eds) pp. 172-186. JIBP Synthesis vol. 16. University of Tokyo Press.

Japan	ca.36°47'N 139°56'E 940 m Tochigi Prefecture, near Yaita city *Fagus crenata* (Kawahara *et al.* 1979)	35°20'N 135°45'E 680 m Kyoto Prefecture, Yura River *Fagus crenata* (72%)[a] *Carpinus tschonoskii,* *Acer tschonoskii et al.* (Ogino 1977)
Age (years)	Mature	ca. 150
Trees/ha	844	785
Tree height (m)	11.5 (top height 23.9)	14.3
Basal area (m²/ha)	30.9	30.7
Leaf area index		4.5
Stem volume (m³/ha)	288	318

Dry biomass (t/ha)			
	Stem wood	} 163.5	} 194.3
	Stem bark		
	Branches	81.2	95.1
	Fruits etc.		
	Foliage	3.0	3.0
	Root estimate	81.6	64.6

CAI (m³/ha/yr)			4.0
Net production (t/ha/yr)	Stem wood	} 9.80 + 0.05[b] } + 0.62[b]	} 2.88
	Stem bark		
	Branches		2.07 + 1.38[c]
	Fruits etc.	0.04[b]	
	Foliage	3.18[b]	0.03 + 3.45[d]
	Root estimate		0.77 + 0.69[e]

Ten trees were sampled at Yaita and stand biomass values for a 0.25 ha plot were derived from regressions on D²H. Root biomass was assumed to be one third of aboveground biomass. Fourteen trees were sampled at Yura, and 8 root systems were excavated; stand biomass values for six 100 m² plots for trees at least 4.5 cm D were derived from regressions on D²H.

a. Percentage of the total tree number.

b. Litterfall, where 0.62 t/ha/yr was miscellaneous litterfall.

c. Woody litterfall, assumed to be 40% of foliage litterfall value.

d. Foliage litterfall; the total foliage production of 3.48 t/ha/yr was equal to foliage biomass in August plus 15%.

e. Root losses, assumed to be 20% of foliage litterfall value.

Katagiri, S. and Tsutsumi, T. (1975). The relationship between site condition and circulation of nutrients in forest ecosystems. III Aboveground biomass and nutrient contents of stands. *J. Jap. For. Soc.* <u>57</u>, 412-419.

Katagiri, S. and Tsutsumi, T. (1976). (as above) IV The amount of mineral nutrient returned to forest floor. *J. Jap. For. Soc.* <u>58</u>, 79-85.

Katagiri, S. and Tsutsumi, T. (1978). (as above) V The differences in nutrient circulation between stands located in upper part of slope and lower part of slope. *J. Jap. For. Soc.* <u>60</u>, 195-202.

35°18'N 135°43'E   (alt. given below)   Japan, Kyoto Prefecture, Ashiu Forest.

	Fagus crenata, Cornus controversa, Aesculus turbinata, Styrax obassia, et al.		*Quercus serrata, Quercus crispula, Hydrangea paniculata, Hamamelis japonica, et al.*	
	Ridge top 715-725 m		Lower slope 680-695 m	
Age (years)				
Trees/ha	2737	3808	433	561
Tree height (m)	7.9	6.0	12.5	15.1
Basal area (m²/ha)	19.9	25.6	36.4	39.8
Leaf area index				
Stem volume (m³/ha)				
Dry biomass (t/ha)				
Stem wood	} 56.0	} 56.5	} 143.8	} 170.7
Stem bark				
Branches	19.9	18.7	104.6	102.0
Fruits etc.				
Foliage	3.2	3.3	2.5	4.2
Root estimate				
CAI (m³/ha/yr)				
Net production (t/ha/yr)				
Stem wood	}	}	}	}
Stem bark	} 0.90^a	} $3.29+0.90^a$	} 1.28^a	} $8.95+2.11^a$
Branches	}	}	}	}
Fruits etc.	} 3.26^b	} 3.19^b	} 3.25^b	} 4.80^b
Foliage	}	}	}	}
Root estimate				

Thirty-two trees were sampled. Stand biomass values for the above four 25 m² plots were derived from regressions on D²H. Nutrient contents were determined.

a. Big wood litterfall; note that stem and branch increments are given for only two plots.

b. Foliage, twig and miscellaneous litterfall.

Satoo, T., Kunugi, R. and Kumekawa, A. (1956). Materials for the studies of growth in stands. III Amount of leaves and production of wood in an aspen (*Populus davidiana*) second growth in Hokkaido. *Bull. Tokyo Univ. For.* 52, 33-51.

Satoo, T. (1970a). A synthesis of studies by the harvest method: primary production relations in the temperate deciduous forests of Japan. In: "Analysis of Temperate Forest Ecosystems" (D.E. Reichle, ed.) pp. 55-72. Springer-Verlag, New York, Heidelberg and Berlin.

---

ca.43°13'N 142°25'E   230-260 m   Japan, Hokkaido, near Mount Asibetu.

Populus davidiana (82%)[a]

Magnolia obovata et al.

Age (years)		25-40
Trees/ha		1244
Tree height (m)		14-21
Basal area (m²/ha)		26.7
Leaf area index		2.4
Stem volume (m³/ha)		274
Dry biomass (t/ha)	Stem wood	} 105.0
	Stem bark	
	Branches	25.4
	Fruits etc.	
	Foliage	2.2
	Root estimate	
CAI (m³/ha/yr)		14.7
Net production (t/ha/yr)	Stem wood	} 5.62[b]
	Stem bark	
	Branches	0.88[b]
	Fruits etc.	
	Foliage	2.17
	Root estimate	

Fourteen trees were sampled and stand biomass values for a 0.16 ha plot were derived from regressions on D.
a. Percentage of the total tree number; stand values were calculated for all trees using regressions for *P. davidiana*, as if it were a pure stand.
b. Excluding woody litterfall and any mortality.

Kan, M., Saito, H. and Shidei, T. (1965). Studies of the productivity of evergreen broadleaved forests. *Bull. Kyoto Univ. For.* <u>37</u>, 55-75.

ca.34°30'N 136-137°E   300-400 m   Japan, Mie Prefecture.

Quercus glauca, Quercus salicina, Quercus acuta,
Camellia japonica et al.

Evergreen broadleaved two-storeyed forest

Age (years)	70	70	70
Trees/ha	8300 + 5300	1200 + 4300	10800 + 4800
Tree height (m)	2.7 9.9	2.2 10.4	3.5 10.0
Basal area (m²/ha)	12 + 36	3 + 38	11 + 33
Leaf area index	1.4 + 6.1	0.5 + 6.5	1.4 + 5.4
Stem volume (m³/ha)	53 + 198	14 + 214	46 + 179

Dry biomass (t/ha)

Stem wood	} 37 + 136	} 10 + 147	} 32 + 123
Stem bark			
Branches	8 + 40	3 + 46	6 + 31
Fruits etc.			
Foliage	2.3 + 4.8	0.7 + 5.2	2.0 + 4.3
Root estimate			

CAI (m³/ha/yr)	2.7 + 10.3	0.6 + 11.4	2.6 + 9.4

Net production (t/ha/yr)

Stem wood	} 1.7^a + 7.3^a	} 0.4^a + 8.1^a	} 1.8^a + 6.6^a
Stem bark			
Branches	0.3^a	0.4^a	0.3^a
Fruits etc.			
Foliage	2.3^b + 4.8^b	0.7^b + 5.2^b	2.0^b + 4.3^b
Root estimate			

Twenty trees were sampled in August. Stand biomass values for plots of 90 m² (left and centre columns) or 25 m² (right column) were derived from regressions on D²H. Values are given above for the lower storey plus the upper storey (left and right in each column), except for branch production.
a. Excluding woody litterfall and any mortality.
b. Assumed to be equal to the foliage biomass in August.

Kan, M. Saito, H. and Shidei, T. (1965). Studies of the productivity of evergreen broadleaved forests. *Bull. Kyoto Univ. For.* **37**, 55-75.

33-34°N 133-134°E   --   Japan, Kochi Prefecture.

Quercus phillyraeoides and

Rapanaea neriifolia (syn. *Mysine neriifolia*)

Evergreen broadleaved two-storeyed forest.

---

		ca. 80	ca. 80	ca. 80
Age (years)		ca. 80	ca. 80	ca. 80
Trees/ha		14600 + 4000	13400 + 3700	9800 + 5200
Tree height (m)		5.7 9.6	6.0 9.7	6.0 9.6
Basal area (m²/ha)		11 + 29	16 + 33	11 + 37
Leaf area index		2.7 + 4.2	4.4 + 5.0	2.8 + 4.9
Stem volume (m³/ha)		40 + 150	59 + 170	39 + 195
Dry biomass (t/ha)	Stem wood	} 34 + 138	} 50 + 155	} 33 + 178
	Stem bark			
	Branches	6 + 31	11 + 37	7 + 36
	Fruits etc.			
	Foliage	2.8 + 6.0	4.4 + 7.6	2.8 + 7.1
	Root estimate			
CAI (m³/ha/yr)		3.8 + 4.6	4.6 + 4.7	3.2 + 6.0
Net production (t/ha/yr)	Stem wood	} 3.2^a+ 4.2^a	} 3.9^a+ 4.4^a	} 2.7^a+ 5.5^a
	Stem bark			
	Branches	2.1^a	2.6^a	2.5^a
	Fruits etc.			
	Foliage	0.5^c+ 8.8^b	0.7^c+12.0^b	0.6^c+ 9.9^b
	Root estimate			

About 20 trees were sampled in November. Stand biomass values for plots of 300, 164 and 9 m² (in columns left to right) were derived from regressions on D²H. Values are given above for the lower storey plus the upper storey (left and right in each column), except for branch and foliage production.
a. Excluding woody litterfall and any mortality.
b. Total foliage litterfall of both storeys, assumed to be equal to foliage biomass in November.
c. Foliage increment.

Oshima, Y. (1961a). Ecological studies of Sasa communities. I Productive structure of some of the Sasa communities in Japan. *Bot. Mag., Tokyo* 74, 199-210.

Oshima, Y. (1961b). Ecological studies of Sasa communities. II Seasonal variations of productive structure and annual net production in Sasa communities. *Bot. Mag., Tokyo* 74, 280-290.

Japan Black-brown loams.		ca.35°55'N 138°10'E Nagano Prefecture		ca.37°00'N 139°20'E	ca.43°00'N 140°55'E
	Evergreen bamboos	*Sasa nipponica*	*Sasa nikkoensis*	*Sasa oseana*	*Sasa kurilensis*
	Altitude:	1700 m	2150 m	1400 m	550 m
Age (years)		2	3-5	3-5	9.2
Trees/ha		3,890,000[a]	2,600,000[a]	2,160,000[a]	275,000[a]
Tree height (m)		0.9	1.2	1.2	3.3
Basal area (m²/ha)					
Leaf area index		4.7	5.2	4.7	5.3
Stem volume (m³/ha)					
Dry biomass (t/ha)	Stem wood / Stem bark / Branches / Fruits etc.	}5.5[b]	}15.7[b]	}13.7[b]	}7.5[b]
	Foliage	2.7	3.2	2.8	4.7
	Root estimate	7.4[c]	13.1[c]	11.4[c]	31.0[c]
CAI (m³/ha/yr)					
Net production (t/ha/yr)	Stem wood / Stem bark / Branches / Fruits etc.				}8.6[d]
	Foliage				3.1
	Root estimate				4.3[c]

In the *S. kurilensis* stand all bamboos were harvested within two 4 m² plots; in each of the other stands all bamboos were harvested within two 0.25 m² plots.

a. Numbers of culms per hectare.
b. Culms and branches.
c. Rhizomes and roots.
d. New growth plus estimated losses.

Satoo, T., Negisi, K. and Senda, M. (1959). Materials for the studies of growth in stands. V Amount of leaves and growth in plantations of *Zelkova serrata* applied with crown thinning. *Bull. Tokyo Univ. For.* <u>55</u>, 101–123.

35°56'N   138°51'E   ca. 1000 m   Japan, Titibu.

Plantations.   *Zelkova serrata*

Deciduous broadleaved species

		Thinned	Unthinned
Age (years)		42	42
Trees/ha		290^a + 983^a	2600
Tree height (m)			
Basal area (m²/ha)		19.5	33.2
Leaf area index		2.2	4.0
Stem volume (m³/ha)		157	248
Dry biomass (t/ha)	Stem wood		
	Stem bark		
	Branches	24.8	48.7
	Fruits etc.		
	Foliage	1.5	2.8
	Root estimate		
CAI (m³/ha/yr)		3.5	8.7
Net production (t/ha/yr)	Stem wood		
	Stem bark		
	Branches		
	Fruits etc.		
	Foliage	1.52^b	2.75^b
	Root estimate		

Ten trees were sampled in each stand, and stand values were derived from regressions on D. No values were reported for stem biomass or wood increment.
a. Understorey (290) and overstorey (983).
b. Foliage biomass.

Furuno, T. and Kawanabe, S. (1967). Investigations on the productivity of Japanese fir (*Abies firma* Sieb. et Zucc.) and hemlock (*Tsuga sieboldii* Carr.) stands in Kyoto University Forest in Wakayama. I On the growth of Japanese fir stands. *Bull. Kyoto Univ. For.* **39**, 9–26.
Furuno, T. (1971). Investigations on the productivity of Japanese fir (*Abies firma* Sieb. et Zucc.) and hemlock (*Tsuga sieboldii* Carr.) stands in Kyoto University Forest in Wakayama. II On the mixed stand of Japanese fir and hemlock. *Bull. Kyoto Univ. For.* **42**, 128–142.

34°04'N   135°30'E   ca. 700 m   Japan, Wakayama Prefecture, Kyoto University Forest.

	Abies firma (99%)[a] (Furuno and Kawanabe 1967)	*Abies firma* and *Tsuga sieboldii* (Furuno 1971) (53%)[b]	(48%)[b]
Age (years)			
Trees/ha	3730	1995	2471
Tree height (m)	ca. 7.6	12–15	
Basal area (m²/ha)	36.1	47.7	26.2
Leaf area index			
Stem volume (m³/ha)		445	157
Dry biomass (t/ha) Stem wood	}76.9	}189.1	}68.6
Stem bark			
Branches	30.4	28.3	13.4
Fruits etc.			
Foliage	20.8	14.6	7.7
Root estimate			
CAI (m³/ha/yr)			
Net production (t/ha/yr) Stem wood	}12.4[c]	}11.0[c]	
Stem bark			
Branches		6.5[c]	
Fruits etc.			
Foliage	2.4[d]	2.5	1.1
Root estimate			

Furuno and Kawanabe (1967): eight trees were sampled and stand biomass values for a 70 m² plot were derived from regressions on D. (The authors also reported values for 3 plots with many overstorey shade trees.)
Furuno (1971): ten *A. firma* and twenty *T. sieboldii* were sampled and stand biomass values for the above plots of 381 and 85 m² (left and right columns, respectively) were derived from various regressions on D and H.
a. Percentage of the total basal area.
b. Percentage of the total biomass accounted for by *A. firma*.
c. Excluding woody litterfall and any mortality.
d. Assuming foliage production to be 11.5% of the foliage biomass.

Furuno, T., Uenishi, S. and Uenishi, K. (1979). Investigations on the productivity of Japanese fir (*Abies firma* Sieb. et Zucc.) and hemlock (*Tsuga sieboldii* Carr.) stands in Kyoto University Forest in Wakayama. V Biomass of upperground parts and litterfall in fir-hemlock stands. *Bull. Kyoto Univ. For.* 51, 58-70.

Furuno, T. and Yamada, K. (1974). (as above) III Seasonal variation of litterfall and primary consumption by herbivorous insects in the mixed fir and hemlock stand. *Bull. Kyoto Univ. For.* 46, 7-22.

30°04'N 135°30'E   700 m   Japan, Wakayama Prefecture, Kyoto University Forest.

Abies firma and *Tsuga sieboldii*

	76%[a]	67%[a]	56%[a]	76%[a]	100%[a]	81%[a]	64%[a]
Age (years)							
Trees/ha	854	1249	1087	1034	353	1123	1220
Tree height (m)	ca.26	ca.26	ca.26	ca.26	ca.26	ca.26	ca.26
Basal area (m²/ha)	57.4	56.0	49.1	46.3	56.6	47.3	59.0
Leaf area index							
Stem volume (m³/ha)	653	583	515	443	704	480	605
Dry biomass (t/ha)							
Stem wood / Stem bark	}274.1	}246.0	}217.5	}188.0	}294.6	}203.2	}255.9
Branches	56.6	51.4	46.9	40.4	58.1	38.4	55.0
Fruits etc.	1.2	1.0	0.7	0.8	1.7	1.0	0.9
Foliage	21.9	20.7	18.2	17.1	22.3	17.7	21.8
Root estimate							
CAI (m³/ha/yr)							
Net production (t/ha/yr)							
Stem wood							
Stem bark							
Branches	1.87[b]	1.87[b]	1.87[b]	1.87[b]	1.87[b]	1.87[b]	1.87[b]
Fruits etc.	0.25[b]	0.25[b]	0.25[b]	0.25[b]	0.25[b]	0.25[b]	0.25[b]
Foliage	3.63[c]	3.39[c]	3.16[c]	2.70[c]	3.55[c]	2.80[c]	3.64[c]
Root estimate							

Data from 6 *A. firma* and 5 *T. sieboldii* that were sampled in September were pooled with values for 7 *A. firma* and 12 *T. sieboldii* sampled earlier to calculate regressions on D from which stand values were derived for the above seven sample plots of 474 to 1427 m². In each plot there was about 0.8 t/ha of parasitic plants and lianes. Nutrient contents were determined.
a. Percentage of the total basal area accounted for by *A. firma*.
b. Litterfall only; the average of all plots.
c. New foliage biomass, mean foliage litterfall was 2.77 t/ha/yr.

Ando, T., Chiba, K., Nishimura, T. and Tanimoto, T. (1977). Temperate fir and hemlock forests in Shikoku. In: "Primary Productivity in Japanese Forests" (T. Shidei and T. Kira, eds) pp. 213–245. JIBP Synthesis vol. 16. University of Tokyo Press.

---

ca.33°20'N 133°00'E   420 m   Japan, Kochi Prefecture, Yusuhara district.

Deep,
fertile
soil.

Abies firma with an understorey of evergreen and deciduous broadleaved species including *Cyclobalanopsis* spp., *Actinodaphne lancifolia et al.*

Age (years)	97–145
Trees/ha	$288 + 1789^{a}$
Tree height (m)	$29.1 \quad 12.1^{a}$
Basal area (m²/ha)	$60.7 + 21.0^{a}$
Leaf area index	$5.4^{b} + 1.3^{a}$
Stem volume (m³/ha)	$783 + 137^{a}$

Dry biomass (t/ha)

Stem wood	$\left.\begin{array}{l}\\ \\\end{array}\right\}303.8 + 90.4^{a}$
Stem bark	
Branches	$57.0 + 26.5^{a}$
Fruits etc.	
Foliage	$15.0 + 2.5^{a}$
Root estimate	$114.9 + 28.6^{a}$

CAI (m³/ha/yr)

Net production (t/ha/yr)

Stem wood	$\left.\begin{array}{l}\\ \\\end{array}\right\}1.66 + 0.94^{a} \text{ (or } 3.9 + 1.3^{a})^{c}$
Stem bark	
Branches	$0.44 + 1.62^{d} + 0.27^{a} + 0.36^{ad} \text{ (or } 1.3 + 2.2^{a})^{c}$
Fruits etc.	$1.21^{d} + 0.02^{ad}$
Foliage	$0.08 + 2.03^{d} + 0.02^{a} + 1.10^{ad} \text{ (or } 3.2 + 1.0^{a})^{c}$
Root estimate	$0.9 + 0.3^{a} \text{ (or } 2.0 + 3.8^{a})^{c}$

Twelve *A. firma* and 8 understorey trees were sampled, and the roots of 8 *A. firma* trees were excavated. Stand biomass values for a 0.12 ha plot were derived from regressions on D, except for the biomass of understorey roots which was estimated assuming top/root ratios to be 4.2. Biomass values given above are the means over 4 years. Nutrient contents were determined.

a. Understorey values.
b. All-sided LAI was 12.1.
c. Alternative values (in the brackets) estimated as the new growth over the previous year.
d. Litterfall measured over 1 to 3 years.

Yamamoto, T. and Sanada, E. (1970). Nutrients uptake by planted Todo-fir (*Abies sachalinensis* Mast.), nutrient circulation and a change of soil in forest land. *Bull. Govt Forest Exp. Stn Tokyo* 229, 93-121.

43°33'N   142°10'E   100-160 m   Japan, Hokkaido, Takikawa Prefecture,
                                    Tokiwa city.

Plantations.
                                    Abies sachalinensis

Age (years)	8	12	23	29	35
Trees/ha	2870	2726	1849	1427	1178
Tree height (m)	1.0	3.8	8.2	10.4	16.2
Basal area (m²/ha)	1.8	5.3	17.6	25.2	48.9
Leaf area index					
Stem volume (m³/ha)					
Dry biomass (t/ha) — Stem wood	} 0.3	} 8.5	} 33.1	} 48.1	} 123.6
Stem bark					
Branches	0.3	2.6	8.6	15.8	50.6
Fruits etc.					
Foliage	0.4	4.2	11.3	13.4	22.0
Root estimate					
CAI (m³/ha/yr)					
Net production (t/ha/yr) — Stem wood					
Stem bark					
Branches					
Fruits etc.					
Foliage					
Root estimate					

Stand biomass values for the above five 500 m² plots were estimated using regression methods. Nutrient contents were determined.

Satoo, T. (1973a). Materials for the studies of growth of forest stands. XI Primary
production relations in a young plantation of *Abies sachalinensis* in Hokkaido.
Bull. Tokyo Univ. For. **66**, 127–137.

---

ca.43°13'N   142°23'E   230 m   Japan, Hokkaido, Tokyo University Forest.

Plantation.
Brown forest   *Abies sachalinensis*
soil.

Age (years)	26
Trees/ha	2400
Tree height (m)	10.9
Basal area (m²/ha)	32.6
Leaf area index	
Stem volume (m³/ha)	

Dry biomass (t/ha)

Stem wood	$\left.\begin{array}{l}\\\\\end{array}\right\}$ 62.6 (or 64.1, 65.2)^a
Stem bark	
Branches	16.3 (or 18.8, 13.9)^a
Fruits etc.	
Foliage	14.6 (or 15.2, 11.7)^a
Root estimate	

CAI (m³/ha/yr)

Net production (t/ha/yr)

Stem wood	$\left.\begin{array}{l}\\\\\end{array}\right\}$ 7.3^b (or 7.9, 6.1)^{ab}
Stem bark	
Branches	4.0^b (or 4.6, 3.5)^{ab}
Fruits etc.	
Foliage	3.4 (or 4.1, 2.7)^a
Root estimate	

Seven trees were sampled and stand biomass values for a 700 m² plot were derived
from regressions on D. Production was estimated over the previous one year.
a. Alternative values, estimated by proportional basal area allocation or by
multiplying mean tree values by the number of trees per hectare (written left
and right, respectively, within the brackets).
b. Excluding woody litterfall and any mortality.

Ueda, S. (1974). Investigation on the nutrients circulation in the mixed natural forest of Todo-matsu (*Abies sachalinensis* Mast.) and broadleaved trees. *Bull. Kyoto Univ. For.* <u>46</u>, 23-39.

---

ca.44°N 141-144°E   200 m   Japan, northern Hokkaido.

| Clay loam soil, low in phosphate. | *Abies sachalinensis* (29%)[a]
Quercus crispula (24%)[a]
Acer mono var. *marmoratum* (17%)[a] | *A. sachalinensis* (62%)[a]
Betula ermanii (18%)[a]
Q. crispula (13%)[a] |

Both stands with *Sasa kurilensis* understorey

Age (years)		
Trees/ha	130 + 418	343 + 208
Tree height (m)		
Basal area (m²/ha)		
Leaf area index		
Stem volume (m³/ha)		

Dry biomass (t/ha)

Stem wood	} 25.5 + 62.8 + 13.2	} 53.1 + 32.9 + 8.5
Stem bark		
Branches	7.4 + 23.7 + 0.0	20.4 + 10.5 + 0.0
Fruits etc.		
Foliage	3.7 + 2.3 + 3.1	5.1 + 0.9 + 2.4
Root estimate		

Net production (t/ha/yr)

CAI (m³/ha/yr)		
Stem wood		
Stem bark		
Branches		
Fruits etc.		
Foliage	$0.74^b + 2.24^b + 0.70^b$	$1.00^b + 0.80^b + 0.48^b$
Root estimate		

Eight trees were sampled in each stand. Stand biomass values for plots of 1.00 ha (left column) and 0.16 ha (right column) were derived from regressions on D^2H. Values are given above separately for *A. sachalinensis* plus broadleaved trees plus the bamboo understorey (written left to right; the understorey stem number is not given). Nutrient contents were determined.
a. Percentage of the total overstorey biomass.
b. Foliage litterfall, including bamboo understorey culms.

Tadaki, Y., Hatiya, K. and Miyauchi, H. (1967). Studies on the productivity of *Abies veitchii* in the natural forests at Mount Fuji. *J. Jap. For. Soc.* <u>49</u>, 421–428.

ca.35°30'N 138°40'E	(alt. given below)	Japan, Yamanashi Prefecture, NNW slope of Mount Fuji.		
Black volcanic ash soils.		*Abies veitchii*		
	Planted in a nursery	Naturally seeded in an old nursery		
	1530 m	1530 m	1700 m	1640 m
Age (years)	4	20	40–70	60–90
Trees/ha	10 million	19500	9700	3179
Tree height (m)	0.5	4.5	6.8	14.6
Basal area (m²/ha)		33.7	56.8	64.8
Leaf area index	5.5	8.1	9.7	8.2
Stem volume (m³/ha)	12	114	285	516
Dry biomass (t/ha) Stem wood	} 4.9	} 45.7	} 107.6	} 190.2
Stem bark				
Branches	2.2	9.3	16.3	16.5
Fruits etc.				
Foliage	5.5	14.0	17.6	16.7
Root estimate	4.3	16.3	36.9	61.8
CAI (m³/ha/yr)		23.1	21.7	18.8
Net production (t/ha/yr) Stem wood	} 4.1[a]	} 9.6[a]	} 8.2[a]	} 6.9[a]
Stem bark				
Branches		2.5[a]	2.1[a]	1.7[a]
Fruits etc.				
Foliage	1.6[b]	2.9[b]	3.4[b]	3.3[b]
Root estimate	2.7	3.6	3.1	2.6

Five to ten trees were sampled per stand, and stand biomass values for the above four 20–100 m² plots were derived by proportional basal area allocation. Several root systems were excavated in each stand. Projected leaf areas were measured. Stem increments were estimated over one year; branch increments were assumed to be the same as stem increments within the crowns; roots were assumed to grow at the same relative rates as above-ground woody parts.
a. Excluding woody litterfall and mortality.
b. New foliage biomass.

Tadaki, Y., Hatiya, K., Tochiaki, K., Miyauchi, H. and Matsuda, U. (1970). Studies on the production structure of forest. XVI Primary productivity of *Abies veitchii* forests in the subalpine zone of Mount Fuji. *Bull. Govt Forest Exp. Stn Tokyo* <u>229</u>, 1-22.

35°30'N 138°40'E   (alt. given below)   Japan, Yamanashi Prefecture, NNW slope of Mount Fuji.

Abies veitchii

	Planted in a nursery	Naturally seeded in an old nursery	Plantation		
	1500 m	1530 m	1530 m	1700 m	1660 m
Age (years)	5	ca.25	23	40-45	45-130
Trees/ha	630000	12106	2076	3814	1204
Tree height (m)	0.7	5.3	8.5	10.1	16.3
Basal area (m²/ha)		33.4	29.3	58.0	63.4
Leaf area index	7.7	10.9	12.8	7.8	10.6
Stem volume (m³/ha)	26	117	138	341	568
Dry biomass (t/ha)					
Stem wood	} 9.3	} 42.1	} 45.0	} 129.2	} 205.7
Stem bark					
Branches	3.5	13.6	17.0	16.9	32.3
Fruits etc.					
Foliage	7.3	18.3	21.3	13.3	18.8
Root estimate	6.1	17.5	25.9	40.6	54.2
CAI (m³/ha/yr)		15.5	22.0	10.6	14.0
Net production (t/ha/yr)					
Stem wood	} 4.1[a]	} 5.3[a]	} 7.2[a]	} 4.1[a]	} 5.0[a]
Stem bark					
Branches		3.2[a]	3.7[a]	1.4[a]	1.8[a]
Fruits etc.					
Foliage	2.5[b]	3.4[b]	4.5[b]	4.1[b]	4.4[b]
Root estimate	2.0	2.7	4.5	1.5	1.6

Eight trees were sampled per stand, and stand biomass values for the above 20-100 m² plots were derived by proportional basal area allocation. Several root systems were excavated. Projected leaf areas were measured. Stem increments were estimated over one year; branch increments were assumed to be the same as stem increments within the crowns; roots were assumed to grow at the same relative rate as above-ground woody parts.

a. Excluding woody litterfall and mortality.
b. New foliage biomass.

Oohata, S. and Oniishi, C. (1974). Some discussions on tree form and dry matter
  production of a fir stand at Tanohara on Mount Ontake. *Bull. Kyoto Univ. For.*
  <u>46</u>, 51–57.

---

ca.36°00'N 137°30'E   2100 m   Japan, Nagano Prefecture, Mount Ontake.

Abies veitchii

Stunted trees near the snow line

Age (years)	Mature	Mature	Mature
Trees/ha	2850	2325	2400
Tree height (m)	2.9	4.2	3.4
Basal area (m²/ha)	33.5	58.4	37.8
Leaf area index			
Stem volume (m³/ha)			
Dry biomass (t/ha)			
Stem wood	} 23.0	} 90.4	} 40.9
Stem bark			
Branches	8.3	31.8	14.9
Fruits etc.			
Foliage	4.2	13.2	7.3
Root estimate			
CAI (m³/ha/yr)			
Net production (t/ha/yr)			
Stem wood	} 0.32[a]	} 1.10[a]	} 0.57[a]
Stem bark			
Branches	0.11[a]	0.37[a]	0.20[a]
Fruits etc.			
Foliage	0.75	2.48	1.33
Root estimate			

Nine trees were sampled and stand values for the above three 400 m² plots were
derived from regressions on D and H.
a. Excluding woody litterfall and any mortality.

Kimura, M., Mototani, I. and Hogetsu, K. (1968). Ecological and physiological studies on the vegetation of Mount Shimagare. VI Growth and dry matter production of young *Abies* stand. *Bot. Mag., Tokyo* 81, 287-296.

Kimura, M. (1969). (as above) VII Analysis of production processes of young stand based on the carbohydrate economy. *Bot. Mag., Tokyo* 82, 6-19.

Kimura, M. (1963). Dynamics of vegetation in relation to soil development in northern Yatsugataki mountains. *Jap. J. Bot.* 18, 255-287.

ca.36°30'N 138°00'E   (alt. given below)   Japan, Yatsugataki Mountains.

	Abies veitchii	*A. veitchii* (58%)[a] *Abies mariesii* (30%)[a] *et al.*
	(Kimura *et al.* 1968, 1969)	(Kimura 1963)
	2340 m	2250 m
Age (years)	15	40-100
Trees/ha		4625
Tree height (m)	0.9	11
Basal area (m²/ha)		61.5
Leaf area index	5.0	
Stem volume (m³/ha)		
Dry biomass (t/ha)		
Stem wood	} 9.8	} 154
Stem bark		
Branches	7.1	28
Fruits etc.		
Foliage	10.2	21
Root estimate	6.0	55[b]
CAI (m³/ha/yr)		
Net production (t/ha/yr)		
Stem wood	} 1.1[c]	} 4.2[c]
Stem bark		
Branches	1.4[d]	1.3[f] } + 0.9[e]
Fruits etc.		
Foliage	3.9	3.9[g]
Root estimate	0.9	1.7[b]

Eleven trees were sampled from the 40- to 100-year-old stand, and stand values for a 400 m² plot were derived from regressions on D. The biomass and production of the 15-year-old stand were estimated by completely harvesting ten 0.5 m² plots at the beginning and end of one growing season.

a. Percentage of the total basal area.
b. Assumed to be one-third the value of above-ground woody parts.
c. Excluding any mortality.
d. Excluding woody litterfall.
e. Litterfall.
f. Assumed to be 30% of the stem increment value.
g. Estimated from the foliage biomass and average foliage longevity.

Tadaki, Y., Sato, A., Sakurai, S., Takeuchi, I. and Kawahara, T. (1977). Studies on the production structure of forest. XVII Structure and primary production in subalpine "dead tree strips" *Abies* forest near Mount Asahi. *Jap. J. Ecol.* <u>27</u>, 83-90.

---

ca.36°N 138°E   2420-2440 m   Japan, Yamaguchi and Nagano Prefectures.

Abies veitchii, *Abies mariesii*, and *Picea jezoensis* var. *hondoensis*

Stands that naturally deteriorate and die at age 100-130 and then regenerate.

	62%[a]	81%[a]	97%[a]	90%[a]	74%[a]	91%[a]	59%[a]
Age (years)	1-5	7-19	18-34	31-42	51-79	73-118	119-133
Trees/ha	730000	380000	233000	96000	20800	10200	4133
Tree height (m)	0.2	0.6	1.0	2.6	5.4	7.4	8.1
Basal area (m²/ha)				9.7	62.2	66.9	52.2
Leaf area index	1.4[b]	2.8[b]	6.5[b]	6.8[b]	5.9[b]	5.4[b]	2.8[b]
Stem volume (m³/ha)		7.8	55	179	230	307	232

Dry biomass (t/ha)

	62%[a]	81%[a]	97%[a]	90%[a]	74%[a]	91%[a]	59%[a]
Stem wood } Stem bark }	1.0	6.4	29.2	74.9	97.2	130.3	92.2
Branches	0.4	2.6	8.1	10.6	12.2	15.7	11.3
Fruits etc.	0.0	0.0	0.0	0.9	1.4	1.1	0.2
Foliage	2.3	4.6	13.6	17.1	13.6	14.3	8.3
Root estimate							

CAI (m³/ha/yr)

Net production (t/ha/yr)

	62%[a]	81%[a]	97%[a]	90%[a]	74%[a]	91%[a]	59%[a]
Stem wood } Stem bark }	1.3[c]	3.8[c]	4.7[c]	3.6[c]	3.7[c]	1.2[c]	
Branches		0.8[c]	1.4[c]	1.5[c]	1.5[c]	1.6[c]	0.4[c]
Fruits etc.		0.0	0.0	0.9	1.4	1.1	0.2
Foliage		1.3[d]	2.2[d]	3.0[d]	2.2[d]	2.0[d]	1.7[d]
Root estimate							

Seven to 20 trees were sampled per stand in the autumn, and stand biomass values were estimated for the above plots of 1.0, 1.0, 1.4, 5.0, 25, 50 and 75 m², in columns left to right.

a. Percentage of the total basal area accounted for by *A. veitchii*.
b. Projected LAI values as measured by the authors; approximate all-sided LAI values can be obtained by multiplying by 2.3
c. Excluding woody litterfall and mortality.
d. New foliage biomass.

Satoo, T. and Senda, M. (1958). Materials for the studies of growth in stands. IV Amount of leaves and production of wood in a young plantation of *Chamaecyparis obtusa*. *Bull. Tokyo Univ. For.* <u>54</u>, 71-100.

Harada, H. Satoo, H., Hotta, I. and Tadaki, Y. (1969). On the amount of nutrient contained in 28-year-old *Cryptomeria* forest (*C. japonica* D. Don) and *Chamaecyparis* forest (*C. obtusa* Sieb. et Zucc.). *J. Jap. For. Soc.* <u>51</u>, 125-133.

Japan Plantations	ca.35°09'N 140°09'E -- Chiba Prefecture *Chamaecyparis obtusa* (Satoo and Senda 1958)			ca.35°N 138°E 680-760 m *C. obtusa* with bamboo understorey (Harada *et al.* 1969)	
				Poor gravelly soil	Fertile loam
Age (years)	28	28	28	28	28
Trees/ha	5941	6747	6234	3483	2004
Tree height (m)	10.5	9.0	9.3	5.3	9.9
Basal area (m²/ha)	39.5	35.6	33.7	24	38
Leaf area index					
Stem volume (m³/ha)				74	189
Dry biomass (t/ha) — Stem wood				$\big\}32.6\big\}$ $+5.7^b\big\}$	$82.0\big\}$ $+1.1^b$
Stem bark					
Branches	15.7	15.0	22.0	8.0	16.4
Fruits etc.					
Foliage	12.4	10.3	18.2	$11.7+2.2^b$	$17.5+0.3^b$
Root estimate				13.6	27.1
CAI (m³/ha/yr)	17.6^a	12.1^a	16.8^a		
Net production (t/ha/yr) — Stem wood	$\big\}7.3^a$	$\big\}4.0^a$	$\big\}7.0^a$		
Stem bark					
Branches					
Fruits etc.					
Foliage					
Root estimate					

Satoo and Senda (1958) sampled 28 trees and derived stand values for the above three plots of 133-182 m² from regressions on D.

Harada *et al.* (1969) sampled 5 trees from the poor site, 8 from the fertile site, and derived stand values for the above two plots of 200-400 m² from regressions on D^2H; nutrient contents were determined.

a. Values derived from the authors' Table 7 which expressed volume production per unit of foliage biomass.

b. Bamboo understorey.

Yamakura, T., Saito, H., Shidei, T. (1972a). Investigations on the primary produc-
tivity and production structure of *Chamaecyparis obtusa* Sieb. et Zucc. stands.
Bull. Kyoto Univ. For. 43, 106-123.
Yamakura, T., Saito, H., Shidei, T. (1972b). Production structure of underground
parts of Hinoki (*C. obtusa*) stand. I Estimation of root production by means of
root analysis. *J. Jap. For. Soc.* 54, 118-125.
Saito, H. (1977). *Chamaecyparis* plantations. In: "Primary Productivity in Japanese
Forest" (T. Shidei and T. Kira, eds) pp.252-268. JIBP Synthesis vol.16.
University of Tokyo Press.

35°00'N 136°20'E   440 m   Japan, Shiga Prefecture, Hino.

Plantations.
Brown forest   *Chamaecyparis obtusa*
soils.

Age (years)	30	40
Trees/ha	3500	1300
Tree height (m)	10.4	15.9
Basal area (m²/ha)	45.6	60.4
Leaf area index	5.8^a	6.3^a
Stem volume (m³/ha)	268	507

Dry biomass (t/ha)			
	Stem wood	} 115.0	} 219.0
	Stem bark		
	Branches	12.0	25.0
	Fruits etc.	0.0	0.2
	Foliage	14.0	19.0
	Root estimate	43.0	76.0

CAI (m³/ha/yr)		20.7	18.1

Net production (t/ha/yr)			
	Stem wood	} 9.8 (or 8.9)^b	} 11.5 (or 7.8)^b
	Stem bark		
	Branches	0.6^c (or 1.7)^b	1.0 + 0.6^d + 0.2^e (or 2.1)^b
	Fruits etc.	0.0	0.6 (or 0.2)^b
	Foliage	0.9 + 2.4^d + 1.0^e (or 3.4)^b	0.1 + 3.8^d + 1.4^e (or 4.6)^b
	Root estimate	3.3 (or 3.0)^b	3.9 (or 2.4)^b

Six trees were sampled per stand in the autumn and roots were excavated. Stand
biomass values for a 150 m² plot in the 30-year-old stand and a 500 m² plot in the
40-year-old stand were derived from regressions on D²H.
a. All-sided LAI values can be obtained by multiplying by 2.3.
b. Alternative values obtained by measuring the new growth (from Saito 1977).
c. Excluding any woody litterfall.
d. Litterfall.
e. Consumption and other losses.

Ogata, N., Nagatomo, Y., Kaminaka, S. and Takeshita, K. (1973). Effects of fertilization and site on the matter production of *Chamaecyparis* stand, and differences
between *Chamaecyparis* and *Cryptomeria* stands. *A. Rep. Kyushu Branch, Govt Forest
Exp. Stn* <u>15</u>, 21-24.

---

ca.33°N 131°E   --   Japan, Kyushu.

Plantations   *Chamaecyparis obtusa*

Age (years)	36	51	51	62	62
Trees/ha	1141	1149	1186	1012	2555
Tree height (m)	12.5	14.0	12.7	21.2	14.4
Basal area (m²/ha)	47.9	47.0	45.0	55.7	63.8
Leaf area index	6.8^a	5.3^a	4.8^a	5.4^a	4.7^a
Stem volume (m³/ha)	312	343	304	554	467
Dry biomass (t/ha) Stem wood	}160.2	}166.8	}142.9	}217.0	}207.7
Stem bark					
Branches	13.7	26.7	20.3	13.5	11.4
Fruits etc.					
Foliage	20.7	16.3	15.3	15.8	14.9
Root estimate					
CAI (m³/ha/yr)	13.9	11.3	8.3	18.9	11.2
Net production (t/ha/yr) Stem wood	}13.4	}10.8	}8.6	}12.3	}9.0
Stem bark					
Branches					
Fruits etc.					
Foliage					
Root estimate					

Stand values for the above 100 m² plot were derived using published regressions on
D and H.
a. All-sided LAI values can be obtained by multiplying by 2.3.

Takeuchi, I., Tadaki, Y., Hatiya, K., Kawahara, T. and Sato, A. (1975). Thinning experiment of 30-year-old plantation of *Chamaecyparis obtusa*. In reference to line thinning. *Bull. Govt Forest Exp. Stn Tokyo* <u>272</u>, 141-155.

Tadaki, Y., Ogata, N., Nagatomo, Y. and Yoshida, T. (1966). Studies on the production structure of forest. X Primary productivity of an unthinned 45-year-old stand of *Chamaecyparis obtusa*. *J. Jap. For. Soc.* <u>48</u>, 387-393.

Japan		ca.33°N 350 m			32°30'N 130°30'E 750 m
Plantations		*Chamaecyparis obtusa*			Kumamoto Prefecture
		Values before thinning (Takeuchi *et al.* 1975)			Unthinned (Tadaki *et al.* 1966)
Age (years)		30	30	30	30	45
Trees/ha		1951	1776	2097	2266	3400
Tree height (m)		14.9	15.2	14.7	14.5	16.0
Basal area (m²/ha)		39.5	39.5	39.7	41.2	70.5
Leaf area index						5.1
Stem volume (m³/ha)		304	297	301	313	560

Dry biomass (t/ha):

	Stem wood	141.1	137.7	139.2	145.4	229.6
	Stem bark					
	Branches	14.4	14.9	13.8	14.1	12.8
	Fruits etc.					0.1
	Foliage	13.3	13.7	12.9	13.2	11.9
	Root estimate	50.7	49.9	49.8	51.9	72.7

CAI (m³/ha/yr)		14.6	14.6	14.4	14.7	15.6

Net production (t/ha/yr):

	Stem wood	6.7^a	6.7^a	6.3^a	6.7^a	8.6^{ab}
	Stem bark					
	Branches	1.8^a	1.9^a	1.7^a	1.8^a	
	Fruits etc.					0.1
	Foliage	4.0	4.1	3.9	4.0	3.0^c
	Root estimate	2.4	2.4	2.3	2.4	3.7^b

Eight trees were sampled per plot. Stand biomass values for the above four 400 m² plots containing the 30-year-old trees were derived from regressions on D²H. Stand biomass values for a 100 m² plot containing the 45-year-old trees were derived by proportional basal area allocation. In all cases root biomass was assumed to be 30% of the above-ground woody biomass. Projected leaf areas were measured in the 45-year-old plot.

a. Excluding woody litterfall and mortality.

b. Tadaki *et al.* (1966) gave the total woody increment (stems, branches and roots); here it is assumed that 30% of that total was roots.

c. Assumed to be 25% of the foliage biomass.

Kawanabe, S., Tamai, S. and Tsutsumi, T. (1975). Effects of thinning on the biomass and the light climate in *Chamaecyparis obtusa* Sieb. et Zucc. stand. *Bull. Kyoto Univ. For.* <u>47</u>, 26-33.

34°04'N 135°30'E   1000 m   Japan, Nara Prefecture.

Plantations.
Deep brown
forest soils.

Chamaecyparis obtusa

	Unthinned				30% thinned at age 45[a]			
Age (years)	45	46	45	46	45	46	45	46
Trees/ha	1152	1152	1040	1040	704	704	592	592
Tree height (m)	13.7	13.9	13.8	14.0	14.3	14.5	15.4	15.6
Basal area (m²/ha)	38.0	39.3	32.4	33.2	27.0	27.8	26.3	27.0
Leaf area index								
Stem volume (m³/ha)	262	275	225	234	193	202	203	212

Dry biomass (t/ha)

Stem wood	}113.1	}118.5	}96.8	}100.7	}83.1	}87.1	}87.5	}91.2
Stem bark								
Branches	16.2	17.3	13.3	14.1	12.7	13.6	14.7	15.6
Fruits etc.								
Foliage	10.6	11.2	8.9	9.4	8.0	8.3	8.7	9.1
Root estimate								

CAI (m³/ha/yr)	12.5		9.0		9.3		8.6	

Net production (t/ha/yr)

Stem wood	}5.4		}3.9		}4.0		}3.7	
Stem bark								
Branches								
Fruits etc.								
Foliage								
Root estimate								

Nine trees were sampled and stand values for the above 625 m² plots were derived from regressions on D²H.
a. The values given here refer to the stands after 30% of the trees had been removed.

Continued from p.138.

Same as p.138.

All stands were thinned at age 45.

	32%[a]		41%[a]		51%[a]		55%[a]	
Age (years)	45	46	45	46	45	46	45	46
Trees/ha	576	576	544	544	448	448	464	464
Tree height (m)	14.8	15.0	15.3	15.5	14.7	14.9	14.7	14.9
Basal area (m²/ha)	23.3	23.9	25.1	25.8	18.1	18.7	18.4	19.1
Leaf area index								
Stem volume (m³/ha)	172	179	193	201	132	138	136	143
Dry biomass (t/ha) — Stem wood / Stem bark	}74.3	}77.2	}83.0	}86.5	}56.8	}59.6	}58.8	}61.7
Dry biomass (t/ha) — Branches	11.7	12.4	14.2	15.1	8.8	9.4	9.3	9.9
Dry biomass (t/ha) — Fruits etc.								
Dry biomass (t/ha) — Foliage	7.2	7.5	8.3	8.7	5.5	5.8	5.7	6.0
Dry biomass (t/ha) — Root estimate								
CAI (m³/ha/yr)		6.7		8.1		6.5		6.7
Net production (t/ha/yr) — Stem wood / Stem bark		}2.9		}3.5		}2.8		}2.9
Net production (t/ha/yr) — Branches								
Net production (t/ha/yr) — Fruits etc.								
Net production (t/ha/yr) — Foliage								
Net production (t/ha/yr) — Root estimate								

Nine trees were sampled and stand values for the above 625 m² plots were derived from regressions on D^2H.

[a]. Percentage of trees removed at age 45; all values in the table refer to the stands after thinning.

Kawahara, T., Tadaki, Y., Takeuchi, I., Sato, A., Higuchi, K. and Kamo, K. (1979). Productivity and cycling of organic matter in natural *Fagus crenata* and two planted *Chamaecyparis obtusa* forests. *Jap. J. Ecol.* <u>29</u>, 387-395.

36°47'N 139°56'E   (alt. given below)   Japan, Tochigi Prefecture, near Yaita city.

Plantations   *Chamaecyparis obtusa*

	600 m	730 m
Age (years)	17	48
Trees/ha	3600	1230
Tree height (m)	7.8	17.6
Basal area (m²/ha)	34.5	49.3
Leaf area index		
Stem volume (m³/ha)	153	488

Dry biomass (t/ha)		600 m	730 m
	Stem wood	} 69.3	} 170.2
	Stem bark		
	Branches	10.8	18.8
	Fruits etc.		
	Foliage	14.3	15.7
	Root estimate	31.5[a]	68.2[a]

Net production (t/ha/yr)	CAI (m³/ha/yr)		
	Stem wood		
	Stem bark	} $17.3 + 0.01^b$	} $14.7 + 0.03^b$
	Branches		
	Fruits etc.	0.02^b	0.06^b
	Foliage	2.42^b	2.67^b
	Root estimate		

Stand biomass values for the above plots of 100 m² and 1600 m² (in columns left and right, respectively) were derived using published regressions on D and D²H.
a. Assumed to be one third the value of above-ground parts.
b. Litterfall, measured over 4 years.

Satoo, T. (1979a). Standing crop and increment of bole in plantations of *Chamae-cyparis obtusa* near an electric power plant in Owase, Mie. *Jap. J. Ecol.* <u>29</u>, 103–109.

Satoo, T. (1979b). Leaf litter production in plantations of *Chamaecyparis obtusa* near an electric power plant in Owase, Mie. *Jap. J. Ecol.* <u>29</u>, 205–208.

Satoo, T. (1979c). Production of reproductive organs in plantations of *Chamae-cyparis obtusa* near an electric power plant in Owase, Mie. *Jap. J. Ecol.* <u>29</u>, 315–321.

ca.34°44'N 136°12'E   400–500 m   Japan, Mie Prefecture, near Owase city.

Plantations.   *Chamaecyparis obtusa*

	Stands exposed to a sulphur polluted atmosphere since age 33.		Unpolluted
Age (years)	40	40	38
Trees/ha	1345	1231	1206
Tree height (m)	17.3	17.1	16.7
Basal area (m²/ha)	38.2	34.1	37.4
Leaf area index			
Stem volume (m³/ha)	300 (or 307)^a	241 (or 242)^a	255 (or 259)^a
Dry biomass (t/ha)			
Stem wood	137.3 (or 144.0)^a	105.6 (106.4)^a	115.8 (or 106.3)^a
Stem bark	10.0 (or 10.0)^a	7.1 (or 6.9)^a	9.0 (or 8.2)^a
Branches	13.3 (or 17.1)^a	13.2 (or 15.7)^a	16.2 (or 14.7)^a
Fruits etc.	0.1 (or 0.1)^a	0.1 (or 0.1)^a	0.1 (or 0.1)^a
Foliage	6.7 (or 7.8)^a	5.9 (or 6.9)^a	10.2 (or 9.8)^a
Root estimate			
CAI (m³/ha/yr)	7.2 (or 6.6)^a	7.9 (or 8.1)^a	12.4 (or 11.9)^a
Net production (t/ha/yr) Stem wood / Stem bark	}3.63^b(or 3.26)^{ab}	}3.68^b(or 2.85)^{ab}	}5.28^b(or 5.10)^{ab}
Branches			
Fruits etc.	0.06^c	0.08^c	0.18^c
Foliage	1.27^c	1.57^c	2.60^c
Root estimate			

Five trees were sampled per stand, and stand values for the above 200 to 400 m²
plots were derived from regressions on D.
a. Alternative values derived by proportional basal area allocation.
b. Excluding any woody litterfall and mortality.
c. Litterfall measured over 17 months.

Miyamoto, M., Tanimoto, T., and Ando, T. (1980). Analysis of the growth of Hinoki (*Chamaecyparis obtusa*) artificial forests in Shikoku district. *Bull. Forestry and Forest Products Research Inst. (Ibaraki, Japan)* 309, 89-107.

34°10'N   133°40'E   150 m   Japan, Kagawa Prefecture, Manno Town.

Plantations.   *Chamaecyparis obtusa*

All stands on 'site class 1'.

Age (years)		20	20	20	20
Trees/ha		2600	3025	3125	3244
Tree height (m)		8.9	9.0	9.2	9.4
Basal area (m²/ha)		30.1	30.8	31.1	34.1
Leaf area index		6.4^a	6.3^a	6.3^a	7.1^a
Stem volume (m³/ha)		139	139	139	157
Dry biomass (t/ha)	Stem wood	}55.2	}55.6	}55.7	}62.2
	Stem bark				
	Branches	7.9	7.5	7.3	8.7
	Fruits etc.				
	Foliage	14.5	14.4	14.3	16.3
	Root estimate	22.2^b	22.1^b	22.1^b	24.9^b
CAI (m³/ha/yr)		18.4	17.7	17.4	20.5
Net production (t/ha/yr)	Stem wood	}7.3^c	}7.1^c	}7.0^c	}8.1^c
	Stem bark				
	Branches	2.2^c	2.1^c	2.0^c	2.4^c
	Fruits etc.				
	Foliage	3.6	3.6	3.6	4.1
	Root estimate	2.9	2.8	2.8	3.2

Fourteen trees were sampled in each stand, and biomass values for each of the above 225 - 400 m² plots were estimated by proportional basal area allocation. Increments were estimated over one year.
a. Approximate all-sided LAI values may be obtained by multiplying by 2.3.
b. Assuming top/root ratios to be 3.5.
c. Excluding any woody litterfall and mortality.

Continued from p.142.

33°30'N　133°30'E　　1250 m　　Japan, Kochi Prefecture, Terakawa National Forest.

Plantations.　　　　　　　　　　*Chamaecyparis obtusa*

All stands on 'site class 1'.

Age (years)	42	42	42	42	42
Trees/ha	1338	1375	1463	1488	1613
Tree height (m)	15.8	16.6	16.0	15.9	14.6
Basal area (m²/ha)	60.7	57.6	61.0	64.4	64.0
Leaf area index	5.6[a]	5.4[a]	5.8[a]	6.0[a]	5.9[a]
Stem volume (m³/ha)	468	482	486	506	464
Dry biomass (t/ha) — Stem wood / Stem bark	}188.3	}178.3	}189.5	}200.0	}199.2
Branches	18.0	17.2	17.9	18.9	18.3
Fruits etc.					
Foliage	16.2	15.4	16.2	17.1	16.8
Root estimate	63.6[b]	60.3[b]	63.9[b]	67.4[b]	66.9[b]
CAI (m³/ha/yr)	25.3	26.2	26.1	27.3	24.8
Net production (t/ha/yr) — Stem wood / Stem bark	}10.2[c]	}9.7[c]	}10.2[c]	}10.8[c]	}10.7[c]
Branches	2.7[c]	2.5[c]	2.6[c]	2.8[c]	2.7[c]
Fruits etc.					
Foliage	4.1	3.9	4.1	4.3	4.2
Root estimate	3.4	3.3	3.4	3.6	3.6

Ten trees were sampled in each stand.

See p.142.

Continued from p.143.

33°10'N 133°10'E   450 m   Japan, Kochi Prefecture, Kubokawa Forest Station.

Plantations.	*Chamaecyparis obtusa*			Fertilizers applied at age 43		
	'Site class 1'			'Site class 3'	'Site class 2'	'Site class 2'
Age (years)	48	50	48	49	49	50
Trees/ha	1325	1434	1600	2854	2470	3192
Tree height (m)	17.2	17.8	17.6	11.3	13.3	13.7
Basal area (m²/ha)	56.5	58.3	59.4	47.6	50.8	59.5
Leaf area index	4.7^a	5.9^a	5.4^a	6.6^a	5.7^a	6.2^a
Stem volume (m³/ha)	455	472	503	264	331	412
Dry biomass (t/ha) Stem wood	}190.9	}179.6	}208.2	}114.2	}140.8	}171.6
Stem bark						
Branches	32.9	20.4	25.7	26.4	20.9	22.2
Fruits etc.						
Foliage	12.1	12.8	13.3	14.7	11.7	11.4
Root estimate	67.4^b	60.8^b	70.6^b	44.4^b	49.5^b	58.6^b
CAI (m³/ha/yr)	11.6	16.9	17.2	16.7	14.1	12.5
Net production (t/ha/yr) Stem wood	}4.8^c	}6.4^c	}7.1^c	}7.2^c	}6.0^c	}5.2^c
Stem bark						
Branches	1.8^c	2.0^c	1.6^c	2.7^c	1.7^c	1.2^c
Fruits etc.						
Foliage	3.0	3.2	3.3	3.7	2.9	2.8
Root estimate	1.7	2.2	2.4	2.8	2.1	1.8

Eight trees were sampled in each stand.

See p.142.

Ando, T., Takeuchi, I., Saito, A. and Watanabe, H. (1969). Some observations of dry-matter production on the artificial two-storied stand. *J. Jap. For. Soc.* 51, 102-107.

		Before pruning or thinning			
ca.33°30'N 132°30'E -- Japan, Ehime Prefecture. *Cryptomeria japonica* (overstorey) with Plantations. *Chamaecyparis obtusa* (understorey)					
		Overstorey	Understorey	Overstorey	Understorey
Age (years)		8	4	68	36
Trees/ha		4100 +	4700	383 +	950
Tree height (m)		6.7	2.1	23.6	9.5
Basal area (m²/ha)		19.8 +	1.0	38.2 +	8.1
Leaf area index					
Stem volume (m³/ha)		70.5 +	3.6	452 +	36
Dry biomass (t/ha)	Stem wood	} 23.4 +	1.6	} 160.3 +	15.2
	Stem bark				
	Branches	3.6 +	0.6	14.5 +	4.5
	Fruits etc.	0.0 +	0.0	1.4 +	0.0
	Foliage	11.0 +	2.1	17.0 +	3.8^a
	Root estimate	10.8^b +	1.2^b	55.2^b +	6.7^b
CAI (m³/ha/yr)		25.6 +	1.5	11.8 +	1.9
Net production (t/ha/yr)	Stem wood	} 8.5^c +	0.6^c	} 4.2^c +	0.8^c
	Stem bark				
	Branches	2.2^c +	0.2^c	0.6^c +	0.5^c
	Fruits etc.	0.0 +	0.0	1.4 +	0.0
	Foliage	5.5 +	1.0	4.2 +	1.0
	Root estimate	3.9 +	0.5	1.4 +	0.4

Stand biomass values were derived by proportional basal area allocation. The younger plot was 100 m², the older one was 600 m².
a. An error in the authors' Table 2 has been corrected.
b. Assuming top/root ratios to be 3.5.
c. Excluding any woody litterfall and mortality.

Tadaki, Y., Ogata, N., Nagatomo, Y., Yoshioka, K. and Miyagawa, Y. (1964). Studies on production structure of forest. VI Productivities of scaffolding producing stands of *Cryptomeria japonica*. *J. Jap. For. Soc.* 46, 246-253.

Tadaki, Y., Ogata, N. and Nagatomo, Y. (1965). The dry matter productivity in several stands of *Cryptomeria japonica* in Kyushu. *Bull. Govt Forest Exp. Stn Tokyo* 173, 45-66.

ca.32°45'N 130°00'E   100 m   Japan, Nagasaki Prefecture, Higashi-Nagasaki.

Plantations.   *Cryptomeria japonica*

Unthinned stands, raised from seed.

Age (years)	11	11	22	22	31
Trees/ha	10000	9500	6400	6300	3600
Tree height (m)	6.0	5.3	9.1	9.4	10.7
Basal area (m²/ha)	27.1	21.3	38.8	38.1	40.1
Leaf area index	4.7[a]	4.7[a]	5.1[a]	5.1[a]	5.9[a]
Stem volume (m³/ha)	98	77	220	215	247

Dry biomass (t/ha)					
Stem wood	} 34.9	} 27.4	} 83.2	} 81.6	} 93.8
Stem bark					
Branches	3.4	2.7	7.7	7.6	12.3
Fruits etc.					
Foliage	18.7	14.7	18.7	18.3	21.8
Root estimate					

	11	11	22	22	31
CAI (m³/ha/yr)	13.2	10.4	17.3	16.9	22.0

Net production (t/ha/yr)					
Stem wood	} 4.97[bd]	} 3.90[bd]	} 7.58[bd]	} 7.43[bd]	} 8.99[bd]
Stem bark					
Branches					
Fruits etc.					
Foliage	4.68[c]	3.68[c]	4.68[c]	4.58[c]	5.45[c]
Root estimate	1.24[d]	0.98[d]	1.90[d]	1.86[d]	2.25[d]

Eight or nine trees were sampled per stand in November, and biomass values for the above plots (25 m² at age 11, 100 m² at age 22 and 510 m² at age 31) were derived by proportional basal area allocation. Increments were estimated over one year.
a. All-sided LAI values can be obtained by multiplying by 2.3.
b. Excluding any woody litterfall and mortality.
c. Assumed to be 25% of the foliage biomass.
d. The authors gave total woody increment (stems, branches and roots); here it is assumed that the roots grew at the same relative rates as above-ground woody parts; alternatively the roots could be assumed to be 20% of the total increment including leaves.

Tadaki, Y., Ogata, N. and Nagatomo, Y. (1965). The dry matter productivity in several stand of *Cryptomeria japonica* in Kyushu. *Bull. Govt Forest Exp. Stn Tokyo* 173, 45-66.

Japan	ca.33°00'N 131°30'E 850 m Oita Prefecture			ca.32°00'N 131°31'E 150 m Miyazaki Prefecture		
Plantations	*Cryptomeria japonica*					
	Clone 'Yabukuguri'			Clone 'Aka'		
Age (years)	34	34	34	24	34	49
Trees/ha	1333	1952	2420	2110	1239	722
Tree height (m)	12.3	11.4	12.9	12.5	14.7	17.3
Basal area (m²/ha)	59.4	58.0	63.3	46.6	42.7	47.6
Leaf area index	4.1[a]	5.3[a]	5.7[a]	6.4[a]	5.0[a]	4.3[a]
Stem volume (m³/ha)	380	377	409	266	311	372
Dry biomass (t/ha) — Stem wood / Stem bark	}115.7	}110.3	}120.3	}95.2	}88.4	}109.0
Branches	9.0	14.6	10.4	15.6	14.7	13.3
Fruits etc.						
Foliage	16.5	21.1	22.7	25.6	20.0	17.3
Root estimate	35.3[b]	36.5[b]	38.3[b]	34.1[b]	30.8[b]	34.9[b]
CAI (m³/ha/yr)	15.2	22.5	25.6	19.4	17.8	8.5
Net production (t/ha/yr) — Stem wood / Stem bark / Branches / Fruits etc.	}3.6[c]	}5.9[c]	}7.1[c]	}5.5[c]	}4.8[c]	}2.8[c]
Foliage	4.1[d]	5.3[d]	5.7[d]	6.4[d]	5.0[d]	4.3[d]
Root estimate	1.9[e]	2.8[e]	3.2[e]	3.2	2.5	1.8

Eight to ten trees of clone 'Yabukuguri' were sampled per stand in March; three to six trees of clone 'Aka' were sampled per stand in November. Stand biomass values for the above 999-1373 m² plots of 'Yabukuguri' and 123-416 m² plots of 'Aka' were derived by proportional basal area allocation. Increments were estimated over one year.

a. All-sided LAI values can be obtained by multiplying by 2.3.
b. Assumed to be 20% of the total biomass.
c. Excluding any woody litterfall and mortality.
d. Assumed to be 25% of the foliage biomass.
e. The authors gave total woody increment (stem, branches and roots); here it is assumed that 20% of that total was roots.

Yoshioka, K. and Miyagawa, Y. (1965). On the productivity with [the] different [of] stand densities in Sugi (*Cryptomeria japonica* D. Don). *Bull. Nagasaki Agric. For. Exp. Stn* 1, 34-43.

---

ca.33°N 130°E   200-500 m   Japan, Nagasaki Prefecture, Higashi-Nagasaki District.

Plantations.   *Cryptomeria japonica*

Only the 28-year-old stand was thinned;
all the other stands were unthinned.

Age (years)	10	10	11	18	21	22	28	31	34
Trees/ha	4240	8000	9930	2540	5420	7540	2780	3660	1200
Tree height (m)	6.3	6.4	5.4	10.1	8.0	8.4	13.9	10.7	13.1
Basal area (m² /ha)	20.3	24.2	23.6	29.7	28.3	38.8	39.2	40.1	32.6
Leaf area index									
Stem volume (m³ /ha)	83	101	85	179	136	221	304	247	226

Dry biomass (t/ha)

	10	10	11	18	21	22	28	31	34
Stem wood	}31.3	}34.5	}30.6	}62.0	}46.6	}80.7	}108.3	}93.8	}71.6
Stem bark									
Branches	4.8	2.2	3.0	6.4	6.5	7.7	7.8	12.3	12.4
Fruits etc.									
Foliage	11.0	11.7	16.3	14.6	20.1	18.6	14.5	21.8	17.7
Root estimate									

CAI (m³ /ha/yr)

Net production (t/ha/yr)

Stem wood									
Stem bark									
Branches									
Fruits etc.									
Foliage									
Root estimate									

About 65 trees were sampled in all. Stand values for the above 500-510 m² plots were derived from regressions on D.

Tadaki, Y., Ogata, N. and Nagatomo, Y. (1967). Studies on production structure of forest. XI Primary productivities of 28-year-old plantations of *Cryptomeria* of cuttings and seedlings origin. *Bull. Govt Forest Exp. Stn Tokyo* 199, 47-65.

Tadaki, Y. and Kawasaki, Y. (1966). Studies on the production structure of forest. IX Primary productivity of a young *Cryptomeria* plantation with excessively high stand density. *J. Jap. For. Soc.* 48, 55-61.

Japan	ca.33°20'N 131°00'E 300 m Kyushu District, nr Hita city		ca.32°50'N 130°42'E 50 m nr Kumamoto city
Plantations.	*Cryptomeria japonica*		
	(Tadaki *et al.* 1967)		Clones
	Raised from cuttings	Raised from seed	Black volcanic nursery soil
Age (years)	28	28	5
Trees/ha	1250	1150	29500
Tree height (m)	18.5	15.0	5.0
Basal area (m²/ha)	54.9	37.0	37.2
Leaf area index	3.8^a	4.3^a	7.4^a
Stem volume (m³/ha)	501	285	137
Dry biomass (t/ha) — Stem wood	} 177.5	} 110.6	} 50.7
Dry biomass (t/ha) — Stem bark			
Dry biomass (t/ha) — Branches	9.8	14.6	1.8
Dry biomass (t/ha) — Fruits etc.	0.1	0.4	
Dry biomass (t/ha) — Foliage	15.0	17.3	26.5
Dry biomass (t/ha) — Root estimate	50.6^b	35.7^b	18.3
CAI (m³/ha/yr)	23.7	13.1	35.3
Net production (t/ha/yr) — Stem wood	} $8.7^e + 2.5^c$	} $5.5^e + 3.7^c$	} 16.6^{de}
Net production (t/ha/yr) — Stem bark			
Net production (t/ha/yr) — Branches			
Net production (t/ha/yr) — Fruits etc.	0.1	0.4	
Net production (t/ha/yr) — Foliage	3.8^c	4.3^c	6.6^c
Net production (t/ha/yr) — Root estimate	3.7^e	2.3^e	5.9^e

Six trees were sampled per plot, and stand biomass values for the above plots of 200 m² at Hita and 20 m² at Kumamoto were derived by proportional basal area allocation. At Hita very similar alternative values were derived using regressions on D^2H. Increments were estimated over one year.

a. All-sided LAI values can be obtained by multiplying by 2.3.

b. Assumed to be 25% of the above-ground biomass value.

c. Foliage increment and branch litterfall were both assumed to be about 25% of their biomass values.   *d*. Excluding woody litterfall and mortality.

e. The authors gave the total woody increment (stems, branches and roots); here it is assumed that 30% of the woody increment at Hita was roots, and that roots at Kumamoto grew at the same relative rate as above-ground woody parts.

Satoo, T. and Senda, M. (1966). Materials for the studies of growth in stands. VI Biomass, dry matter production, and efficiency of leaves in a young *Cryptomeria* plantation. *Bull. Tokyo Univ. For.* 62, 117-146.

Harada, H., Satoo, H., Hotta, I. and Tadaki, Y. (1969). On the amount of nutrient contained in 28-year-old *Cryptomeria* forest (*C. japonica* D. Don) and *Chamaecyparis* forest (*C. obtusa* Sieb. et Zucc.). *J. Jap. For. Soc.* 51, 125-133.

Japan	ca.43°13'N 142°23'E 230 m Hokkaido	ca.35°N 138°E 680-700 m	
Plantations	*Cryptomeria japonica*[a]	*C. japonica* with bamboo understorey	
		(Harada *et al.* 1969)	
	(Satoo and Senda 1966)	Poor gravelly soil	Fertile loam
Age (years)	29	28	28
Trees/ha	3675	2800	1724
Tree height (m)	10	5.8	12.2
Basal area (m²/ha)		27	45
Leaf area index			
Stem volume (m³/ha)	293 (or 210)[b]	86	245

Dry biomass (t/ha)

Stem wood	} 98.9 (or 71.0)[b]	} 37.0		} 98.3	
Stem bark			+ 6.6[e]		+ 1.1[e]
Branches	9.0 (or 6.9)[b]	11.7		12.9	
Fruits etc.				
Foliage	17.9 (or 14.3)[b]	13.0 + 2.2[e]	17.2 + 0.1[e]
Root estimate			30.3

CAI (m³/ha/yr)	9.1 (or 8.4)[b]		

Net production (t/ha/yr)

Stem wood	} 3.07[c] (or 2.84)[bc]		
Stem bark			
Branches	1.12[c] (or 0.98)[bc]		
Fruits etc.			
Foliage	5.38[d] (or 4.28)[bd]		
Root estimate			

Satoo and Senda (1966) sampled 28 trees and derived stand biomass values for a 600 m² plot from regressions on D.

Harada *et al.* (1969) sampled 5 trees from the poor site, 8 from the fertile site, and derived stand values for 200-400 m² plots from regressions on D²H; nutrient contents were determined.

a. *C. japonica* actually comprised only 70% of the total basal area, but calculations were made as if it were a pure stand.

b. Alternative values estimated by proportional basal area allocation.

c. Excluding woody litterfall and any mortality.

d. Assumed to be 30% of the foliage biomass.

e. Bamboo understorey.

Yamada, I. and Shidei, T. (1968). On the root biomass of *Cryptomeria japonica* stands. *Bull. Kyoto Univ. For.* <u>40</u>, 67-92.

33-35°N   131-136°E   --   Japan.

Plantations   *Cryptomeria japonica*

Age (years)	13	14	14	17	21	23	23
Trees/ha	40740	4400	5880	2083	2770	1887	2935
Tree height (m)	4.3	8.2	8.7	11.5	9.6	13.3	9.3
Basal area (m²/ha)							
Leaf area index							
Stem volume (m³/ha)							

Dry biomass (t/ha)

Stem wood	}50.4	}53.0	}76.4	}76.4	}54.3	}101.5	}88.6
Stem bark							
Branches	1.2	5.6	6.5	4.5	3.6	7.1	10.9
Fruits etc.							
Foliage	18.9	21.0	20.5	21.1	11.8	16.4	23.9
Root estimate	14.7	18.0	20.1	29.1	21.5	33.6	33.4

CAI (m³/ha/yr)

Net production (t/ha/yr)

Stem wood	
Stem bark	
Branches	
Fruits etc.	
Foliage	
Root estimate	

Between 5 and 50 trees were sampled per stand at various locations, and many root systems were excavated. Stand values were derived from regressions on D and D²H.

Saito, H., Yamada, I. and Shidei, T. (1967). Studies on the effects of thinning from small diametered trees. II Changes in stand condition after single growing season. *Bull. Kyoto Univ. For.* <u>39</u>, 64-78.

Saito, H., Kan, M. and Shidei, T. (1966). Studies on the effects of thinning from small diametered trees. I Changes in stand conditions before and after thinning. *Bull. Kyoto Univ. For.* <u>38</u>, 50-67.

---

ca.34°24'N 136°05'E   900 m   Japan, Nara Prefecture, Yoshino District.

Plantation.
Soils derived
from shale.

Cryptomeria japonica

Thinning treatments were applied at age 9.

	Unthinned	16%[a] thinned	37%[a] thinned	46%[a] thinned	64%[a] thinned
Age (years)	10	10	10	10	10
Trees/ha	4400	3280	2660	2000	1200
Tree height (m)	8.2	8.3	7.3	7.7	10.0
Basal area (m²/ha)	36.4	29.8	19.2	18.0	16.0
Leaf area index					
Stem volume (m³/ha)					

Dry biomass (t/ha)

	Unthinned	16% thinned	37% thinned	46% thinned	64% thinned
Stem wood	} 53.0	} 42.0	} 25.0	} 23.0	} 26.0
Stem bark					
Branches	5.6	4.5	2.7	2.8	2.7
Fruits etc.					
Foliage	21.0	17.0	11.0	10.0	9.3
Root estimate	18.0	15.0			8.7

CAI (m³/ha/yr)

Net production (t/ha/yr)

	Unthinned	16% thinned	37% thinned	46% thinned	64% thinned
Stem wood	} 8.11	} 5.39	} 5.06	} 3.35	} 4.37
Stem bark					
Branches	0.63	0.61	0.44[c]	0.45[c]	0.59
Fruits etc.	} +5.03[b]	} +2.63[b]			} +0.93[b]
Foliage	1.94	1.86	1.48[c]	1.45[c]	1.59
Root estimate	2.25	1.71			1.42

Twelve trees were sampled, 9 root systems were excavated, and stand biomass values for the above 64-100 m² plots were derived from regressions on D, H, and diameters at the base of the crowns. Increments were calculated for the one year since thinning. See also Saito *et al.* (1968), p.153, and Tamai and Shidei (1971), p.154, who measured this experiment at ages 12 and 14, respectively.
a. Percentage of the trees removed at age 9.
b. Total litterfall; there was no mortality.
c. Excluding litterfall.

Saito, H., Tamai, S., Ogino, K. and Shidei, T. (1968). Studies on the effects of thinning from small diametered trees. III Changes in stand condition after the second growing season. *Bull. Kyoto Univ. For.* 40, 81-92.

---

ca.34°24'N   136°05'E   900 m   Japan, Nara Prefecture, Yoshino District.

Cryptomeria japonica

Plantation.
Soils derived
from shale.

Thinning treatments were applied at age 9.

	Unthinned	16%[a] thinned	37%[a] thinned	46%[a] thinned	64%[a] thinned
Age (years)	12	12	12	12	12
Trees/ha	4400	3280	2660	2000	1200
Tree height (m)	9.0	9.4	8.0	8.5	10.6
Basal area (m²/ha)	41.9	33.9	23.0	22.1	19.2
Leaf area index					
Stem volume (m³/ha)					

Dry biomass (t/ha)

	Unthinned	16% thinned	37% thinned	46% thinned	64% thinned
Stem wood	} 65.4	} 53.3	} 32.2	} 31.4	} 32.3
Stem bark					
Branches	6.5	5.0	3.3	3.9	3.5
Fruits etc.					
Foliage	23.2	18.6	12.6	13.8	11.5
Root estimate	21.2	16.5			11.1

CAI (m³/ha/yr)

Net production (t/ha/yr)

	Unthinned	16% thinned	37% thinned	46% thinned	64% thinned
Stem wood	} 12.3	} 10.8	} 7.1	} 8.8	} 6.3
Stem bark					
Branches	0.9	0.5	0.6[c]	1.1[c]	0.8
Fruits etc.	} +5.4[b]	} +5.7[b]			} +1.3[b]
Foliage	2.2	1.6	1.9[c]	3.4[c]	2.3
Root estimate	3.5	2.0			2.4

Stand biomass values for the above 64-100 m² plots were derived from regressions calculated by Saito *et al.* (1967), p.152. Increments were calculated for the previous one year (i.e. ages 11 to 12). See also Saito *et al.* (1967), p.152, and Tamai and Shidei (1971), p.154, who measured this experiment at ages 10 and 14, respectively.
a. Percentage of the trees removed at age 9.
b. Total litterfall; there was no mortality.
c. Excluding litterfall.

Tamai, S. and Shidei, T. (1971). Studies on the effects of thinning from small diametered trees. IV Changes in stand condition after the fourth growing season. *Bull. Kyoto Univ. For.* <u>42</u>, 163-173.

---

ca.34°24'N 136°05'E   900 m   Japan, Nara Prefecture, Yoshino District.

Plantation.
Soils derived
from shale.

Cryptomeria japonica

Thinning treatments were applied at age 9.

	Unthinned	16%[a] thinned	37%[a] thinned	46%[a] thinned	64%[a] thinned
Age (years)	14	14	14	14	14
Trees/ha	4100	3280	2500	2000	1200
Tree height (m)	10.4	10.4	9.4	9.6	12.6
Basal area (m²/ha)	47.8	38.0	27.7	27.3	24.2
Leaf area index					
Stem volume (m³/ha)					

Dry biomass (t/ha)

	Unthinned	16% thinned	37% thinned	46% thinned	64% thinned
Stem wood	} 78.8	} 66.6	} 41.6	} 43.2	} 47.2
Stem bark					
Branches	6.5	6.2	4.4	5.6	4.7
Fruits etc.					
Foliage	23.5	21.9	15.6	17.3	15.8
Root estimate	25.4	20.5			15.0

CAI (m³/ha/yr)

Net production (t/ha/yr)

	Unthinned	16% thinned	37% thinned	46% thinned	64% thinned
Stem wood	} 2.7	} 8.0	} 3.2	} 5.0	} 7.9
Stem bark					
Branches	-0.2 ⎫	0.5 ⎫	0.3[c]	0.7[c]	0.4 ⎫
Fruits etc.	⎬ +2.8[b]	⎬ +8.2[b]			⎬ +0.8[b]
Foliage	-1.3 ⎭	1.4 ⎭	0.6[c]	0.5[c]	1.8 ⎭
Root estimate	2.2	0.6			1.6

Twelve trees were sampled and their roots were excavated. Stand biomass values for the above 64-100 m² plots were derived from regressions on D, H, and diameters at the base of the crowns. Increments were calculated for the one year prior to sampling (i.e. ages 13 to 14). See also Saito *et al.* (1967), p.152, and (1968), p.153, who measured this experiment at ages 10 and 12, respectively.
a. Percentage of the trees removed at age 9.
b. Total litterfall, but excluding mortality in the stand with 4100 trees/ha.
c. Excluding litterfall, and mortality in the stand with 2500 trees/ha.

Ando, T., Hatiya, K., Doi, K., Kataoka, H., Kato, Y. and Sakaguchi, K. (1968). Studies on the system of density control of Sugi (*Cryptomeria japonica*) stand. *Bull. Govt Forest Exp. Stn Tokyo* <u>209</u>, 1-76.

34°10'N 136°10'E   --   Japan, Nara Prefecture, Yoshino District.

Plantations.   *Cryptomeria japonica*

Seedlings planted at close spacings, given frequent, light thinnings[a], and pruned to 1.2-1.5 m height at age 10.

Age (years)	10	15	19	24	31	45	51	60
Trees/ha	12019	6865	4503	3438	2557	1557	1321	980
Tree height (m)	5.3	7.1	10.5	12.3	14.5	18.6	21.3	21.9
Basal area (m²/ha)	25.1	29.0	36.6	44.8	52.8	61.0	59.1	60.6
Leaf area index								
Stem volume (m³/ha)								

Dry biomass (t/ha)								
Stem wood / Stem bark	} 25.9	} 40.6	} 77.6	} 106.0	} 137.5	} 201.4	} 220.4	} 238.7
Branches	3.5	3.2	7.9	8.2	11.6	13.8	15.3	13.6
Fruits etc.								
Foliage	20.0	15.2	22.7	24.0	21.4	21.9	30.4	21.7
Root estimate								

CAI (m³/ha/yr)	13.4	18.3	22.7	27.7	21.8	24.4	24.1	21.1
Net production (t/ha/yr)								
Stem wood / Stem bark	} 4.4^b	} 6.1^b	} 7.7^b	} 9.5^b	} 7.5^b	} 8.5^b	} 8.4^b	} 7.4^b
Branches	1.8^b	0.9^b	2.0^b	1.9^b	1.6^b	1.5^b	1.8^b	1.5^b
Fruits etc.								
Foliage	5.0^c	3.8^c	5.7^c	6.0^c	5.3^c	5.5^c	7.6^c	5.4^c
Root estimate	2.4^d	2.5^d	3.0^d	3.5^d	2.7^d	2.8^d	2.9^d	2.4^d

Eight trees were sampled in each of the above stands, and stand biomass values were derived by proportional basal area allocation.
a. Values given here refer to the stands before each thinning.
b. Excluding thinnings and any woody litterfall.
c. Assumed to be 25% of the foliage biomass.
d. Assuming top/root ratios to be 3.5.

Continued from p.155.

ca.36°N 139°E   --   Japan, Saitama Prefecture, Nishikawa District, near Hanno city.

Plantations.   *Cryptomeria japonica*

Seedlings planted at close spacings. Lightly thinned[a] and pruned at ages 10-13, 18-20 and 25-26.

Age (years)	10	15	20	25	29	35
Trees/ha	3504	2933	2650	2238	2022	1754
Tree height (m)	6.8	9.5	11.4	14.5	15.5	18.2
Basal area (m²/ha)	19.0	28.0	37.6	39.9	45.1	49.1
Leaf area index						
Stem volume (m³/ha)	68	141	230	319	375	499

Dry biomass (t/ha)		10	15	20	25	29	35
	Stem wood	}24.4	}47.9	}77.4	}106.8	}125.8	}181.1
	Stem bark						
	Branches	3.1	4.7	7.2	5.1	8.7	11.0
	Fruits etc.						
	Foliage	18.9	25.5	25.7	23.1	24.0	18.1
	Root estimate						

Net production (t/ha/yr)		10	15	20	25	29	35
CAI (m³/ha/yr)		16.6	17.7	21.0	22.1	20.8	23.8
	Stem wood	}5.8[b]	}5.8[b]	}6.9[b]	}7.6[b]	}7.0[b]	}8.9[b]
	Stem bark						
	Branches	0.6[b]	1.0[b]	1.5[b]	0.9[b]	1.2[b]	0.8[b]
	Fruits etc.						
	Foliage	4.7[c]	6.1[c]	6.4[c]	5.8[c]	6.0[c]	4.5[c]
	Root estimate	3.2[d]	2.7[d]	2.8[d]	2.6[d]	2.3[d]	3.0[d]

See p.155.

Continued from p.156.

---

ca.37°00'N 140°40'E   --   Japan, Ibaraki-Fukushima Prefectures, Kitakanto-Abukuma District.

Plantations.   *Cryptomeria japonica*

Seedlings planted at intermediate spacings, moderately thinned[a] but not pruned.

Age (years)	9	16	20	26	29	34	45	53
Trees/ha	2210	2652	2378	1723	1528	1189	822	726
Tree height (m)	4.4	7.8	11.4	14.6	15.8	17.6	19.0	22.1
Basal area (m²/ha)	7.0	28.4	38.8	40.3	48.7	52.2	51.7	53.3
Leaf area index								
Stem volume (m³/ha)	24	121	255	342	393	487	490	620

Dry biomass (t/ha)

	9	16	20	26	29	34	45	53
Stem wood / Stem bark	}6.3	}33.8	}72.1	}97.7	}113.5	}143.1	}150.0	}198.3
Branches	2.8	8.1	8.2	6.3	13.6	13.1	14.8	17.6
Fruits etc.								
Foliage	10.1	20.6	22.9	18.6	16.6	18.3	23.1	28.1
Root estimate								

CAI (m³/ha/yr)	6.9	13.8	26.6	22.9	27.7	23.4	19.1	28.2

Net production (t/ha/yr)

	9	16	20	26	29	34	45	53
Stem wood / Stem bark	}2.0[b]	}3.9[b]	}7.7[b]	}6.7[b]	}8.3[b]	}7.2[b]	}6.4[b]	}10.4[b]
Branches	0.8[b]	2.1[b]	2.1[b]	1.5[b]	1.7[b]	1.5[b]	1.2[b]	1.9[b]
Fruits etc.								
Foliage	2.5[c]	5.2[c]	5.7[c]	4.7[c]	4.2[c]	4.6[c]	5.8[c]	7.0[c]
Root estimate	1.7[d]	1.8[d]	3.0[d]	2.4[d]	3.0[d]	2.5[d]	1.8[d]	3.5[d]

See p.155.

Continued from p.157.

31°40'N 131°20'E   ca. 50 m   Japan, Miyazaki Prefecture, Obi District, near Nichinan city.

Plantations.   *Cryptomeria japonica*

Cuttings planted at wide spacings and lightly thinned[a] but not pruned.

Age (years)	10	17	21	25	31	34	39	45
Trees/ha	1500	1541	923	838	673	575	458	435
Tree height (m)	3.7	7.1	10.0	12.7	14.9	17.9	18.2	20.2
Basal area (m²/ha)	4.3	19.6	30.2	35.6	40.4	45.1	47.2	54.0
Leaf area index								
Stem volume (m³/ha)	12	70	151	222	288	376	378	464

Dry biomass (t/ha)		10	17	21	25	31	34	39	45
	Stem wood	}4.2	}24.7	}53.1	}76.6	}97.7	}129.8	}129.7	}159.0
	Stem bark								
	Branches	2.0	7.5	12.9	13.6	12.2	13.1	16.0	18.1
	Fruits etc.								
	Foliage	6.7	14.1	20.5	23.3	16.8	17.2	16.5	21.4
	Root estimate								

Net production (t/ha/yr)		10	17	21	25	31	34	39	45
	CAI (m³/ha/yr)	3.9	7.3	17.6	21.0	12.6	16.7	13.0	16.1
	Stem wood	}1.4[b]	}2.6[b]	}6.2[b]	}7.3[b]	}4.3[b]	}5.8[b]	}4.5[b]	}5.5[b]
	Stem bark								
	Branches	0.7[b]	2.2[b]	3.2[b]	2.8[b]	2.5[b]	2.6[b]	2.5[b]	3.3[b]
	Fruits etc.								
	Foliage	1.7[c]	3.5[c]	5.1[c]	5.8[c]	4.2[c]	4.3[c]	4.1[c]	5.4[c]
	Root estimate	1.3[d]	1.4[d]	2.9[d]	3.1[d]	1.6[d]	2.0[d]	1.6[d]	2.0[d]

See p.155.

Harada, H., Satoo, H., Hotta, I., Hatiya, K. and Tadaki, Y. (1972). Study on the nutrient contents of mature *Cryptomeria* forest. *Bull. Govt Forest Exp. Stn Tokyo* <u>249</u>, 17–74.

---

ca.35°N 138°E (alt. given below) Japan, Shizuoka Prefecture.

Cryptomeria japonica

Plantations.

Soils derived from volcanic ash.

	Hakone		Amagi			Keta		
	760 m	680 m	760 m	700 m	560 m	940 m	1040 m	900 m
Age (years)	28	28	35	38	38	48	48	49
Trees/ha	1720	2800	2470	2050	1410	860	1100	1310
Tree height (m)	12.2	5.8	11.4	14.0	18.3	21.8	18.3	12.9
Basal area (m²/ha)	45	27	41	49	55	66	55	51
Leaf area index								
Stem volume (m³/ha)	260	70	280	410	520	600	490	350

Dry biomass (t/ha)

	760 m	680 m	760 m	700 m	560 m	940 m	1040 m	900 m
Stem wood / Stem bark	} 98	} 37	} 103	} 155	} 175	} 183	} 152	} 115
Branches	16	14	20	18	18	24	20	28
Fruits etc.	0	0	0	0	0	1	1	1
Foliage	11	14	12	10	14	17	15	18
Root estimate	30		32	49	45	81		54

CAI (m³/ha/yr)

Net production (t/ha/yr)

	760 m	680 m	760 m	700 m	560 m	940 m	1040 m	900 m
Stem wood / Stem bark	} 5.87[a]	} 1.91[a]	} 3.93[a]	} 4.39[a]	} 6.26[a]	} 4.15[a]	} 3.94[a]	} 3.10[a]
Branches	1.12[a]	0.36[a]	0.75[a]	0.83[a]	1.19[a]	0.79[a]	0.75[a]	0.59[a]
Fruits etc.								
Foliage	3.50[b]	2.75[b]	3.00[b]	2.50[b]	3.50[b]	4.25[b]	3.75[b]	4.50[b]
Root estimate	1.63	0.53	1.09	1.41	1.62	1.76	1.62	1.23

At Hakone and Amagi five to eight trees were sampled per stand and a total of eleven root systems were excavated. At Keta eight trees were sampled from each stand and a total of four root systems were excavated. Stand biomass values were derived from regressions on D²H. Nutrient contents were determined.
a. Excluding woody litterfall and any mortality.
b. Foliage litterfall.

Continued from p.159.

Japan	36°00'N 139°08'E 300 m Saitama Prefecture, Chichibu			38°13'N 139°28'E 160–180 m Niigata Prefecture, Murakami		
Plantations						
	Cryptomeria japonica					
Age (years)	55	55	55	59	59	59
Trees/ha	2005	2020	928	837	2680	621
Tree height (m)	13.4	15.1	20.9	21.5	7.1	25.8
Basal area (m²/ha)	40	50	53	64	31	68
Leaf area index						
Stem volume (m³/ha)	320	420	520	680	130	770
Dry biomass (t/ha)						
Stem wood	}112	}144	}170	}240	}73	}268
Stem bark						
Branches	12	17	11	24	28	30
Fruits etc.	0	0	0	0	0	0
Foliage	16	22	15	18	14	20
Root estimate	32		75		27	
CAI (m³/ha/yr)						
Net production (t/ha/yr)						
Stem wood	}5.45^a	}6.35^a	}3.12^a	}8.79^a	}3.24^a	}7.74^a
Stem bark						
Branches	1.04^a	1.21^a	0.59^a	1.67^a	0.62^a	1.47^a
Fruits etc.						
Foliage	4.00^b	5.50^b	3.75^b	4.50^b	3.50^b	5.00^b
Root estimate	1.48	2.16	1.43	2.68	0.99	2.36

Five to eight trees were sampled per stand. Four root systems were excavated in all from the 55-year-old stands, and two from the 59-year-old stands. Stand biomass values were derived from regressions on D²H. Nutrient contents were determined.
a. Excluding woody litterfall and any mortality.
b. Foliage litterfall.

Saito, H., Yamada, I. and Shidei, T. (1972). Some discussions on dry matter production of young stands of *Cryptomeria japonica* D. Don with excessively high stand density. *Bull. Kyoto Univ. For.* 44, 121-140.

Japan	34°04'N 131°48'E ca. 50 m Yamaguchi Prefecture, Tokuyama nursery		ca.34°40'N 136°30'E 50 m Mie Prefecture Ichishi nursery
Plantations. Fertile nursery loams.	*Cryptomeria japonica*		
Age (years)	10	10	9
Trees/ha	42600	38900	100000
Tree height (m)	5.3[a]	5.1[a]	3.4[a]
Basal area (m²/ha)	37.1	26.0	14.4
Leaf area index			
Stem volume (m³/ha)	131.0	87.8	27.7
Dry biomass (t/ha) Stem wood / Stem bark	} 60.0	} 39.9	} 12.7
Branches	1.9	0.5	0.8
Fruits etc.	2.0	0.9	0.0
Foliage	18.5	16.5	7.4
Root estimate	15.3	14.1	3.2
CAI (m³/ha/yr)	18.4	19.6	7.2
Net production (t/ha/yr) Stem wood / Stem bark	} 8.5[b]	} 9.0[b]	} 3.3[b]
Branches	2.2[b]	2.5[b]	1.0[b]
Fruits etc.	ca.0.5	ca.0.5	0.0
Foliage	4.6[c]	4.1[c]	1.9[c]
Root estimate	2.6	4.0	1.0

Stand values at Tokuyama were derived by harvesting all trees, including roots, within plots of only about 6 m². Stand values at Ichishi were derived from regressions on D²H, calculated using over 45 sample trees, except for branch biomass values which were estimated by proportional basal area allocation. There were 3700, 7400 and 78600 dead trees per hectare in columns left to right. Increments were calculated for the previous one year.
a. Mean height of the dominant trees.
b. Excluding woody litterfall and mortality.
c. Assumed to be 25% of the foliage biomass.

Saito, H. and Shidei, T. (1973). Studies on the productivity and its estimation methodology in a young stand of *Cryptomeria japonica* D. Don. *J. Jap. For. Soc.* <u>55</u>, 52-62.

34°30'N   136°20'E   400 m   Japan, Mie Prefecture, Kii Peninsula.

Plantations

Cryptomeria japonica

Age (years)		12	12
Trees/ha		6106	5600
Tree height (m)		8.4[a]	9.2[a]
Basal area (m²/ha)		40.8	40.4
Leaf area index			
Stem volume (m³/ha)		180	186
Dry biomass (t/ha)	Stem wood	} 68.3	} 70.5
	Stem bark		
	Branches	5.3	5.8
	Fruits etc.		
	Foliage	19.8	20.5
	Root estimate	18.1	18.7

CAI (m³/ha/yr)		23.8	24.6
Net production (t/ha/yr)	Stem wood	} 9.1 (or 9.0)[c]	} 9.4 (or 10.3)[c]
	Stem bark		
	Branches	1.9[b]	2.2[b]
	Fruits etc.	} (or 6.4)[bc]	} (or 6.9)[bc]
	Foliage	2.9[d]	3.0[d]
	Root estimate	2.4	2.7

Thirteen trees were sampled in all and 8 root systems were excavated. Stand biomass values for the above two 100 m² plots were derived from regressions on D²H.
a. Mean height of the dominant trees.
b. Including woody litterfall of 0.03 and 0.08 t/ha/yr in the left and right columns, respectively.
c. Alternative values, estimated from the difference in biomass measured at ages 10 and 12, rather than from the increase in D²H and the biomass of new foliage.
d. New foliage biomass; foliage litterfall was 0.9 and 1.5 t/ha/yr in the left and right columns, respectively.

Satoo, T. (1979d). Loss of canopy biomass due to thinning - a comparison of two
young stands of *Cryptomeria japonica* of cutting and seedling origins. *J. Jap.
For. Soc.* <u>61</u>, 83-87.

ca.35°N 140°E   --   Japan, Chiba Prefecture, southern Boso Peninsula.

Plantations.   *Cryptomeria japonica*

	Raised from seed		Raised from cuttings	
Age (years)	13	13	13	13
Trees/ha	4800	4700	3700	4300
Tree height (m)	9.0	9.2	8.5	8.9
Basal area (m²/ha)	38.6	41.7	33.5	34.8
Leaf area index				
Stem volume (m³/ha)				
Dry biomass (t/ha) Stem wood	}64.1	}69.2	}57.8	}60.7
Stem bark				
Branches	8.6	8.4	6.1	5.3
Fruits etc.				
Foliage	35.2	30.6	28.5	30.3
Root estimate				
CAI (m³/ha/yr)				
Net production (t/ha/yr) Stem wood	}11.3^a	}9.2^a	}7.2^a	}8.0^a
Stem bark				
Branches				
Fruits etc.				
Foliage				
Root estimate				

Five trees were sampled from each of the above 100 m² plots in early May and stand
values were derived from regressions on D.
a. Excluding woody litterfall and any mortality.

Kawanabe, S., Saito, H. and Shidei, T. (1975). Studies on the effects of thinning small diametered trees. V Changes in stand conditions and biomass of *Cryptomeria japonica* D. Don. stand during six years after thinning. *J. Jap. For. Soc.* <u>57</u>, 215-223.

34°24'N 136°05'E   900 m   Japan, Nara Prefecture, Yoshino District.

Plantations.
Soils derived
from shale.

Cryptomeria japonica

All plots were unthinned.

Age (years)	10	11	12	13	14	15	16
Trees/ha	4400	4400	4400	4200[a]	4200	4100[a]	4100
Tree height (m)	7.4	8.2	9.0	9.8	10.4	10.6	11.1
Basal area (m²/ha)	33.1	36.4	41.9	44.6	47.8	48.9	50.5
Leaf area index							
Stem volume (m³/ha)							

Dry biomass (t/ha)

	10	11	12	13	14	15	16
Stem wood	}44.6	}53.0	}65.4	}76.1	}81.0	}86.9	}93.1
Stem bark							
Branches	5.0	5.6	6.5	6.7	6.5	6.0	6.2
Fruits etc.							
Foliage	18.9	20.8	23.2	24.8	23.5	21.7	22.5
Root estimate	15.5	17.6	21.2	23.2	25.4	26.5	27.8

CAI (m³/ha/yr)

Net production (t/ha/yr)

	10	11	12	13	14	15	16
Stem wood	}8.4	}12.4	}10.7	}4.9	}5.9	}6.2	
Stem bark							
Branches							
Fruits etc.							
Foliage							
Root estimate							

Twelve trees were sampled at age 10, twenty at age 16, and stand values for the above 64-100 m² plots were derived from regressions on D and D²H.
a. The densities of these stands were decreased by snow breakage.

Continued from p.164.

Same as p.164.

All plots were thinned from 3910 trees/ha at age 10.

Age (years)	10	11	12	13	14	15	16
Trees/ha	3280	3280	3280	3280	3280	3280	3280
Tree height (m)	7.8	8.3	9.4	9.9	10.4	11.0	11.3
Basal area (m²/ha)	27.3	29.8	33.9	37.4	38.7	41.1	45.5
Leaf area index							
Stem volume (m³/ha)							
Dry biomass (t/ha)							
Stem wood	}36.9	}42.7	}53.3	}61.8	}66.6	}73.7	}81.1
Stem bark							
Branches	3.9	4.5	5.0	5.7	6.2	5.2	5.4
Fruits etc.							
Foliage	14.9	16.7	18.6	20.5	21.9	18.8	19.4
Root estimate	12.9	14.5	16.5	19.6	20.5	22.3	24.5
CAI (m³/ha/yr)							
Net production (t/ha/yr)							
Stem wood		}5.8	}10.6	}8.5	}4.8	}7.1	}7.4
Stem bark							
Branches							
Fruits etc.							
Foliage							
Root estimate							

Twelve trees were sampled at age 10, twenty at age 16, and stand values for the above 64–100 m² plots were derived from regressions on D and D²H.

Continued from p.165.

Same as p.164.

All plots were thinned from 3300 trees/ha at age 10.

Age (years)	10	11	12	13	14	15	16
Trees/ha	1200	1200	1200	1200	1200	1200	1200
Tree height (m)	9.3	10.0	10.6	11.3	12.1	12.9	13.5
Basal area (m²/ha)	14.0	16.0	19.2	22.3	24.2	26.6	29.5
Leaf area index							
Stem volume (m³/ha)							

Dry biomass (t/ha)		10	11	12	13	14	15	16
	Stem wood	}21.6	}26.0	}32.3	}39.3	}45.2	}52.3	}59.9
	Stem bark							
	Branches	2.1	2.7	3.5	4.3	4.7	5.4	6.2
	Fruits etc.							
	Foliage	7.6	9.2	11.5	13.9	15.8	16.6	19.0
	Root estimate	7.3	8.7	11.1	13.4	15.0	17.0	19.5

CAI (m³/ha/yr)

Net production (t/ha/yr)								
	Stem wood	}4.4	}6.3	}7.0	}5.9	}7.1	}7.6	
	Stem bark							
	Branches							
	Fruits etc.							
	Foliage							
	Root estimate							

Twelve trees were sampled at age 10, twenty at age 16, and stand values for the above 64-100 m² plots were derived from regressions on D and D²H.

(The authors present similar data for two other heavily thinned plots.)

Satoo, T. (1970b). Primary production in a plantation of Japanese larch, *Larix leptolepis*: a summarized report of JPTF-66 KOIWAI. *J. Jap. For. Soc.* 52, 154-158.

Satoo, T. (1977). Larch plantations. In: "Primary Productivity of Japanese Forests" (T. Shidei and T. Kira, eds) pp. 169-172. JIBP Synthesis vol. 16. University of Tokyo Press.

Satoo, T. (1973b). Materials for the studies of growth in stands. X Primary production relations in a plantation of *Larix leptolepis* in Hokkaido. *Bull. Tokyo Univ. For.* 66, 119-126.

Japan	39°45'N 141°08'E 360 m Iwate Prefecture, Morioka city	ca.43°13'N 142°33'E 230 m Hokkaido, near Mount Asibetu
Plantations	*Larix leptolepis* with understorey broadleaved trees. Deep volcanic ash (Satoo 1970b, 1977)	*L. leptolepis* with understorey shrubs. Brown forest soil (Satoo 1973b)
Age (years)	39	21
Trees/ha	$1155 + 2113^a$	1240
Tree height (m)	$19.4 \quad 3.7^a$	15.3
Basal area (m²/ha)	$36.0 + 1.3^a$	22.4
Leaf area index	$4.2^b + 0.9^a$	
Stem volume (m³/ha)		169

Dry biomass (t/ha)	Stem wood	$\left.\begin{array}{l} \\ \end{array}\right\} 145.4 + 1.9^a$	$\left.\begin{array}{l} \\ \end{array}\right\} 69.2$ (or 66.8, 66.5)^d
	Stem bark		
	Branches	$15.5 + 1.0^a$	12.2 (or 12.0, 12.3)^d $\left.\begin{array}{l} \\ \\ \\ \\ \end{array}\right\} +1.3^a$
	Fruits etc.		
	Foliage	$3.6 + 0.3^a$	4.9 (or 4.3, 4.4)^d
	Root estimate	$34.8 + 0.8^a$	

CAI (m³/ha/yr) .... 169 *(value reference)*

Net production (t/ha/yr)	Stem wood	$\left.\begin{array}{l} \\ \end{array}\right\} 5.80^c + 0.35^{ac}$	$\left.\begin{array}{l} \\ \end{array}\right\} 6.7^c$ (or 6.7, 7.1)^{cd}
	Stem bark		
	Branches	$3.26^c + 0.22^{ac}$	3.0^c (or 3.1, 3.4)^{cd} $\left.\begin{array}{l} \\ \\ \\ \\ \end{array}\right\} +0.3^a$
	Fruits etc.		
	Foliage	$3.59 + 0.31^a$	4.9 (or 4.3, 4.4)^d
	Root estimate	$1.96 + 0.18^a$	

In both studies 10 trees were sampled. Roots were excavated at Morioka. Stand biomass values were derived for a 407 m² plot at Morioka by proportional basal area allocation, and for a 600 m² plot at Asibetu from regressions on D. Increments were estimated for the previous one year. Roots at Morioka were assumed to grow at the same relative rates as above-ground woody parts.

a. Understorey trees and shrubs.

b. Approximate all-sided LAI value can be obtained by multiplying by 2.3.

c. Excluding woody litterfall and any mortality.

d. Alternative values, estimated by proportional basal area allocation and by multiplying mean tree values by the numbers of trees per hectare (written left and right, respectively, within the brackets).

Saito, H., Kawahara, T., Shidei, T. and Tsutsumi, T. (1970). Productivity of young stands of *Metasequoia glyptostroboides*. *Bull. Kyoto Univ. For.* <u>41</u>, 80-95.

Satoo, T. (1974b). Materials for the studies of growth in forest stands. XIII Primary production relations of a young stand of *Metasequoia glyptostroboides* planted in Tokyo. *Bull. Tokyo Univ. For.* <u>66</u>, 153-164.

Japan	ca.34°10'N 131°30'E 30 m Yamaguchi Prefecture		35°45'N 139°30'E ca.50 m W of Tokyo, Tanasi Expt. Field
Plantations	*Metasequoia glyptostroboides*		
			2 clone mixture
	(Saito *et al*. 1970)		(Satoo 1974b)
Age (years)	9	9	17
Trees/ha	6180	12900	753
Tree height (m)	8.9	8.0	14.7
Basal area (m²/ha)	24.3	22.6	23.7
Leaf area index			8.5^e (or 8.5)^{de}
Stem volume (m³/ha)	125	123	180 (or 174)^d
Dry biomass (t/ha) — Stem wood	}40.4	}40.0	}57.7 (or 54.3)^d
Dry biomass (t/ha) — Stem bark			
Dry biomass (t/ha) — Branches	7.5	7.0	12.7 (or 12.2)^d
Dry biomass (t/ha) — Fruits etc.			
Dry biomass (t/ha) — Foliage	5.1^a	5.0^a	4.3 (or 4.3)^d
Dry biomass (t/ha) — Root estimate	8.9		16.4 (or 17.1)^d
CAI (m³/ha/yr)	25.4	24.7	23.3 (or 23.4)^d
Net production (t/ha/yr) — Stem wood	}8.2^b	}8.0^b	}6.94^b (or 7.09)^{bd}
Net production (t/ha/yr) — Stem bark			
Net production (t/ha/yr) — Branches	3.8^b		4.29^b (or 4.15)^{bd}
Net production (t/ha/yr) — Fruits etc.			
Net production (t/ha/yr) — Foliage	$5.1^c + 1.0^c$		4.97^f (or 5.01)^{df}
Net production (t/ha/yr) — Root estimate			

At Yamaguchi, 17 trees were sampled, roots were excavated, and stand biomass values were derived from regressions on D^2H; increments were estimated for the previous one year. At Tanasi, 5 trees were sampled in October, roots were excavated and stand biomass values for a 600 m² plot were derived from regressions on D.
a. Including about 1.0 t/ha of short branchlets.
b. Excluding woody litterfall and any mortality.
c. New foliage biomass (5.1 t/ha/yr) and estimated losses (1.0 t/ha/yr).
d. Alternative values derived by proportional basal area allocation.
e. Approximate all-sided LAI value can be obtained by multiplying by 2.3.
f. Foliage biomass plus estimated leaf fall prior to the October sampling.

Satoo, T. (1971). Materials for the studies of growth in stands. VIII Primary production relations in plantations of Norway spruce in Japan. *Bull. Tokyo Univ. For.* <u>65</u>, 125–142.

43°13'N   142°26'E   360 m   Japan, Hokkaido

Picea abies

Plantations.
Brown forest soils.   All stands were thinned at ages 8 and 30.

	Fertile site	Sites of average fertility		Infertile site
Age (years)	47	46	46	45
Trees/ha	488	756	756	914
Tree height (m)	23.5	18.3	18.3	15.6
Basal area (m²/ha)	33.8	24.9	26.1	21.4
Leaf area index				
Stem volume (m³/ha)	427	244	252	193
Dry biomass (t/ha)				
Stem wood	}208.6	}119.3	}123.2	}94.1
Stem bark				
Branches	16.7	14.1	18.5	12.2 } +0.4^a
Fruits etc.				
Foliage	18.6	14.4	14.7	16.9
Root estimate				
CAI (m³/ha/yr)	16.5	11.7	11.7	8.8
Net production (t/ha/yr)				
Stem wood	}8.04^b	}5.70^b	}5.67^b	}4.31^b + 0.13^{ba}
Stem bark				
Branches	0.98^b	1.17^b	1.29^b	0.84^b + 0.01^{ba}
Fruits etc.				
Foliage	3.37	4.54	4.69	2.19 + 0.03^a
Root estimate				

Five trees were sampled per stand and biomass values for the above four 0.16 ha plots were derived by proportional basal area allocation. Increments were estimated for the previous one year.
a. Understorey shrubs.
b. Excluding woody litterfall and any mortality.

Satoo, T. (1971). Materials for the studies of growth in stands. VIII Primary production relations of Norway spruce in Japan. *Bull. Tokyo Univ. For.* 65, 125–142.

Yoshimura, K. (1967). Growth and biomass of Norway spruce forest in Ashu experimental forest. *Bull. Kyoto Univ. For.* 39, 27–34.

Japan	35°56'N 138°51'E 1030 m Titibu	Ashu, Kyoto University Forest
Plantations. Brown forest soils.	*Picea abies* Unthinned stand (Satoo 1971)	*Picea abies* (Yoshimura 1967)
Age (years)	39	30
Trees/ha	2240[a]	1072
Tree height (m)	16.5	15.3
Basal area (m²/ha)	52.6	41.7
Leaf area index		
Stem volume (m³/ha)	386	269
Dry biomass (t/ha) — Stem wood	} 188.5	} 120.2
Stem bark		
Branches	8.6	31.3
Fruits etc.		
Foliage	23.9	24.6
Root estimate		
CAI (m³/ha/yr)		
Net production (t/ha/yr) — Stem wood	} 8.5[b]	
Stem bark		
Branches	0.9[b]	
Fruits etc.		
Foliage	4.5	
Root estimate		

At Titibu 6 trees were sampled and stand biomass values for a 0.10 ha plot were
derived by proportional basal area allocation; increments were estimated for the
previous one year. At Ashu, 9 windfall trees were sampled (i.e. biassed towards
large trees) and stand values for a 0.17 ha plot were derived from regressions on D
and D²H.

a. There were also 570 dead trees per hectare, not included here.
b. Excluding woody litterfall and any mortality.

Ando, T., Sakaguchi, K., Narita, T. and Satoo, S. (1962). Growth analysis on the natural stands of Japanese red pine (*Pinus densiflora* Sieb. et Zucc.). I Effects of improvement cutting and relative growth. *Bull. Govt Forest Exp. Stn Tokyo* 144, 1-30.

Ando, T. (1962). Growth analysis on the natural stands of Japanese red pine (*Pinus densiflora* Sieb. et Zucc.). II Analysis of stand density and growth. *Bull. Govt Forest Exp. Stn Tokyo* 147, 45-77.

36°00'N 140°00'E   --   Japan, Tochigi Prefecture, Mashiko town National Forest

Pinus densiflora

	Heavily thinned at age 8	Lightly thinned at age 8	Unthinned
Age (years)	18	18	18
Trees/ha	5167	9633	36933
Tree height (m)	4.3	4.5	3.9
Basal area (m²/ha)	13.3	18.7	25.1
Leaf area index			
Stem volume (m³/ha)	41.1	72.1	93.2

Dry biomass (t/ha)

Stem wood	}19.7	}31.1	}44.1
Stem bark			
Branches	6.2	8.7	7.7
Fruits etc.	0.0	0.0	0.0
Foliage	5.3	6.5	6.3
Root estimate			

CAI (m³/ha/yr)

Net production (t/ha/yr)

Stem wood			
Stem bark			
Branches			
Fruits etc.			
Foliage			
Root estimate			

Twenty to thirty trees were sampled per treatment, and stand values were estimated by proportional basal area allocation. Values given in each column above are the means of 3 replicate 100 m² plots.

Hatiya, K., Fujimori, T., Tochiaki, K. and Ando, T. (1966). Studies on the seasonal variations of leaf and leaf-fall amount in Japanese red pine (*Pinus densiflora*) stands. *Bull. Govt Forest Exp. Stn Tokyo* 191, 101-113.

Japan	35°40'N　139°20'E　100 m Kanagawa Prefecture, Hachioji, Asakawa nursery			ca.36°40'N　140°00'E　-- Tochigi Prefecture, Mashiko	
	Pinus densiflora				
	Plantations			Natural stands	
Age (years)	7	7	7	13	13
Trees/ha	145000	62500	20400	69300	14900
Tree height (m)	1.3	1.1	1.2	3.2	6.5
Basal area (m²/ha)					
Leaf area index					
Stem volume (m³/ha)					
Dry biomass (t/ha) — Stem wood	}12.40	}9.00	}5.92		
Stem bark					
Branches	6.35	7.13	8.39		
Fruits etc.					
Foliage	9.83	10.63	11.48	6.1	9.9
Root estimate					
CAI (m³/ha/yr)					
Net production (t/ha/yr) — Stem wood					
Stem bark					
Branches					
Fruits etc.					
Foliage	8.42[a]	7.70[a]	4.90[a]	3.73[a]	5.60[a]
Root estimate					

Trees were sampled on several occasions during the year; values given above are for June at Asakawa and August at Mashiko, when foliage biomass values were greatest. Stand values were derived by proportional basal area allocation for plots of 16, 26, 54, 16 and 77 m² in columns left to right.
a. Foliage litterfall measured during one year.

Hatiya, K., Doi, K. and Kobayashi, R. (1965). Analysis of the growth of Japanese red pine (*Pinus densiflora*) stands. A report on the matured plantation in Iwate Prefecture. *Bull. Govt Forest Exp. Stn Tokyo* 176, 75-88.

41°30'N 139°30'E   --   Japan, Iwate Prefecture

Plantations   *Pinus densiflora*

These stands, with different numbers of trees/ha, were growing on sites of similar quality.

Age (years)	46	44	43	46	33
Trees/ha	370	750	1009	1310	2340
Tree height (m)	19.5	18.5	18.2	19.2	15.7
Basal area (m²/ha)	23.7	32.3	38.5	46.4	45.6
Leaf area index					
Stem volume (m³/ha)	215	325	375	473	409
Dry biomass (t/ha)					
Stem wood	}83.9	}126.0	}153.6	}198.5	}163.4
Stem bark					
Branches	14.1	13.4	15.6	16.6	14.6
Fruits etc.					
Foliage	4.0	5.1	6.4	7.0	6.9
Root estimate	29.1[a]	41.3[a]	50.2[a]	63.4[a]	52.8[a]
CAI (m³/ha/yr)	9.5	13.3	13.7	14.1	21.8
Net production (t/ha/yr)					
Stem wood	}3.7[b]	}5.2[b]	}5.6[b]	}5.9[b]	}8.7[b]
Stem bark					
Branches	1.8[b]	1.7[b]	1.9[b]	2.1[b]	2.1[b]
Fruits etc.					
Foliage	2.2[c]	2.8[c]	3.4[c]	3.8[c]	3.4[c]
Root estimate	1.3	1.7	1.8	1.9	2.8

Three to ten trees were sampled per stand. Stand biomass values were estimated by proportional basal area allocation for plots of 216, 400, 308, 397 and 180 m² in columns left to right.
a. Assuming top/root ratios to be 3.5.
b. Excluding woody litterfall and any mortality.
c. New foliage biomass.

Satoo, T. (1968). Primary production relations in woodlands of *Pinus densiflora*.
   In: "Symposium on Primary Productivity and Mineral Cycling in Natural Ecosystems"
   pp. 52-80. University of Maine, Orono, U.S.A.
Satoo, T. (1981). In: "Dynamic Properties of Forest Ecosystems" (D.E. Reichle, ed.)
   p. 602. Cambridge University Press, Cambridge, London, New York, Melbourne.
Yuasa, Y. and Kamio, K. (1973). Leaf biomass and leaf-fall of young stands of
   Japanese red pine (*Pinus densiflora*) and Japanese black pine (*Pinus thunbergii*).
   Bull. Shizuoka Univ. For. **2**, 25-33.

Japan		39°02'N 141°21'E 300 m		
		Okita		Shizuoka University Forest
Brown forest soils, pH 5.4		*Pinus densiflora*		*P. densiflora*
		(Satoo 1968)	(Satoo 1981)	(Yuasa and Kamio 1973)
Age (years)		17	20	16
Trees/ha			6600	8390
Tree height (m)		8	10.2	5.4
Basal area (m²/ha)			32.3	24.0
Leaf area index				
Stem volume (m³/ha)				
Dry biomass (t/ha)	Stem wood	} 41.9	} 70.9	} 34.5
	Stem bark			
	Branches	6.3	15.6	8.0
	Fruits etc.			
	Foliage	4.6	6.8	5.1
	Root estimate	$7.6^a + 3.3^b$	22.9	
CAI (m³/ha/yr)				
Net production (t/ha/yr)	Stem wood	} 7.50	} $11.33 + 0.35^d$	
	Stem bark			
	Branches	2.73^c		
	Fruits etc.			
	Foliage	4.24	$1.25 + 3.44^d + 0.12^e$	2.9^f
	Root estimate	$0.92^a + 0.40^b$		

Satoo (1968, 1981) sampled all the trees, including the roots, in a 20 m² plot, and
derived stand biomass values by proportional basal area allocation. Yuasa and Kamio
(1973) sampled 11 trees and derived stand values from regressions on D and D²H.
a. Stumps.
b. Roots.
c. Excluding woody litterfall.
d. Litterfall.
e. Frass litterfall.
f. New foliage biomass; foliage litterfall was 2.4 t/ha/yr.

Akai, T., Ueda, J. and Furuno, T. (1970). Mechanisms related to matter production in young slash pine forest. *Bull. Kyoto Univ. For.* <u>41</u>, 56-79.

---

ca.33°40'N 135°30'E   50 m   Japan, Wakayama Prefecture, Shirahama, Kyoto University Forest

Plantation.
Sandy infertile soil   *Pinus elliottii*

	100^a	200^a	50^a	100^a	0^a	25^a	50^a
Age (years)	8	8	8	8	8	8	8
Trees/ha	2200	2000	2800	4000	4000	5400	5800
Tree height (m)	6.8	6.9	6.5	6.2	1.9	5.8	7.4
Basal area (m²/ha)	25.8	23.5	24.1	29.3		27.5	43.6
Leaf area index							
Stem volume (m³/ha)	102	91	95	111	8	105	196

Dry biomass (t/ha)		100^a	200^a	50^a	100^a	0^a	25^a	50^a
	Stem wood	}37.6	}33.7	}34.9	}40.6	}3.8	}38.2	}72.1
	Stem bark							
	Branches	7.8	6.9	6.6	6.9	0.4	4.1	9.3
	Fruits etc.	0.0	0.0	0.0	0.0	0.0	0.0	0.0
	Foliage	11.7	10.5	10.6	12.2	1.9	8.7	17.6
	Root estimate							

CAI (m³/ha/yr)

Net production (t/ha/yr)		100^a	200^a	50^a	100^a	0^a	25^a	50^a
	Stem wood	}6.9	}6.4	}6.5	}8.8		}5.2	}10.1
	Stem bark							
	Branches							
	Fruits etc.							
	Foliage							
	Root estimate							

Twenty-four trees were sampled. Stand values for the above seven 500 m² plots were derived from regressions on D²H. Nutrient contents were determined.
a. Grammes of NPK (15:13:12) applied per tree in the first and second years after planting.

Shidei, T. (1963). Productivity of Haimatsu (*Pinus pumila*) community growing in Alpine zone of Tateyama-Range. *J. Jap. For. Soc.* 45, 169-173.

ca.36°30'N 137°00'E   2200-2800 m   Japan, near Mount Tsurugi

Pinus pumila

Dwarf stand in the 'alpine' zone.

Age (years)		22	22	40	45
Trees/ha		400000	880000	360000	80000
Tree height (m)		0.65	0.45	0.55	1.3
Basal area (m²/ha)					
Leaf area index					
Stem volume (m³/ha)					
Dry biomass (t/ha)	Stem wood	}40.6	}34.2	}24.5	}57.3
	Stem bark				
	Branches	15.9	16.1	17.4	46.9
	Fruits etc.				
	Foliage	22.6	25.3	17.1	21.6
	Root estimate				
CAI (m³/ha/yr)					
Net production (t/ha/yr)	Stem wood	}3.14[a]	}2.83[a]	}1.36[a]	}3.69[a]
	Stem bark				
	Branches	1.60[a]	1.33[a]	0.96[a]	3.02[a]
	Fruits etc.				
	Foliage	7.04[b]	7.92[b]	5.52[b]	6.76[b]
	Root estimate				

All trees were harvested within the above four plots of only 0.5 x 0.5 m and biomass values were expressed per hectare. Creeping stems and roots were not included.
a. Excluding woody litterfall and mortality.
b. New foliage biomass.

Akai, T., Ueda, S. and Furuno, T. (1971). Mechanisms related to matter production in a young white pine forest. *Bull. Kyoto Univ. For.* <u>42</u>, 143-162.

35°32'N 137°48'E   1100 m   Japan, Nagano Prefecture, near Iida City, Achi
                                       National Forest.

Plantation.
Sandy loam   *Pinus strobus*
soil.

Age (years)	11	11
Trees/ha	3000	3400
Tree height (m)	5.4	5.2
Basal area (m²/ha)	12.9	13.4
Leaf area index		
Stem volume (m³/ha)	45.6	45.3

Dry biomass (t/ha)

Stem wood	} 14.9	} 14.9
Stem bark		
Branches	7.5	7.3
Fruits etc.		
Foliage	2.8	2.7
Root estimate		

CAI (m³/ha/yr)	12.2	12.2

Net production (t/ha/yr)

Stem wood	} 4.0	} 4.0
Stem bark		
Branches		
Fruits etc.		
Foliage	2.14[a]	2.08[a]
Root estimate		

Fourteen trees were sampled and stand biomass values for the above two 200 m² plots were derived from regressions on $D^2 H$. Stem increments were estimated for the previous one year. Branch increments were not estimated. Nutrient contents were determined.
a. New foliage biomass.

Akai, T., Furuno, T., Ueda, S. and Sano, S. (1968). Mechanisms of matter production in young loblolly pine forest. *Bull. Kyoto Univ. For.* <u>40</u>, 26-49.

ca.33°40'N 135°30'E   50 m   Japan, Wakayama Prefecture, Shirahama, Kyoto University Forest.

Plantations.
Infertile, sandy loam.

Pinus taeda

	100 g/tree of NPK (15:8:8) applied each year for three years after planting						No fertilizer applied
Age (years)	7	7	7	7	7	7	7
Trees/ha	2066	2151	3835	3765	6536	6543	3750
Tree height (m)	7.0	6.7	5.1	6.2	7.3	6.3	1.6
Basal area (m²/ha)	18.4	18.5	16.0	23.0	36.5	34.6	
Leaf area index							
Stem volume (m³/ha)	69.0	68.9	53.3	81.9	149.4	138.3	3.3

Dry biomass (t/ha)

Stem wood	}27.2	}25.9	}21.5	}32.7	}59.4	}55.1	}1.3
Stem bark							
Branches	9.1	8.7	6.0	10.0	14.4	12.5	0.5
Fruits etc.	0.0	0.0	0.0	0.0	0.0	0.0	0.0
Foliage	8.8	8.4	7.6	11.0	13.9	12.6	0.9
Root estimate							

CAI (m³/ha/yr)

Net production (t/ha/yr)

Stem wood	}8.2	}8.2	}6.2	}9.7	}14.4	}12.9	
Stem bark							
Branches							
Fruits etc.							
Foliage							
Root estimate							

Twenty-three trees were sampled, and stand values for the above seven 100 m² plots were derived from regressions on D^2H. Nutrient contents were determined.

Akai, T. and Furuno, T. (1971). Amount of litterfall and grazing in young loblolly pine forest. *Bull. Kyoto Univ. For.* 42, 83-95.

Akai, T., Ueda, S., Furuno, T. and Saito, H. (1972). Mechanisms related to matter production in a thrifty loblolly pine forest. *Bull. Kyoto Univ. For.* 43, 85-105.

Japan Plantations	ca.33°40'N 135°30'E 50 m Wakayama Prefecture Shirahama Kyoto Univ. Forest			ca.32°30'N 130°50'E 200 m Kumamoto Prefecture Nishikihara Nat. Forest	
	Pinus taeda A spacing experiment (Akai and Furuno 1971)			Clay loam soil (Akai *et al.* 1972)	
Age (years)	10	10	10	34	34
Trees/ha	2151	3765	6543	696	700
Tree height (m)	6.7	6.2	6.3	21.1	20.3
Basal area (m²/ha)	18.5	23.0	34.6	42.8	38.7
Leaf area index					
Stem volume (m³/ha)				337	302
Dry biomass (t/ha) Stem wood				}168.0	}151.0
Stem bark					
Branches	8.7	10.0	12.5	22.0	19.7
Fruits etc.				0.3	0.4
Foliage	8.4	11.0	12.6	9.5	8.5
Root estimate					
CAI (m³/ha/yr)				ca.14	ca.14
Net production (t/ha/yr) Stem wood				}ca.7.4	}ca.7.4
Stem bark					
Branches	0.1[a]	0.1[a]	0.1[a]		
Fruits etc.					
Foliage	6.5[a]	7.1[a]	8.0[a]		
Root estimate					

Akai and Furuno (1971) used published regression equations to derive stand biomass values for the above three 500 m² plots from regressions on D²H.
Akai *et al.* (1972) sampled 13 trees and derived stand values for the above two 0.20 ha plots from regressions on D² and D²H; foliage biomass was estimated in October; there was about 6.6 t/ha of dead branches in each plot, and 0.1 to 0.9 t/ha of dead cones; nutrient contents were determined.
a. Litterfall only; excluding insect consumption estimated to be 0.05 to 0.09 t/ha/ yr.

Kabaya, H., Ikusima, I. and Numata, M. (1964). Growth and thinning of *Pinus thunbergii* stand - ecological studies of coastal pine forest. *Bull. Marine Lab. Chiba Univ.* **6**, 1-26.

---

ca.35°30'N   140-141°E   5-50 m   Japan, Chiba Prefecture, Futtsu.

Coastal sites

Pinus thunbergii

Dwarf unthinned stands.

Age (years)	8	9	10	11	12	13
Trees/ha	12400	11938	12329	12384	12353	11860
Tree height (m)	1.2	1.5	1.8	2.2	2.5	2.8
Basal area (m²/ha)						
Leaf area index						
Stem volume (m³/ha)						

Dry biomass (t/ha)

	Stem wood						
	Stem bark	12.4	19.1	27.0	37.4	46.2	51.0
	Branches						
	Fruits etc.						
	Foliage						
	Root estimate	3.3^a	5.2^a	7.3^a	10.1^a	12.5^a	13.8^a

CAI (m³/ha/yr)

Net production (t/ha/yr)

	Stem wood						
	Stem bark	$4.7+0.0^b$	$6.7+0.0^b$	$7.9+0.0^b$	$10.4+1.0^b$	$8.8+2.0^b$	$5.6+2.6^b$
	Branches						
	Fruits etc.						
	Foliage	2.3^b	4.0^b	5.4^b	6.5^b	7.4^b	8.0^b
	Root estimate						

Twenty trees were sampled and 11 root systems were excavated. Stand values for the above six 19 m² plots were derived from regressions on stem basal diameter.
a. Assumed to be 27% of the value of above-ground woody biomass.
b. Litterfall.

Ando, T. (1965). Estimation of dry matter and growth analysis of the young stand of Japanese black pine (*Pinus thunbergii*). *Advg Front. Pl. Sci., New Delhi* <u>10</u>, 1-10.

35°00'N 139°00'E -- Japan, Shinuoka Prefecture, near Ito city.

Plantations.   *Pinus thunbergii*

	good[a]	good[a]	good[a]	moderate[a]	moderate[a]	poor[a]
Age (years)	10	10	10	10	10	10
Trees/ha	6863	7231	3245	4573	10204	9824
Tree height (m)	4.5	4.5	4.7	3.7	3.2	2.6
Basal area (m² /ha)	21.1	24.9	17.8	8.5	13.9	7.0
Leaf area index						
Stem volume (m³ /ha)	72.3	81.1	54.4	23.9	40.1	20.9
Dry biomass (t/ha)						
Stem wood	}33.4	}38.1	}23.4	}11.3	}18.7	}10.7
Stem bark						
Branches	11.2	15.9	10.6	5.6	8.5	4.1
Fruits etc.	0.0	0.0	0.0	0.0	0.0	0.0
Foliage	11.9	13.8	10.1	5.7	10.7	6.2
Root estimate						
CAI (m³ /ha/yr)						
Net production (t/ha/yr)						
Stem wood	}9.5	}8.1	}6.9	}3.9	}6.4	}3.5
Stem bark						
Branches	4.2[b]	4.7[b]	3.4[b]	2.2[b]	2.9[b]	1.5[b]
Fruits etc.						
Foliage						
Root estimate						

Five to eight trees were sampled in each of the above 60 m² plots, and stand biomass values were estimated by proportional basal area allocation.
a. The site quality.
b. Assumed to be equal to the increment of the stems within the crowns; excluding any woody litterfall.

Uenaka, K., Haya, K., Nasu, T. and Akai, T. (1972). Primary production of young
stands of *Pinus thunbergii* in various planting densities. *Rep. Kyoto Univ. For.*
10, 53-59.

Yuasa, Y. and Kamio, K. (1973). Leaf biomass and leaf-fall of young stands of
Japanese red pine (*Pinus densiflora*) and Japanese black pine (*Pinus thunbergii*).
Bull. Shizuoka Univ. For. 2, 25-33.

ca.35°N -- Japan Plantations.	Kyoto University Forest *Pinus thunbergii* A spacing experiment. (Uenaka *et al.* 1972)				Shizuoka University Forest *P. thunbergii* (Yuasa and Kamio 1973)
Age (years)	14	14	14	14	16
Trees/ha	2554	9938	26605	27649	9786
Tree height (m)	4.2	4.7	6.5	4.7	6.6
Basal area (m²/ha)	5.2	13.2	38.2	20.6	35.4
Leaf area index					
Stem volume (m³/ha)	16.6	42.2	159	76	
Dry biomass (t/ha) Stem wood	}9.1	}23.9	}90.7	}43.8	}58.1
Stem bark					
Branches	4.5	3.9	11.3	4.5	9.9
Fruits etc.					
Foliage	3.2	3.4	9.9	4.0	8.0
Root estimate					
CAI (m³/ha/yr)					
Net production (t/ha/yr) Stem wood					
Stem bark					
Branches					
Fruits etc.					
Foliage					4.10^{a}
Root estimate					

At Kyoto, 24 trees were sampled and stand values were derived from regressions on
D² and on D²H.
At Shizuoka, 11 trees were sampled and stand values were derived from regressions
on D²H.
a. New foliage biomass; foliage litterfall was 3.73 t/ha/yr.

Yasui, H. and Narita, T. (1972). Studies on the selection forest of Ate (*Thujopsis dolabrata* Sieb. et Zucc. var. *hondai* Makino). 3. Biomass of Maate (a cv. of Ate) selection forest. *Bull. Fac. Agr. Univ. Shimane* <u>6</u>, 39-44.

---

ca.35°N   132°E   --   Japan, Shimane Prefecture

Thujopsis dolabrata **var.** *hondai*

Natural regeneration with trees of different ages and heights.

Age (years)						
Trees/ha						
Tree height (m)	ca.13	ca.15	ca.17	ca.17	ca.18	ca.23
Basal area (m²/ha)	26.5	32.4	45.6	42.8	39.8	29.4
Leaf area index						
Stem volume (m³/ha)	147	213	293	321	343	291
Dry biomass (t/ha) — Stem wood / Stem bark	}66.3	}95.8	}131.8	}144.6	}154.5	}131.0
Branches	12.6	27.8	21.5	22.3	19.7	12.5
Fruits etc.						
Foliage	14.6	21.6	24.9	24.9	17.0	11.8
Root estimate						
CAI (m³/ha/yr)	10.3	11.1	15.9	15.5	15.2	13.6
Net production (t/ha/yr) — Stem wood						
Stem bark						
Branches						
Fruits etc.						
Foliage						
Root estimate						

Twelve to 25 trees were sampled at each of the above six sites, and stand values were derived from regressions on D²H. The sites were at Koonosu, Koizumi, Futamata, Ishiyasumiba, Yamamoto and Hosoya in columns left to right.

Satoo, T., Negisi, K. and Yagi, K. (1974). Materials for the studies of growth in forest stands. XII Primary production relations in plantations of *Thujopsis dolabrata* in the Noto Peninsula. *Bull. Tokyo Univ. For.* <u>66</u>, 139-151.

ca.37°30'N 136°50'E   (alt. given below)   Japan, Noto Peninsula.

Clonal
plantations   *Thujopsis dolabrata*

	150 m	120 m	280 m
Age (years)	24-31	23-27	35-42
Trees/ha	5584	6490	2760
Tree height (m)	8.7	7.4	12.2
Basal area (m²/ha)	42.2	47.3	56.0
Leaf area index	12.6^b (or 13.3)^{ab}	17.5^b (or 18.3)^{ab}	12.7^b (or 13.2)^{ab}
Stem volume (m³/ha)	261 (or 265)^a	233 (or 222)^a	415 (or 417)^a

		150 m	120 m	280 m
Dry biomass (t/ha)	Stem wood / Stem bark	}103.8 (or 105.1)^a	}94.3 (or 95.3)^a	}157.3 (or 156.9)^a
	Branches	13.0 (or 13.0)^a	20.2 (or 19.8)^a	24.1 (or 24.6)^a
	Fruits etc.			
	Foliage	30.1 (or 32.2)^a	43.6 (or 44.5)^a	31.7 (or 32.1)^a
	Root estimate			

CAI (m³/ha/yr)

		150 m	120 m	280 m
Net production (t/ha/yr)	Stem wood / Stem bark	}5.66^c (or 6.74)^{ac}	}13.29^c (or 13.79)^{ac}	}6.14^c (or 6.23)^{ac}
	Branches	1.36^c (or 1.44)^{ac}	1.82^c (or 1.71)^{ac}	
	Fruits etc.			
	Foliage	3.81 (or 3.58)^a	3.95 (or 3.80)^a	
	Root estimate			

Ten trees were sampled at each of the above three sites and stand values were derived from regressions on D.
a. Alternative values derived by proportional basal area allocation.
b. All-sided LAI values can be obtained by multiplying by 2.3.
c. Excluding woody litterfall and any mortality.

Kitazawa, Y. (1981). In: "Dynamic Properties of Forest Ecosystems" (D.E. Reichle, ed.) p. 603. Cambridge University Press, Cambridge, London, New York and Melbourne.

36°40'N   138°30'E   1790 m   Japan, Shigayama.

Wet podzols,
pH 3.6-4.2

Tsuga diversifolia, Abies mariesii with
Betula ermanii.

Age (years)		290
Trees/ha		1199
Tree height (m)		18
Basal area (m²/ha)		53.1
Leaf area index		6.8
Stem volume (m³/ha)		

Dry biomass (t/ha)	Stem wood	}139.9
	Stem bark	
	Branches	51.7
	Fruits etc.	
	Foliage	9.9
	Root estimate	

CAI (m³/ha/yr)

Net production (t/ha/yr)	Stem wood	}$1.80 + 0.02^a$
	Stem bark	
	Branches	$1.79 + 0.93^a$
	Fruits etc.	0.06^c } $+ 0.13^b$
	Foliage	2.12^c
	Root estimate	

---

a. Woody litterfall, excluding any mortality.
b. Miscellaneous litterfall.
c. Litterfall.

Ando, T., Chiba, K., Nishimura, T. and Tanimoto, T. (1977). Temperate fir and hem-
lock forests in Shikoku. In: "Primary Productivity in Japanese Forests"
(T. Shidei and T. Kira, eds) pp. 213-245. JIBP Synthesis vol. 16. University of
Tokyo Press.

---

ca.33°20'N 133°00'E   720 m   Japan, Kochi Prefecture, Yusuhara district.

Shallow,   *Tsuga sieboldii* with a few *Pinus densiflora* and
infertile,
soil.   *Chamaecyparis obtusa* and deciduous and evergreen

                            broadleaved understorey trees.

Age (years)	120-443	
Trees/ha	$475 + 1473^a$	
Tree height (m)	$24.0 \quad 10.2^a$	
Basal area (m²/ha)	$70.3 + 17.7^a$	
Leaf area index	$4.3^b + 1.0^a$	
Stem volume (m³/ha)	$793 + 129^a$	

Dry biomass (t/ha)	Stem wood	$\left.\begin{array}{c} \\ \\ \end{array}\right\}347.1 + 85.6^a$
	Stem bark	
	Branches	$91.8 + 25.5^a$
	Fruits etc.	
	Foliage	$7.8 + 1.9^a$
	Root estimate	$136.6 + 27.1^{ac}$

CAI (m³/ha/yr)

Net production (t/ha/yr)	Stem wood	$\left.\begin{array}{c} \\ \\ \end{array}\right\}1.45 + 0.73^a \text{ (or } 1.9 + 1.2^a)^d$
	Stem bark	
	Branches	$0.51 + 1.23^e + 0.23^a + 0.38^{ae} \text{ (or } 0.8 + 1.5^a)^d$
	Fruits etc.	$0.12 + 0.00^a$
	Foliage	$0.04 + 2.24^e + 0.01^a + 0.62^{ae} \text{ (or } 2.4 + 0.8^a)^d$
	Root estimate	$0.6 + 0.2^a \text{ (or } 0.8 + 0.4^a)^d$

Seven *T. sieboldii* and 18 understorey trees were sampled, and roots of the seven *T. sieboldii* trees were excavated. Stand biomass values for a 0.12 ha plot were derived from regressions on D. Biomass values given above are the means over 4 years. Nutrient contents were determined.

a. Understorey values.
b. All-sided LAI was 9.9.
c. Understorey root biomass was estimated assuming the top/root ratio to be 4.2.
d. Alternative production values (in the brackets), estimated as the biomass of new growth made during the previous year.
e. Litterfall measured over 1 to 3 years.

Hozumi, K., Yoda, K., Kokawa, S. and Kira, T. (1969). Production ecology of tropical rainforests in southwestern Cambodia. I Plant biomass. *Nature and Life in S.E. Asia* 6, 1-51.

Hozumi, K., Yoda, K. and Kira, T. (1969). Production ecology of tropical rainforests in southwestern Cambodia. II Photosynthetic production in an evergreen seasonal forest. *Nature and Life in S.E. Asia* 6, 57-81.

10°56'N   103°24'E   ca.100 m   Kampuchea, Koh Kong province, Chékô.

Deep reddish yellow, sandy latosol.   *Dipterocarpus* spp. (35%)[a], *Anisoptera* sp., *Myristica* sp., *Beilschmiedia* sp. *et al.*

Evergreen seasonal tropical forest.

Age (years)		
Trees/ha		
Tree height (m)	(20, 40-45)[b]	(20, 40-45)[b]
Basal area (m²/ha)	31-34	31-34
Leaf area index	5.8 + 1.6[c]	5.8 + 1.6[c]
Stem volume (m³/ha)		
Dry biomass (t/ha)		
Stem wood	}225.4 + 4.2[c]	}197.1 + 4.2[c]
Stem bark		
Branches	107.5 + 0.6[c]	88.1 + 0.6[c]
Fruits etc.		
Foliage	6.5 + 0.8[c]	6.4 + 0.8[c]
Root estimate	69.2 + 0.9[c]	49.9 + 0.9[c]
CAI (m³/ha/yr)		
Net production (t/ha/yr)		
Stem wood		
Stem bark		
Branches		
Fruits etc.		
Foliage		
Root estimate		

One hundred and thirty trees were sampled and 8 root systems were excavated. Stand values for the above two 0.25 ha plots were derived from regressions on D and D²H.
a. Percentage of the total basal area.
b. Middle and upper storeys.
c. Understorey shrubs less than 4.5 cm D.

Hozumi, K., Yoda, K., Kokawa, S. and Kira, T. (1969). Production ecology of tropical rainforests in southwestern Cambodia. I Plant biomass. *Nature and Life in S.E. Asia* <u>6</u>, 1-51.

10°56'N 103°24'E   50 m   Kampuchea, Koh Kong province, Chékô.

	Melaleuca leucadendron Swamp. Deep sand with red-brown pan at 1 m depth.	*Dacrydium* sp., *Tristania* sp *et al.* Heath, evergreen forest. Light reddish yellow sand.
Age (years)		
Trees/ha		
Tree height (m)	10	(3-10, 20)[a]
Basal area (m²/ha)	0.4	23.9
Leaf area index	0.4	5.5
Stem volume (m³/ha)		

Dry biomass (t/ha)

Stem wood	} 7.4	} 106.3
Stem bark		
Branches	3.9	34.0
Fruits etc.		
Foliage	0.8	6.8
Root estimate	2.6	18.0

CAI (m³/ha/yr)

Net production (t/ha/yr)

Stem wood		
Stem bark		
Branches		
Fruits etc.		
Foliage		
Root estimate		

Several trees of *M. leucadendron*, *Dacrydium* sp. and *Tristania* sp. were sampled. Stand values for the above two 0.1 ha plots were derived from regressions on D²H. Values given in the right column refer to trees at least 4.5 cm D.
a. Middle and upper storeys.

Magambo, M.J.S. and Cannell, M.G.R. (1981). Dry matter production and partition in relation to yield of tea. *Expl Agric.* <u>17</u>, 33-38.

0°22'S 35°21'E   2178 m   Kenya, Kericho.

Camellia sinensis (tea)

Plantation. Fertile lateritic red loam.	Coppiced at age 6 and pruned to form a 'plucking table'.	
	Plucked	Unplucked
Age (years)	7-8	7-8
Trees/ha	10766	10766
Tree height (m)	ca.1.0	ca.1.5
Basal area (m²/ha)		
Leaf area index	5-6	5-6
Stem volume (m³/ha)		

Dry biomass (t/ha)

	Plucked	Unplucked
Stem wood		
Stem bark	} 19.5	} 25.6
Branches		
Fruits etc.		
Foliage	9.0	12.2
Root estimate	17.9	19.3

CAI (m³/ha/yr)

Net production (t/ha/yr)

	Plucked	Unplucked
Stem wood		
Stem bark	} 7.4	} 13.4
Branches		
Fruits etc.		
Foliage	$2.5 + 1.4^a + 1.4^b$	$4.9 + 2.3^a$
Root estimate	4.2	5.7

Three bushes were sampled within a closed stand on each of 5 occasions during one year, and stand values were obtained by multiplying mean tree values by the numbers of bushes per hectare. Biomass values given above were those estimated at the last sampling.
a. Leaf litterfall.
b. Leaves removed as pluckings.

Ng, S.K., Thamboo, S. and de Souza, P. (1968). Nutrient contents of oil palms in Malaya. II Nutrients in vegetative tissues. *Malay. agric. J.* <u>46</u>, 332-390.

---

ca.3°20'N   101°20'E   10-50 m   Malaysia, Selangor, Bantung.

Plantations.
Marine clay, with   *Elaeis guineensis* (oil palm)
fine sandy loams.

High-yielding fertilized stands.

Age (years)	3.3	4.3	5.3	6.3	7.5	8.6
Trees/ha	148	148	148	148	148	148
Tree height (m)	1.3	1.4	2.0	2.3	2.6	3.0
Basal area (m²/ha)						
Leaf area index						
Stem volume (m³/ha)						

Dry biomass (t/ha)

Stem wood	} 8.5	} 13.7	} 17.9	} 21.3	} 31.5	} 32.8
Stem bark						
Branches	} 17.5^a	} 22.6^a	} 22.3^a	} 25.7^a	} 32.1^a	} 33.6^a
Fruits etc.						
Foliage						
Root estimate				10.3		

CAI (m³/ha/yr)

Net production (t/ha/yr)

Stem wood	
Stem bark	
Branches	
Fruits etc.	
Foliage	
Root estimate	

One or two palms of average size were sampled from each stand and parts of the root systems of 4 palms were excavated. Stand values were obtained by multiplying mean palm values by the numbers of palms per hectare. Nutrient contents were determined.
a. Including pinnae, raches, unexpanded leaves and 6.0 t/ha of fruits at age 3.3 increasing to 14.0 t/ha at age 8.6.

Continued from p.190.

Same as p.190.

	9.6	10.6	11.4	13.2	14.4	15.3
Age (years)	9.6	10.6	11.4	13.2	14.4	15.3
Trees/ha	148	148	148	148	148	148
Tree height (m)	3.5	4.0	4.8	6.0	7.1	8.0
Basal area (m²/ha)						
Leaf area index						
Stem volume (m³/ha)						

Dry biomass (t/ha)

	9.6	10.6	11.4	13.2	14.4	15.3
Stem wood / Stem bark	43.2	48.5	55.6	75.9	67.8	96.3
Branches / Fruits etc. / Foliage	37.2^a	32.7^a	37.2^a	39.2^a	32.3^a	36.3^a
Root estimate					16.3	

CAI (m³/ha/yr)

Net production (t/ha/yr)

Stem wood
Stem bark
Branches 12.9 to 17.0^b
Fruits etc. $+ 14.3^c$
Foliage
Root estimate 3.5–4.5

One or two palms of average size were sampled from each stand and parts of the root systems of 3 palms were excavated. Stand values were obtained by multiplying mean palm values by the numbers of palms per hectare. Nutrient contents were determined.
a. Including pinnae, raches, unexpanded leaves and about 14.0 t/ha of fruits.
b. Mean annual increment of vegetative parts, including litterfall, between ages 8 and 15.
c. Fruit production, assuming a yield of 25 t/ha/yr fresh weight.

Shorrocks, V.M. (1965). Mineral nutrition, growth and nutrient cycle of *Hevea braziliensis*. I Growth and nutrient cycle. *J. Rubb. Res. Inst. Malaysia* 19, 32-47.

---

ca.3°10'N   101°40'E   50-200 m   Malaysia, near Kuala Lumpur.

Plantations.   *Hevea braziliensis*
Fertile 'Malacca'
series soils.   Grafted clones RRIM 501 and Tjir 1.

                              Stands were tapped after age 6.

Age (years)	1	2	3	4	5	6
Trees/ha	445	445	445	408	371	346
Tree height (m)	4.6	7.0	7.3		13.1	
Basal area (m²/ha)	0.5	1.9	2.9	5.2	6.9	7.5
Leaf area index	0.1	1.1	2.6	3.4	4.1	6.3

Stem volume (m³/ha)

Dry biomass (t/ha)

	1	2	3	4	5	6
Stem wood / Stem bark	}0.5	}3.1	}5.6	}13.6	}12.8	}18.8
Branches	0.2	2.6	5.4	23.1	31.9	79.1
Fruits etc.						
Foliage	0.1	0.9	2.3	3.3	3.8	4.7
Root estimate	0.4	1.8	4.0	7.5	8.7	12.9

CAI (m³/ha/yr)

Net production (t/ha/yr)

Stem wood			
Stem bark			
Branches	}6.3	}8.4	}ca.22.0[a]
Fruits etc.			
Foliage			
Root estimate			

One to four trees of average size were sampled in each stand and their roots were excavated (except for the stand aged 8). Stand biomass values were obtained by multiplying mean tree values by the numbers of trees per hectare. Basal areas given above were derived from mean girths and numbers of trees per hectare. The high branch/stem biomass ratios may be attributed to tapping and the wide spacings. Nutrient contents were determined.

a. Production value estimated from graphs of cumulative biomass with age, assuming no branch litterfall and a leaf longevity of only one year; also assuming that trees that were culled were of average size, and (on p.193) that 8300 fruits/ha were produced per year after age 6.

Same as p.192.

Age (years)	8	10	11	24	33
Trees/ha	321	296	296	247	267
Tree height (m)		18.9	20.1	23.2	24.1
Basal area (m²/ha)	11.5	17.4	14.6	20.5	29.2
Leaf area index	5.3	14.0	9.9	8.9	5.4
Stem volume (m³/ha)					

Dry biomass (t/ha)

	8	10	11	24	33
Stem wood	} 16.3	} 30.2	} 28.4	} 61.3	} 84.6
Stem bark					
Branches	54.3	219.6	169.9	161.7	391.8
Fruits etc.					
Foliage	3.9	11.0	7.8	7.0	4.4
Root estimate	16.2	43.0	36.4	40.6	84.9

CAI (m³/ha/yr)

Net production (t/ha/yr)

Stem wood			
Stem bark			
Branches	ca.35.0[a]	ca.37.0[a]	ca.19.0[a]
Fruits etc.			
Foliage			
Root estimate			

Templeton, J.K. (1968). Growth studies in *Hevea braziliensis*. I Growth analysis up to seven years after budgrafting. *J. Rubb. Res. Inst. Malaysia* 20, 136-146.

ca.3°10'N   101°40'E   50-200 m   Malaysia, near Kuala Lumpur.

Plantations.
Deep sandy   *Hevea braziliensis*
clay loam

                          Grafted clones RRIM 501 and RRIM 513.

Age (years)	1.25	2.25	3.25	4.25	5.25	6.25
Trees/ha	445	445	445	445	445	445
Tree height (m)	0.3	1.4	3.9	5.2	5.8	5.8
Basal area (m²/ha)						
Leaf area index						
Stem volume (m³/ha)						

Dry biomass (t/ha)

	Stem wood						
	Stem bark						
	Branches	2	9	34	63	95	125
	Fruits etc.						
	Foliage						
	Root estimate						

CAI (m³/ha/yr)

Net production (t/ha/yr)

	Stem wood						
	Stem bark						
	Branches	7[a]	25[a]	29[a]	32[a]	30[a]
	Fruits etc.						
	Foliage						
	Root estimate						

Nine or ten trees were sampled 9, 15, 21, 27, 39, 55, 63 and 81 months after grafting, and roots of half of the sample trees were excavated. Stand biomass values were derived from regressions on tree girths. About 20% of the total biomass consisted of roots. Values for ages 4.25 and 6.25 were interpolated from the author's Figure 8 (correcting an error on the ordinate axis).

a. Increments between ages 1.25 and 2.25, 2.25 and 3.25, 3.25 and 4.25, 4.25 and 5.25, and 5.25 and 6.25 years in columns left to right; the author estimated peak production to be about 35.5 t/ha/yr at about age 4.5.

Bullock, J.A. (1981). In: "Dynamic Properties of Forest Ecosystems" (D.E. Reichle, ed.) p.606. Cambridge University Press, Cambridge, London, New York and Melbourne.

2°59'N 102°18'E   100 m   Malaysia, Negiri Sembilan, Pasoh Forest Reserve.

Sandy clay loam, pH 4.3-4.8

Shorea spp., *Dipterocarpus* spp., *Koompassia* spp. *et al.*

Tropical rainforest.

Age (years)	Mature
Trees/ha	
Tree height (m)	35-40
Basal area (m²/ha)	27.3
Leaf area index	8 to 10
Stem volume (m³/ha)	

Dry biomass (t/ha)

Stem wood	$\Big\}$ 287.1 + 10.4^a
Stem bark	
Branches	59.4 + 2.0^a
Fruits etc.	
Foliage	4.8 + 1.1^a
Root estimate	29.6

CAI (m³/ha/yr)

Net production (t/ha/yr)

Stem wood	$\Big\}$ -1.59 + 5.97^b
Stem bark	
Branches	-0.29 + 3.10^b
Fruits etc.	0.47^b
Foliage	-0.04 + 7.49^b + 0.27^c
Root estimate	

There was also 13.9 t/ha of standing dead wood.
a. Understorey vegetation.
b. Litterfall.
c. Consumption.

(See also p.196.)

Kato, R., Tadaki, Y. and Ogawa, H. (1978). Plant biomass and growth increment studies in Pasoh Forest. *Malay Nat. J.* 30, 211-224.
Kira, T. (1978). Primary productivity of Pasoh Forest - a synthesis. *Malay Nat. J.* 30, 291-297.
Kira, T. (1978). Community architecture and organic matter dynamics in tropical lowland rainforests of southeastern Asia with special reference to Pasoh Forest, west Malaysia. In: "Tropical Trees as Living Systems" (P.B. Tomlinson and M.H. Zimmermann, eds) pp. 561-590. Cambridge University Press, Cambridge, London.

2°59'N 102°18'E   100 m   Malaysia, Negiri Sembilan, Pasoh Forest Reserve.

Sandy clay loam, pH 4.3-4.8   *Shorea* spp., *Dipterocarpus* spp., *Koompassia* spp. *et al.*

Tropical rainforest.

Age (years)	Mature	Mature	Mature
Trees/ha			
Tree height (m)	35-40	35-40	35-40
Basal area (m²/ha)			
Leaf area index	$6.2 + 1.0^a$	$7.1 + 0.9^a$	6.87
Stem volume (m³/ha)			

Dry biomass (t/ha)

Stem wood	} 522.2	} 367.5	} 346.0
Stem bark			
Branches	125.4	90.1	77.9
Fruits etc.			
Foliage	$7.8 + 1.2^a$	$8.0 + 0.2^a$	7.77
Root estimate			

CAI (m³/ha/yr)

Net production (t/ha/yr)

Stem wood		} $5.1 + 3.7^b + 0.7^d$	} $4.24 + 3.30^b + 0.37^d$
Stem bark			
Branches		$1.2 + 2.7^c + 1.2^d$	$0.98 + 2.43^c + 1.04^d$
Fruits etc.			1.36^e
Foliage		$7.8^c + 0.1^d$	$7.03^c + 0.31^d$
Root estimate		4.4	$5.00^f + 0.53^g$

Seventy-three trees were sampled in 1971 and 83 trees were sampled in 1973. Stand biomass values were estimated from regressions on D, H and D²H for trees at least 4.5 cm D in a 0.1 ha plot containing big trees (left column), a randomly chosen 0.20 ha plot (centre column) and all trees in a 0.8 ha area (right column).
a. Ground vegetation.
b. 'Big wood' litterfall.
c. Litterfall.
d. Decay loss.
e. Miscellaneous litterfall.
f. One quarter of the fine root biomass.
g. Estimated increment of thick roots.

Yoda, K. (1968). A preliminary survey of the forest vegetation of eastern Nepal. III Plant biomass in the sample plots chosen from different vegetation zones. *J. Coll. Arts Sci. Chiba Univ.* 5, 277-302.

Yoda, K. (1967). A preliminary survey of the forest vegetation of eastern Nepal. II General description, structure and floristic composition of the sample plots chosen from different vegetation zones. *J. Coll. Arts Sci. Chiba Univ.* 5, 99-140.

27°30-45'N 86°20' to 84°40'E (alt. given below) Nepal, 100 km E of Kathmandu.	
Quercus sp.?, *Quercus glauca*, *Quercus lamellosa* and *Quercus spicata* (21%)[a] *Machilus edulis* (17%)[a] *Cinnamomum* sp. (17%)[a] *et al.*	*Quercus glauca* (27%)[a] *Cinnamomum* sp. (22%)[a] *Buddleja* sp. (13%)[a] *et al.*
2270 m	2390 m
Age (years)	
Trees/ha — 745	600
Tree height (m) — 16.6	22.9
Basal area (m²/ha) — 63.3	65.8
Leaf area index — 4.5	6.3
Stem volume (m³/ha)	
Stem wood / Stem bark } 319	} 338
Branches — 130	131
Fruits etc.	
Foliage — 5.8	8.1
Root estimate — 91[b]	96[b]
CAI (m³/ha/yr)	
Stem wood	
Stem bark	
Branches	
Fruits etc.	
Foliage	
Root estimate	

(Left margin labels: Dry biomass (t/ha); Net production (t/ha/yr))

Up to 16 trees were sampled per plot. Stand values for the above 0.20 and 0.08 ha plots (left and right columns, respectively) were derived from regressions on D and H. Only trees over 5 cm D were included.
a. Percentage of the total basal area (21% in the left column refers to all *Quercus* species).
b. Assumed to be 20% of the above-ground biomass value.

Same as p.197.

27°30-45'N 86°20' to 84°40'E (alt. given below) Nepal, 100 km E of Kathmandu.		
Rhododendron *lanatum* (53%)[a] *Juniperus* *wallichiana?* (47%)[a]	*Juniperus* *squamata* (87%)[b] *Rhododendron* *anthopogon* (13%)[b]	*Rh. anthopogon* *Rhododendron setosum* *Rhododendron nivale* (94%)[b]
3830 m	3870 m	4050 m

Age (years)			
Trees/ha	3320	201600	1350000
Tree height (m)	8.3	1.2	1.2
Basal area (m²/ha)	39.2		
Leaf area index			
Stem volume (m³/ha)			

Dry biomass (t/ha)

Stem wood	} 67	} 1.1	} 65
Stem bark			
Branches	27	2.2	
Fruits etc.			
Foliage	10.4	4.7	103.5
Root estimate	23[c]	6.7[c]	242[c]

CAI (m³/ha/yr)

Net production (t/ha/yr)

Stem wood
Stem bark
Branches
Fruits etc.
Foliage
Root estimate

Up to 16 trees were sampled per plot. Stand values for the above 256, 25 and 10 m² plots (in columns left to right) were derived from regressions on D and H.
a. Percentage of the total basal area.
b. Percentage of the total biomass; 94% in the right column refers to all *Rhododendron* species.
c. Assuming top/root biomass ratios to be 5 for broadleaved trees and 4 for conifers.

Same as p.197.

27°30-45'N 86°20' to 84°40'E   (alt. given below)   Nepal, 100 km E of Kathmandu.
Abies spectabilis, with understorey of *Rhododendron* spp. *et al.*

	19%[a] *Tsuga dumosa* (67%)[b]	(99%)[b]	(98%)[b]	(98%)[b]	(100%)[b]	(79%)[b]
	2920 m	3120 m	3280 m	3420 m	3530 m	3680 m
Age (years)						
Trees/ha	913	563	713	488	275	1450
Tree height (m)	23.6	23.1	22.1	ca.20	21.9	11.6
Basal area (m²/ha)	72.5	73.3	59.6	59.6	49.9	44.6
Leaf area index						
Stem volume (m³/ha)						

Dry biomass (t/ha)

Stem wood	} 346	} 339	} 271	} 289	} 231	} 127
Stem bark						
Branches	51	41	40	36	28	22
Fruits etc.						
Foliage	20.2	20.1	17.5	14.2	10.7	9.4
Root estimate	102[c]	100[c]	80[c]	85[c]	67[c]	38[c]

CAI (m³/ha/yr)

Net production (t/ha/yr)

Stem wood						
Stem bark						
Branches						
Fruits etc.						
Foliage						
Root estimate						

Up to 16 trees were sampled per plot. Stand values for the above 0.23, 0.16, 0.16, 0.80, 0.16 and 0.08 ha plots (in columns left to right) were derived from regressions on D and H. Only trees over 5 cm D were included.
a. Percentage of the total basal area; *T. dumosa* syn. *brunoniana*.
b. Percentage of the total basal area accounted for by *A. spectabilis*.
c. Assuming top/root biomass ratios to be 5 for broadleaved trees and 4 for *A. spectabilis*.

Same as p.197.

27°30-45'N 86°20' to 84°49'E (alt. given below)	Nepal, 100 km E of Kathmandu.
Tsuga dumosa (56%)[a] *Quercus semicarpifolia* (39%)[a] *Rhododendron* spp. (2%)[a] *et al.* 2720 m	*T. dumosa* (97%)[a] *Rhododendron* spp. (2%)[a] 2760 m

		2720 m	2760 m
Age (years)			
Trees/ha		220	280
Tree height (m)		37.1	30.6
Basal area (m²/ha)		67.5	75.6
Leaf area index			
Stem volume (m³/ha)			
Dry biomass (t/ha)	Stem wood	} 461	} 429
	Stem bark		
	Branches	90	70
	Fruits etc.		
	Foliage	5.5	12.0
	Root estimate	125[b]	127[b]
CAI (m³/ha/yr)			
Net production (t/ha/yr)	Stem wood		
	Stem bark		
	Branches		
	Fruits etc.		
	Foliage		
	Root estimate		

Up to 16 trees were sampled per plot. Stand values for the above 0.20 and 0.25 ha plots (left and right columns, respectively) were derived from regressions on D and H. Only trees over 5 cm D were included.

a. Percentage of the total basal area; *Tsuga dumosa* syn. *brunoniana*.

b. Assuming top/root biomass ratios to be 5 for broadleaved trees and 4 for *T. dumosa*.

Drift, J. van der (1974). "Project Meerdink: Production and Decomposition of Organic Matter in an Oak Woodland." Final report 1966-71 of the Netherlands contribution to the IBP. North Holland Publication Co., Amsterdam.

Drift, J. van der (1981). In: "Dynamic Properties of Forest Ecosystems" (D.E. Reichle, ed.) p.607. Cambridge University Press, Cambridge, London.

51°55'N 6°42'E   45 m   Netherlands, Meerdink.

Humus infiltrated sands overlying heavy tertiary loam.	*Quercus petraea* with *Fagus sylvatica* and an understorey of *Sorbus aucuparia* and *Frangula alnus*

Age (years)		140
Trees/ha		300
Tree height (m)		27.2
Basal area (m²/ha)		33.8
Leaf area index		5.2
Stem volume (m³/ha)		

Dry biomass (t/ha)

Stem wood	$\Big\}\ 236.5 + 7.8^{a}$
Stem bark	
Branches	$28.5 + 0.9^{a}$
Fruits etc.	0.8
Foliage	$3.3 + 0.3^{a}$
Root estimate	41.8

CAI (m³/ha/yr)

Net production (t/ha/yr)

Stem wood	$\left.\begin{array}{l}\Big\}\ 3.50^{b} + 0.55^{a} + 0.08^{c}\\[4pt] 0.80\ \ + 0.15^{a} + 1.07^{c}\end{array}\right\} + 0.70^{d}$
Stem bark	
Branches	
Fruits etc.	$0.79^{c} + 0.70^{c}$
Foliage	$3.67^{c} + 0.71^{e}$
Root estimate	2.20^{f}

Many trees were sampled and stand biomass values were derived from regressions on D and H.

a. Understorey shrubs.
b. Including boughs over 7 cm diameter.
c. Litterfall (0.79 t/ha/yr was acorn production).
d. Woody increment of *F. sylvatica*.
e. Foliage consumption and pre-fall losses.
f. Assumed to be 20% of the above-ground increment value.

Minderman, G. (1967). The production of organic matter and the utilization of solar energy by a forest plantation of *Pinus nigra* var. *austriaca*. *Pedobiologia* 7, 11–22.

52°34'N 4°39'E   4 m   Netherlands, North Holland, Bakkum.

Plantation.   *Pinus nigra* var. *austriaca* syn. var. *nigra*.
Dune sand.

Thinned at ages 13, 18 and 21.

Age (years)	22
Trees/ha	4800
Tree height (m)	
Basal area (m²/ha)	
Leaf area index	
Stem volume (m³/ha)	163[a]

Dry biomass (t/ha)		
Stem wood	}	45.5
Stem bark		
Branches		27.7
Fruits etc.		
Foliage		9.4
Root estimate		16.4

CAI (m³/ha/yr)

Net production (t/ha/yr)		
Stem wood	}	4.1
Stem bark		
Branches		2.0[b]
Fruits etc.		
Foliage		4.0[c]
Root estimate		

Five trees were sampled and roots were excavated. Stand biomass values for a 625 m² plot were obtained by multiplying mean tree values by the numbers of trees per hectare. Nitrogen contents were determined.

a. Basal area at 1.3 m times height.
b. Excluding woody litterfall.
c. Assumed to be 43% of the foliage biomass.

Madgwick, H.A.I., Beets, P. and Gallagher, S. (1981). Dry matter accumulation, nutrient and energy content of the above-ground portion of 4-year-old stands of *Eucalyptus nitens* and *E. fastigata*. *N.Z.J. For. Sci.* 11, 53-59.

38°00'S 176°20'E   200-500 m   New Zealand, near Rotorua, Rotoehu State Forest.

Plantations.
Soils derived from
Kaharoa ash.

		Eucalyptus nitens	*Eucalyptus fastigata*
Age (years)		4	4
Trees/ha		6470	7250[a]
Tree height (m)		9.5	7.3
Basal area (m²/ha)		31.1	20.6
Leaf area index			
Stem volume (m³/ha)			
Dry biomass (t/ha)	Stem wood	52.1	33.6
	Stem bark	7.6	4.6
	Branches	9.1	11.2
	Fruits etc.	0.0	0.0
	Foliage	9.4	10.6
	Root estimate		
CAI (m³/ha/yr)			
Net production (t/ha/yr)	Stem wood		
	Stem bark		
	Branches		
	Fruits etc.		
	Foliage		
	Root estimate		

Seven trees of each species were sampled in autumn (April) and stand values for a 82 m² plot in each stand were derived from regressions on D. Nutrient contents were determined. There was 2.1 and 1.9 t/ha of dead branches in the left and right columns, respectively.
a. There were a further 380 dead stems per hectare.

Beets, P.N. (1980). Amount and distribution of dry matter in a mature beech / podocarp community. *N.Z.J. For. Sci.* 10, 395-418.

Miller, R.B. (1963). Plant nutrients in hard beech. I The immobilization of nutrients. III The cycle of nutrients. *N.Z. Jl Sci.* 6, 365-377; 388-413.

New Zealand	42°08'S 171°45'E 100-150 m Tawhai State Forest, Maimai	41°20'S 175°00'E -- Nr Wellington, Western Hut Hills
Shallow, infertile, yellow-brown earths.	*Nothofagus truncata* (47%)[a] *Weinmannia racemosa* (21%)[a] et al.	*Nothofagus truncata*
	(Beets 1980)	(Miller 1963)

Age (years)	up to 300	110
Trees/ha	1346[b]	484
Tree height (m)	$(5, 7\text{-}19, 20\text{-}36)^{c}$	
Basal area (m²/ha)	ca.50	
Leaf area index		
Stem volume (m³/ha)		

Dry biomass (t/ha)

Stem wood	$200.2 + 0.7^{d}$	194
Stem bark	$22.9 + 0.1^{d}$	34
Branches	$69.2 + 0.4^{d}$	44
Fruits etc.		
Foliage	$5.24 + 0.41^{d}$	2.8
Root estimate		40

CAI (m³/ha/yr)

Net production (t/ha/yr)

Stem wood		
Stem bark		
Branches		1.7^{e}
Fruits etc.		
Foliage		4.3^{e}
Root estimate		

Beets (1980) sampled 14 trees at least 25 cm D during the summer (December), and derived stand values for a catchment of 4.14 ha from regressions on D; smaller trees and understorey shrubs were sampled in twenty 400 m² plots; there was 6.5 t/ha of dead branches and 21.5 t/ha of standing dead trees.
Miller (1963) sampled only one tree of average size, plus part of its root system, and derived stand values by proportional basal area allocation; nutrient contents were determined.
a. Percentage of the total above-ground biomass.
b. Canopy and subcanopy trees.
c. Lower, middle and upper storeys.   *d*. Understorey shrubs.
e. Litterfall only, measured over 7 years.

Orman, H.R. and Will, G.M. (1960). The nutrient content of *Pinus radiata* trees. *N.Z. Jl Sci.* <u>3</u>, 510-522.

Will, G.M. (1964). Dry matter production and nutrient uptake by *Pinus radiata* in New Zealand. *Commonw. For. Rev.* <u>43</u>, 57-70.

Will, G.M. (1959). Nutrient return in litter and rainfall under some exotic conifer stands in New Zealand. *N.Z. Jl agric. Res.* <u>2</u>, 719-734.

New Zealand.	38°24'S 176°36'E 518 m Kaingaroa Forest	ca.38°07'S 176°17'E 500 m Whakarewarewa Forest
Fine pumice sands overlying pumice gravel.	*Pinus radiata* Plantation (Orman and Will 1960)	*Pinus radiata* Unthinned natural regeneration. (Will 1959, 1964)
Age (years)	26	12
Trees/ha	301	2637
Tree height (m)	35.4	46.7
Basal area (m²/ha)	40.4	
Leaf area index		
Stem volume (m³/ha)	483	368

		Dry biomass (t/ha)	
	Stem wood	196.6	} 132.8
	Stem bark	28.4	
	Branches	11.7	49.0
	Fruits etc.		
	Foliage	4.5	9.0
	Root estimate		

		Net production (t/ha/yr)	
CAI (m³/ha/yr)			
	Stem wood		
	Stem bark		
	Branches		
	Fruits etc.		
	Foliage		2.3[a]
	Root estimate		

Orman and Will (1960) sampled 8 trees and presented mean tree values; stand values were obtained by the compiler by multiplying mean tree values by the numbers of trees per hectare.

Will (1959, 1964) sampled 24 trees and derived stand values for a 0.16 ha plot, also by multiplying mean tree values by the numbers of trees per hectare; there was 8.2 t/ha of dead stems and 6.7 t/ha of dead branches.

Nutrient contents were determined in both studies.

a. Foliage litterfall.

Madgwick, H.A.I., Jackson, D.S. and Knight, P.J. (1977a). Above-ground dry matter, energy and nutrient contents of trees in an age series of *Pinus radiata* plantations. *N.Z.J. For. Sci.* 7, 445-468.

Madgwick, H.A.I., Jackson, D.S. and Knight, P.J. (1977b). "Dry Matter and Nutrient Data on *Pinus radiata* Trees and Stands." N.Z. Forest Service, Forest Research Institute, Production Forestry Division, Report No.84.

38°18'S   176°44'E   ca.500 m   New Zealand, Kaingaroa Forest.

Plantations. Volcanic ash, Matahina gravels and hill soils.	*Pinus radiata*			Pruned to 2.4 m height at age 6	Stands thinned at age 8 and pruned to 4.3-6.1 m height between ages 4 and 8.				
Age (years)	2	4	6	8	8	9	10	17	22
Trees/ha	2496	2347	2224	1507	544	544	544	855[a]	544
Tree height (m)	1.1	3.9	7.1	11.0	11.7	17.0	16.7	26.1	30.7
Basal area (m²/ha)		7.4	20.1	29.4	14.0	12.3	20.8	51.3	53.9
Leaf area index									
Stem volume (m³/ha)									

Dry biomass (t/ha)

	2	4	6	8	8	9	10	17	22
Stem wood	0.2	7.2	22.6	46.0	25.9	29.1	49.6	214.8	243.5
Stem bark	0.1	1.3	2.9	5.3	2.9	3.6	4.8	21.9	27.4
Branches	0.1	6.6	15.0	23.8	4.8	5.5	11.4	22.0	27.4
Fruits etc.	0.0	0.0	0.0	0.0	0.1	0.6	0.2	5.0	3.2
Foliage	0.4	7.2	11.6	5.9	2.5	3.4	5.9	10.8	9.3
Root estimate									

CAI (m³/ha/yr)

Net production (t/ha/yr)

	2	4	6	8-22
Stem wood		3.5	7.7	11.7
Stem bark		0.7	0.9	1.2
Branches		3.3[b]	4.4[b]	5.1[b]
Fruits etc.		0.0	0.0	0.0
Foliage		3.6[d]	5.7[d]	5.7[d]
Root estimate				

(Net production, ages 10–22: >24.6[c])

Five to seven trees were sampled in each stand, and biomass values for plots of about 400 m² were derived from regressions on D and D²H. Nutrient contents were determined.

a. Note that this plot was less heavily thinned than the others.
b. Including estimated woody litterfall and prunings.
c. Estimated minimum annual production between ages 10 and 22.
d. New foliage biomass.

Rees, A.R. and Tinker, P.B.H. (1963). Dry matter production and nutrient content of plantation oil palms in Nigeria. I Growth and dry matter production. *Pl. Soil* <u>19</u>, 19–32.

Tinker, P.B.H. and Smilde, K.W. (1963). Dry matter production and nutrient content of oil palms in Nigeria. II Nutrient content. *Pl. Soil* <u>19</u>, 350–364.

---

ca.6°20'N 5°40'E   50 m   Nigeria, near Benin city.

Plantations.
Deep, uniform red,   *Elaeis guineensis* (oil palm)
clayey sands.

		7	10	14	17	20
Age (years)		7	10	14	17	20
Trees/ha		147	147	147	147	147
Tree height (m)		0.9	2.9	3.8	6.2	9.3
Basal area (m²/ha)						
Leaf area index		3.0	3.7	3.8	4.4	5.6
Stem volume (m³/ha)						
Dry biomass (t/ha)	Stem wood	}12.1	}25.1	}35.1	}41.2	}64.5
	Stem bark					
	Branches	6.3[a]	8.8[a]	8.0[a]	8.7[a]	14.9[a]
	Fruits etc.	5.6	4.1	2.6	2.4	3.3
	Foliage	4.3	5.4	5.3	6.8	8.5
	Root estimate				18.8	
CAI (m³/ha/yr)						
Net production (t/ha/yr)	Stem wood		}4.6	}2.8	}3.2	}2.6
	Stem bark					
	Branches	5.1[a]	6.0[a]	5.1[a]	6.0[a]	8.5[a]
	Fruits etc.	3.7	5.1	4.3	7.1	4.4
	Foliage	2.9[b]	3.1[b]	2.8[b]	3.7[b]	4.1[b]
	Root estimate					

Two or three palms of average size were sampled in each stand, and the roots of two 17-year-old palms were excavated. Stand biomass values were obtained by multiplying mean palm values by the numbers of palms per hectare. Nutrient contents were determined.
a. The raches, including estimated litterfall.
b. Estimated from the numbers of leaves produced per year.

Egunjobi, J.K. and Fasehun, F. (1972). Preliminary observations on litterfall and nutrient content of *Pinus caribaea* litter. *Niger. J. Sci.* 6, 37-65.

Egunjobi, J.K. (1975). Dry matter production by an immature stand of *Pinus caribaea* in Nigeria. *Oikos* 26, 80-85.

Egunjobi, J.K. (1976). An evaluation of five methods for estimating biomass of an even-aged plantation of *Pinus caribaea* L. *Oecol. Plant.* 11, 109-116.

7°23'N   3°56'E   200 m   Nigeria, Ibadan.

Plantation.
Well-drained   *Pinus caribaea* var. *hondurensis*
soil
(see 1972 paper).   Pruned to 2.4 m height at age 5.

Age (years)	6
Trees/ha	2634
Tree height (m)	
Basal area (m² /ha)	
Leaf area index	
Stem volume (m³ /ha)	

Dry biomass (t/ha)		
Stem wood	} 47.7 (or 42.4)[a]
Stem bark	
Branches	9.4 (or 7.9)[a]
Fruits etc.	0.0
Foliage	11.5 (or 9.4)[a]
Root estimate	17.1 (or 15.1)[a]

Net production (t/ha/yr)	
CAI (m³ /ha/yr)	
Stem wood	
Stem bark	
Branches	
Fruits etc.	
Foliage	2.0[b]
Root estimate	

Forty trees were sampled, including the roots, and stand biomass values were derived from regressions on D. Nutrient contents were determined.
a. Alternative values derived using regressions on D²H.
b. Foliage litterfall measured over 3 years.

Golley, F.B., McGinnis, J.T., Clements, R.G., Child, G.I. and Duever, M.J. (1975). "Mineral Cycling in a Tropical Moist Forest Ecosystem." University of Georgia Press, Athens, Georgia, U.S.A.

Golley, F.B., McGinnis, J.T., Clements, R.G. (1971). La biomasa y la estructura mineral de algunos bosques de Darién, Panama. *Turrialba* 21, 189-196.

8°35'N 70°00'W   0-2 m   Panama, Darién Province, near Santa Fe.

Mangrove
swamp   *Rhizophora brevistyla*
                        et al.

Age (years)
Trees/ha   712
Tree height (m)   41[a]
Basal area (m²/ha)   13.6
Leaf area index
Stem volume (m³/ha)

Dry biomass (t/ha):
  Stem wood
  Stem bark   } 159 + 116[b]
  Branches
  Fruits etc.
  Foliage   3.5
  Root estimate   189.8

CAI (m³/ha/yr)

Net production (t/ha/yr):
  Stem wood
  Stem bark
  Branches
  Fruits etc.
  Foliage
  Root estimate

Seven size classes of trees were sampled and stand biomass values for a 0.25 ha plot were derived from regressions on D. Nutrient contents were determined.
a. Top height.
b. Above-ground prop roots.

Ewel, J. (1971). Biomass changes in early tropical succession. *Turrialba* 21, 110-112.

Golley, F.B., McGinnis, J.T., Clements, R.G., Child, G.I. and Duever, M.J. (1975). "Mineral Cycling in a Tropical Moist Forest Ecosystem." University of Georgia Press, Athens, Georgia, U.S.A.

8°39'N 78°09'W   200 m   Panama, Darién province, near Santa Fe.

Cecropia spp., with *Ochroma* sp., *Trema* sp., *Spondias* sp., *Persea* sp. *et al.*
Tropical moist forest regenerating after clear-felling.

	Poorly drained alluvium		Imperfectly drained upland terrace	
Age (years)	2	2	4	6
Trees/ha				
Tree height (m)	10.1^a	7.2^a	12.3^a	13.6^a
Basal area (m²/ha)				
Leaf area index	6.9	7.5	11.6	16.5
Stem volume (m³/ha)				
Dry biomass (t/ha)				
Stem wood				
Stem bark	} $18.0 + 3.3^b$	} $4.3 + 5.1^b$	} $22.7 + 9.3^b$	} $27.1 + 8.8^b$
Branches				
Fruits etc.	$0.0 + 0.0^b$	$0.0 + 0.0^b$	$0.0 + 0.1^b$	$0.1 + 0.0^b$
Foliage	$2.2 + 0.8^b$	$1.2 + 2.4^b$	$3.3 + 2.6^b$	$4.8 + 1.7^b$
Root estimate	2.6	4.1	4.5	14.2
CAI (m³/ha/yr)				
Net production (t/ha/yr)				
Stem wood				
Stem bark				
Branches				
Fruits etc.				
Foliage				
Root estimate				

All vegetation was harvested in two 625 m² plots per stand in October (far left column) or July (the other columns), and roots were excavated in three 1 m² pits.
a. Heights of the tallest trees.
b. Understorey vegetation; LAI and root biomass values include the understorey.

Golley, F.B., McGinnis, J.T., Clements, R.G., Child, G.I. and Duever, M.J. (1975). "Mineral Cycling in a Tropical Moist Forest Ecosystem." University of Georgia Press, Athens, Georgia, U.S.A.

Golley, F.B., McGinnis, J.T., Clements, R.G. (1971). La biomasa y la estructura mineral de algunos bosques de Darién, Panama. *Turrialba* <u>21</u>, 189-196.

8°30-40'N 78°00'W (alt. given below) Panama, Darién Province, near Santa Fe.

	Tropical moist forest. (395 spp.) Grey-black soil. Below 250 m.		Riverine forest *Prioria copaifera et al.* (218 spp.)	Premontane wet forest (382 spp.) Red latosols.
	Lara[a]	Sabana[a]	ca.200 m	250-600 m
Age (years)				
Trees/ha	4348[b]	6048[b]	3792	6904
Tree height (m)	40[c]	40[c]	49[c]	31[c]
Basal area (m²/ha)	45.2	26.4	59.6	33.1
Leaf area index	10.6	22.4		
Stem volume (m³/ha)				

Dry biomass (t/ha)

	Lara	Sabana	ca.200 m	250-600 m
Stem wood } Stem bark } Branches }	$252.1 + 3.3^{d}$	$354.7 + 1.1^{d}$	$163.8 + 0.6^{d}$	$258.4 + 0.6^{d}$
Fruits etc.	0.0	0.1	0.0	0.0
Foliage	$7.3 + 0.7^{d}$	$11.4 + 0.6^{d}$	$11.4 + 0.7^{d}$	$10.6 + 0.3^{d}$
Root estimate	12.6	9.9	12.2	12.7

CAI (m³/ha/yr)

Net production (t/ha/yr)

	Lara	Sabana		
Stem wood } Stem bark } Branches } Fruits etc.	9.82^{e}	9.82^{e}		
Foliage	1.54^{e}	1.54^{e}		
Root estimate				

Seven size classes of trees were sampled and stand biomass values for 0.25 ha plots were derived from regressions on D. Nutrient contents were determined.

a. Lara was sampled in the dry season (February) and Sabana in the wet season (September).

b. There were 19132 and 19016 stems (as opposed to trees) per hectare at Lara and Sabana, respectively.

c. Top heights.

d. Understorey vegetation.

e. Litterfall only.

Edwards, P.J. and Grubb, P.J. (1977). Studies of mineral cycling in a montane rain-forest in New Guinea. I The distribution of organic matter in the vegetation and soil. *J. Ecol.* <u>65</u>, 943-969.

Edwards, P.J. (1977). Studies of mineral cycling in a montane rainforest in New Guinea. II The production and disappearance of litter. *J. Ecol.* <u>65</u>, 971-992.

6°00'S 145°11'E   2400-2500 m   Papua New Guinea, E of Mount Kerigomma.

Humic brown,
deeply weathered clay.   *Dacrycarpus cinctus, Podocarpus archboldii et al.*
Surface pH 5.5-6.3.

Montane rainforest (119 tree species).

Age (years)					All stands in mature or late-building phases.				
Trees/ha							1350^{a}	950^{a}	950^{a}
Tree height (m)	30	30	30	30	30	30	30	30	30
Basal area (m²/ha)	70	61	28	27	57	37	117	140	98
Leaf area index	$3.3+1.4^{b}$								

Stem volume (m³/ha)

Dry biomass (t/ha)

	Stem wood	⎫ 385 ⎫								
	Stem bark	⎬ +8.9^{b}								
	Branches	98 ⎭ 383	166	158	358	218	740	925	665
	Fruits etc.								
	Foliage	$6.9+1.5^{b}$								
	Root estimate	62.6								

CAI (m³/ha/yr)

Net production (t/ha/yr)

	Stem wood			
	Stem bark			
	Branches	$ca.1.9^{c}$	$ca.1.9^{c}$	$ca.1.9^{c}$
	Fruits etc.			
	Foliage	7.2^{d}	6.8^{d}	6.8^{d}
	Root estimate			

Many trees were sampled and stand biomass values for the above 200 m² plots were derived from regressions on D, except for the far left column which refers to a 200 m² plot that was completely harvested. Roots were excavated in two areas of 50 m². Nutrient contents were determined.
a. Trees greater than 20 cm D.
b. Shrubs, climbers and trees less than 30 cm D.
c. Woody litterfall (1.2 t/ha/yr) measured over one year, plus estimated pre-fall losses.
d. Leaf litterfall measured over one year, including estimated pre-fall losses of 10%.

Petrusewicz, K. (1967). "Secondary Productivity of Terrestrial Ecosystems (Principles and Methods)." Vol.1. Inst. Ecol., Polish Acad. Sci. (Panstowowe Wydawnictwo Naukowe), Warsaw and Krakow.

Medwecka-Kornas, A., Lomnicki, A. and Bandola-Ciolczyk, E. (1974). Energy flow in the oak-hornbeam forest. *Bull. Acad. pol. Sci. Cl. II Sér. biol.* 22, 563-567.

Medwecka-Kornas, A. and Bandola-Ciolczyk, E. (1981). In: "Dynamic Properties of Forest Ecosystems" (D.E. Reichle, ed.) p.609. Cambridge University Press.

	Ojców *Fagus sylvatica* Shallow rendzina soil on N-facing slope. (Petrusewicz *et al.* 1967)	Ispina *Quercus robur, Carpinus betulus, Tilia cordata* with understorey shrubs Leached brown loam, pH 5.2. (Medwecka-Kornas *et al.* 1974, 1981)
50°06'N 20°01'E 180 m Poland, near Krakow.		
Age (years)	70-80	100
Trees/ha		297
Tree height (m)	7-26	25
Basal area (m²/ha)		23.6
Leaf area index		4.7
Stem volume (m³/ha)		
Dry biomass (t/ha)		
Stem wood		$206.4 + 18.6^a$
Stem bark		
Branches	247	$17.5 + 15.1^a$
Fruits etc.		
Foliage		$2.0 + 0.9^a$
Root estimate	50	50.0^b
CAI (m³/ha/yr)		
Net production (t/ha/yr)		
Stem wood		$4.27^c + 0.37^{ac}$
Stem bark	7.10^c	
Branches		$0.30^c + 0.34^{ac}$
Fruits etc.	0.12^d	0.08^d
Foliage	2.91^d	$2.92^d + 0.91^e$
Root estimate		

In both studies stand biomass values for plots of over 0.5 ha were derived from regressions on D and H. The *F. sylvatica* stand was fully stocked, with 94% canopy cover.
a. Understorey shrubs and saplings.
b. Assumed to be 19% of the above-ground biomass value.
c. Excluding woody litterfall and mortality.
d. Litterfall, measured over 2 years at Ispina.
e. Estimated leaf consumption and other losses.

Zajaczkowski, J. and Lech, A. (1981). The effect of different initial growth space on above-ground biomass of Scots pine thicket. In: "Kyoto Biomass Studies", pp.163-171. School of Forestry and Natural Resources, University of Maine, Orono, U.S.A.

Traczk, T. (1981). In: "Dynamic Properties of Forest Ecosystems" (D.E. Reichle, ed.) p. 610. Cambridge University Press, Cambridge, London, New York, Melbourne.

Poland At Kampinos: podzol, pH 3.7-4.7	ca.52°N 20'E -- near Warsaw *Pinus sylvestris* (Zajaczkowski and Lech 1981)				52°20'N 20°50'E 105 m Kampinos National Park *Pinus sylvestris*, with *Quercus robur* and *Betula verrucosa* (Traczk 1981)
Age (years)	17	17	17	17	85
Trees/ha	a	a	a	a	1020
Tree height (m)	5.4	5.5	5.7	5.6	25
Basal area (m²/ha)					
Leaf area index					
Stem volume (m³/ha)					
Dry biomass (t/ha) Stem wood	}28.6	}30.4	}27.1	}33.2	}114.6
Stem bark					
Branches	14.2^b	14.2^b	17.8^b	13.1^b	17.3
Fruits etc.					
Foliage	7.3	5.7	7.7	7.8	2.3
Root estimate					173.6
CAI (m³/ha/yr)					
Net production (t/ha/yr) Stem wood					
Stem bark					
Branches					0.23^c
Fruits etc.					0.14^c
Foliage					$1.84^c + 0.02^d$
Root estimate					

Zajaczkowski and Lech (1981) sampled 24 trees and derived stand biomass values from regressions on D.
Traczk (1981) derived stand biomass values for a plot of over 0.5 ha from regressions on D and H.
a. Initial plant spacings were 0.8 × 0.8 m, 1.0 × 1.0 m, 1.2 × 1.2 m and 1.2 × 0.6 m in columns left to right.
b. Including dead branches; the biomass of living and dead branches excluding young shoots was 10.5, 11.2, 13.4 and 9.3 t/ha in columns left to right.
c. Litterfall only.
d. Frass litterfall.

Golley, F.B., Odum, H.T. and Wilson, R.F. (1962). The structure and metabolism of a Puerto Rican red mangrove forest in May. *Ecology* <u>43</u>, 9-19.

18°00'N　67°05'W　0-1 m　Puerto Rico, near Mayagüez.

Mangrove swamp
in 0.8-1.2 m　　　　　*Rhizophora mangle*
deep peat.

Age (years)		
Trees/ha	1100[a]	
Tree height (m)	7-8	
Basal area (m²/ha)		
Leaf area index	4.4	
Stem volume (m³/ha)		

Dry biomass (t/ha)

Stem wood	⎫ 28.0 + 14.4[b]
Stem bark	⎭
Branches	12.7
Fruits etc.	
Foliage	7.8
Root estimate	50.0[c]

CAI (m³/ha/yr)

Net production (t/ha/yr)

Stem wood	⎫
Stem bark	⎬ 3.07[d]
Branches	⎭
Fruits etc.	
Foliage	4.75[d]
Root estimate	

Ten trees were sampled, roots were sampled in soil cores, and stand biomass values for a 100 m² plot were derived from regressions on D.

a. Trees over 5 m tall; there were 12500 smaller trees per hectare.
b. Above-ground prop roots.
c. Comprised of 40 t/ha of roots 5-10 mm in diameter, and 10 t/ha of larger roots.
d. Litterfall only, expressed by the authors as 0.84 and 1.30 g/m²/day of wood and leaves, respectively.

Ovington, J.D. and Olson, J.S. (1970). Biomass and chemical content of El Verde lower montane forest plants. In: "A Tropical Rainforest" (H.T. Odum and R.F. Pigeon, eds) pp. H 53-61. US Atomic Energy Commission (Nat. Tech. Inf. Service, US Dept. Commerce, Springfield, Va., USA).

Odum, H.T. (1970). Summary: an emerging view of the ecological system at El Verde. *Ibid*. pp. I 191-289.

Odum, H.T., Abbott, W., Selander, R.K., Golley, F.B. and Wilson, R.F. (1970). Estimates of chlorophyll and biomass of the Tabonuco forest of Puerto Rico. *Ib*. p. I 3-19.

18°19'N 65°45'W   510 m   Puerto Rico, Luquillo National Forest.

Dacryodes excelsa (ca.35%)[a], *Sloanea berteriana,*
Manilkara bidentata et al.

Acid clays

Lower montane rainforest.

	Radiation center before irradiation	South control center	North cut center
Age (years)	Mature	Mature	Mature
Trees/ha	ca.800	ca.800	ca.800
Tree height (m)	18–30	18–30	18–30
Basal area (m²/ha)	37–41	36–37	ca.36
Leaf area index	4.5–6.4[b]	4.5–6.4[b]	4.5–6.4[b]
Stem volume (m³/ha)	240	145	202
Dry biomass (t/ha)			
Stem wood	} 191.6	} 111.6	139.5
Stem bark			13.8
Branches	45.1	27.4	36.8
Fruits etc.			
Foliage	9.2–10.2[b]	ca.7.5[b]	7.9–9.4[b]
Root estimate	78.1[c]	54.4[c]	72.3[c]
CAI (m³/ha/yr)			
Net production (t/ha/yr)			
Stem wood			
Stem bark			
Branches			
Fruits etc.			
Foliage	4.9[d]	4.8[d]	
Root estimate			

One hundred and nine trees were sampled, representing 28 species, and roots were excavated. Stand biomass values were derived from regressions on D. The left and centre columns refer to 2862 m² plots, the right column refers to a 1256 m² plot. The values given above exclude palms, which weighed 0.6 to 0.8 t/ha. Nutrient contents were determined.

a. Percentage of the total tree number.
b. Estimates of LAI and foliage biomass depended on the time of sampling and the method used; LAI values up to 12.6 were recorded on some ridges.
c. Roots over 5 mm diameter only.
d. Leaf litterfall, measured over 2 years.

Crow, T.R. (1980). A rainforest chronicle: a 30-year record of change in structure and composition at El Verde, Puerto Rico. *Biotropica* 12, 42-55.

18°19'N 65°45'W   500 m   Puerto Rico, Luquillo National Forest.

Deep acid clays on ridges, stoney clays in drainages.	*Dacryodes excelsa, Sloanea berteriana, Manilkara bidentata et al.*			
	$(36\%)^a$	$(35\%)^a$	$(34\%)^a$	$(41\%)^a$
	1943	1946	1951	1976
Age (years)				
Trees/ha	1326	1728	1717	1410
Tree height (m)				10.5
Basal area (m²/ha)	24.6	28.3	33.8	35.7
Leaf area index				
Stem volume (m³/ha)				
Dry biomass (t/ha)				
Stem wood				
Stem bark				
Branches	} 167	} 188	} 233	} 239
Fruits etc.				
Foliage				
Root estimate	69	84	111	141
CAI (m³/ha/yr)				
Net production (t/ha/yr)				
Stem wood				
Stem bark				
Branches				
Fruits etc.				
Foliage				
Root estimate				

Stand biomass values for a 0.72 ha plot were estimated for the years 1943, 1946, 1951 and 1976 (in columns left to right) from regressions on D taken from Ovington and Olson (1970) (see p.216). Only trees over 4.0 cm D were included. The stand was damaged by a severe hurricane in 1932.

a. Percentage of the total basal area accounted for by *D. excelsa*.

Lugo, A.E., Gonzalez-Liboy, J.A., Cintrón, B. and Dugger, K. (1978). Structure, productivity and transpiration of a subtropical dry forest in Puerto Rico. *Biotropica* 10, 278-291.

17°59'N 67°00'W 145 m Puerto Rico, Guánica Forest.

Sandy, clay soils, about 1 m deep.

Bucida buceras (20%)[a], *Gymnanthes lucida* (14%)[a], *Bursera simaruba* (13%)[a] et al.

Subtropical dry forest.

Age (years)	
Trees/ha	2160[b]
Tree height (m)	7.8
Basal area (m²/ha)	10.7[b]
Leaf area index	1.3-4.2[c]
Stem volume (m³/ha)	

Dry biomass (t/ha)

Stem wood	⎫
Stem bark	⎬ 36.9
Branches	⎭
Fruits etc.	
Foliage	2.3
Root estimate	

CAI (m³/ha/yr)

Net production (t/ha/yr)

Stem wood	⎫
Stem bark	⎬ 0.3[d]
Branches	⎭
Fruits etc.	⎫ 1.7[d]
Foliage	⎬
Root estimate	

Biomass values for a 250 m² plot were derived from estimates of wood volumes, wood densities and samples of the foliage.

a. Percentage of the total basal area; *B. buceras* syn. *Terminalia catappa*.
b. Trees over 5 cm D; there were 7880 trees/ha over 2.5 cm D with a basal area of 16.4 m²/ha.
c. LAI range during one year.
d. Litterfall only, averaged over the deciduous and scrub parts of the Guánica Forest.

Decei, I. (1981). Biomass of high productivity trees and young beech stands (*Fagus sylvatica* L.). In: "Kyoto Biomass Studies", pp.125-128. School of Forestry and Natural Resources, University of Maine, Orono, USA.

Donita, N., Bindiu, C. and Mocanu, V. (1981). In: "Dynamic Properties of Forest Ecosystems" (D.E. Reichle, ed.) pp.611-612. Cambridge University Press, Cambridge, London, New York, Melbourne.

Rumania	ca.45°N 25°E -- *Fagus sylvatica* (Decei 1981)	44°54'N 28°43'E *Quercus pubescens* syn. *lanuginosa* Rendzina pH 7.0	ca.175 m Babadag *Quercus pedunculiflora* Leached chernozem pH 6.4
		(Donita *et al.* 1981)	
Age (years)	21-40	37	37
Trees/ha		2730	1970
Tree height (m)		6.8	8.6
Basal area (m²/ha)		22.7	19.1
Leaf area index		2.5	3.4
Stem volume (m³/ha)			
Dry biomass (t/ha)			
Stem wood	106.0	34.4	26.0
Stem bark	6.6	15.2	19.5
Branches	14.9	7.7	3.3
Fruits etc.			
Foliage	3.0	2.4	3.3
Root estimate	27.0		
CAI (m³/ha/yr)			
Net production (t/ha/yr)			
Stem wood		0.90^a	1.01^a
Stem bark		0.40^a	0.75^a
Branches		0.21 } + 0.40^b	0.28 } + 0.96^b
Fruits etc.			
Foliage		2.38^b	3.26^b
Root estimate			

Decei sampled trees in 5 stands, including the roots, and derived stand values from regressions on D.
a. Excluding any mortality.
b. Litterfall.

Bindiu, C. (1973). Unpublished Doctoral Thesis, Academia de Stunte Agricole si Silvice, Bucharest, Rumania.

Bindiu, C., Popescu-Zeletin, I. and Mocanu, V. (1981). In: "Dynamic Properties of Forest Ecosystems" (D.E. Reichle, ed.) pp.613-614. Cambridge University Press, Cambridge, London, New York, Melbourne.

45°23'N 23°15'E   950-1010 m   Rumania, Sinaia.

Well-drained, brown forest soils. pH 6.1-6.8.	*Fagus sylvatica* (65%)[a] *Abies alba* (35%)[a]	*A. alba*

Age (years)		up to 450	110
Trees/ha		842	485
Tree height (m)		38.5-42.5	36.0
Basal area (m²/ha)		42.0	76.6
Leaf area index		9.7	12.0
Stem volume (m³/ha)			
Dry biomass (t/ha)	Stem wood	243.1	365.0
	Stem bark	23.2	40.6
	Branches	14.3	37.8
	Fruits etc.		
	Foliage	14.1	26.7
	Root estimate		

CAI (m³/ha/yr)			
Net production (t/ha/yr)	Stem wood	4.18[b]	5.46[b]
	Stem bark	0.37[b]	0.60[b]
	Branches	0.75 ⎫	1.90[c]
	Fruits etc.	⎬ + 0.20[d]	
	Foliage	3.75[d]	3.35[d]
	Root estimate		

---

a. Percentage of the total tree number.
b. Excluding any mortality.
c. Excluding woody litterfall.
d. Litterfall.

Alvera, B. (1973). Estudios en bosques de coniferas del Pirineo Central. Serie A: Pinar con acebo de San Juan de la Peña. I Produccion de hojarsca. *Pirineos* 109, 17-29.

Alvera, B. (1981). In: "Dynamic Properties of Forest Ecosystems" (D.E. Reichle, ed.) p.615. Cambridge University Press, Cambridge, London, New York, Melbourne.

42°30'N　　0°40'W　　1230 m　　Spain, San Juan de las Peña, near Jaca.

| Well-drained soil overlying calcareous conglomerate. | *Pinus sylvestris,* with an understorey of *Ilex aquifolium.* |

Age (years)		80	50-60[a]
Trees/ha		832 +	4532[a]
Tree height (m)		17-19	6-7[a]
Basal area (m²/ha)		47.9 +	3.6[a]
Leaf area index		11.9[b]	
Stem volume (m³/ha)			

Dry biomass (t/ha)	Stem wood	130.4 ⎫	
	Stem bark	15.7 ⎬ + 9.5[a]	
	Branches	35.0 ⎪	
	Fruits etc.	0.0 ⎭	
	Foliage	11.5 + 1.8[a]	
	Root estimate		

CAI (m³/ha/yr)		
Net production (t/ha/yr)	Stem wood ⎫	4.26 + 0.04[c]
	Stem bark ⎭	
	Branches	4.83 + 1.34[c]
	Fruits etc.	0.00 + 1.05[c]
	Foliage	2.14 + 2.45[c]
	Root estimate	

Stand biomass values for fifteen 25 m² plots were derived using regression methods. There was 19.9 t/ha of standing dead wood.
a. Values for *I. aquifolium*. Production values include *I. aquifolium*.
b. Including *I. aquifolium*; all-sided LAI was probably in the range 18-24.
c. Litterfall of *P. sylvestris* and *I. aquifolium*.

Hytteborn, H. (1975). Deciduous woodland at Andersby, eastern Sweden. Above-ground tree and shrub production. *Acta phytogeogr. suec.* <u>61</u>, 96 pp.

Persson, H. (1975). Deciduous woodland at Andersby, eastern Sweden: field-layer and below-ground production. *Acta phytogeogr. suec.* <u>62</u>, 71 pp.

60°09'N 17°49'E 25-35 m Sweden, Uppland Province, near Lake Dannemora.		
Betula spp. (88%)[a] *Quercus robur* (11%)[a] 20% ground cover only. Clayey till, partly flooded.	*Acer platanoides* (34%)[a] *Tilia cordata* (33%)[a] *Q. robur* (29%)[a] 56% ground cover. Sandy soil.	*Q. robur* (74%)[a] *Sorbus aucuparia* (10%)[a] *Betula* spp. and *T. cordata.* 48% ground cover. Stoney till.
Age (years)		
70-80	55-177	43-200
Trees/ha		
76	212	452
Tree height (m)		
19.5	6-24	5-23
Basal area (m²/ha)		
6.5	12.2	12.3
Leaf area index		
$1.4 + 0.2^b$	$3.0 + 0.5^b$	$3.1 + 1.6^b$

Stem volume (m³/ha)

Dry biomass (t/ha)

Stem wood	28.3 ⎫	38.9 ⎫	48.7 ⎫
Stem bark	4.7 ⎬ $+ 1.9^b$	10.4 ⎬ $+ 2.3^b$	6.8 ⎬ $+ 6.5^b$
Branches	16.2 ⎭	15.7 ⎭	20.3 ⎭
Fruits etc.	0.9	0.6	0.3
Foliage	$0.9 + 0.1^b$	$1.6 + 0.6^b$	$1.6 + 0.4^b$
Root estimate		ca.7.0	

CAI (m³/ha/yr)

Net production (t/ha/yr)

Stem wood	0.41^c ⎫	0.67^c ⎫	0.77^c ⎫
Stem bark	0.07^c ⎬ $+0.19^b$	0.16^c ⎬ $+0.94^b$	0.14^c ⎬ $+0.99^b$
Branches	$0.99 + 0.33^d$ ⎭	$1.14 + 0.75^d$ ⎭	$1.32 + 1.32^d$ ⎭
Fruits etc.	0.44^d	0.61^d	0.30^d
Foliage	$0.91^d + 0.05^b$	$1.66^d + 0.58^b$	$1.61^d + 0.43^b$
Root estimate			

Thirty-eight trees were sampled, roots were excavated, and stand biomass values for the above 0.25 ha plots were derived by proportional basal area allocation.

a. Percentage of the total basal area.

b. Understorey shrubs and coppice, comprised mainly of *Corylus avellana* in the columns on the left and right and suckers of *Populus tremula* in the centre column.

c. Excluding any mortality.

d. Overstorey litterfall, measured over 3 years.

Andersson, F. (1981). In: "Dynamic Properties of Forest Ecosystems" (D.E. Reichle, ed.) p.620. Cambridge University Press, Cambridge, London, New York, Melbourne.

Andersson, F. (1970). Ecological studies in a Scandinavian woodland and meadow area, southern Sweden. II Plant biomass, primary production and turnover of organic matter. *Bot. Notiser* 123, 8-51.

Andersson, F. (1971). Methods and preliminary results of estimation of biomass and primary production in a south Swedish mixed deciduous woodland. In: "Productivity of Forest Ecosystems" (P. Duvigneaud, ed.) pp.281-287. UNESCO, Paris.

55°44'N 13°18'E   60 m   Sweden, near Lund, Linnebjar Wood.

Brown
forest gley,
pH 4.4-5.0.

Quercus robur $(70\%)^a$, *Tilia cordata* $(14\%)^a$ with understorey
of *Sorbus aucuparia* and coppiced *Corylus avellana* $(14\%)^a$

Age (years)	125-190
Trees/ha	4725
Tree height (m)	23^b
Basal area (m²/ha)	31.4 (21.9 *Quercus*)
Leaf area index	5.4
Stem volume (m³/ha)	323 (260 *Quercus*)

Dry biomass (t/ha)

Stem wood	$100.0 + 26.0^c$
Stem bark	$11.0 + 5.0^c$
Branches	$41.0 + 12.0^c$
Fruits etc.	
Foliage	$3.0 + 2.1^c$
Root estimate	39.2 (21.9 *Quercus*)

CAI (m³/ha/yr)

Net production (t/ha/yr)

Stem wood	} 2.00	
Stem bark		} + 3.00^c
Branches	$-1.30 + 1.07^d$ }	} (or 6.1 + 5.3^d)^e
Fruits etc.	0.51^d	
Foliage	3.26^d	
Root estimate	1.9	(or 2.3)^e

Ten trees were sampled per tree species, 15 per shrub species, and 10 root systems were excavated including 3 *Quercus* and 2 *Tilia*. Stand biomass values for three 0.16 ha plots were derived from regressions on D²H. The data given are from Andersson (1981).

a. Percentage of the total basal area.
b. Stand height.
c. Understorey shrubs and coppice.
d. Litterfall, measured over 3 years.
e. Alternative values taken from Andersson (1970).

Nihlgard, B. (1972). Plant biomass, primary production and distribution of chemical elements in a beech and a planted spruce forest in south Sweden. *Oikos* 23, 69-81.

Nihlgard, B. and Lindgren, L. (1977). Plant biomass, primary production and bio-elements of three mature beech forests in south Sweden. *Oikos* 28, 95-104.

Nihlgard, B. and Lindgren, L. (1981). In: "Dynamic Properties of Forest Ecosystems" (D.E. Reichle, ed.) pp.617-621. Cambridge University Press, Cambridge, London.

55°42-59'N 13°10-55'E (alt. given below) Sweden.

	120 m Kongalund		60 m Oved	150 m Langarod
	Brown loams, pH 4.0-4.5.		*F. sylvatica*	*F. sylvatica*
	Picea abies Plantation.	*Fagus sylvatica*	Brown earth gley, pH 6.0-7.5	Podzol pH 4.0-4.5
Age (years)	55	90 (45-130)	80-100	80-120
Trees/ha	880	240	180	320
Tree height (m)	25	25	28	22
Basal area (m²/ha)	55.6	31.4	31.1	29.7
Leaf area index	11.5[a]	3.4[b]	4.3[b]	3.2[b]
Stem volume (m³/ha)	802	553	452	301
Dry biomass (t/ha) Stem wood	240	212.0	234	158
Stem bark	22	9.0	11	8
Branches	28	99.0	} 69	} 59
Fruits etc.	3.3	3.9		
Foliage	18.0	3.6		
Root estimate	58	50.0	41	34
CAI (m³/ha/yr)	23	8		
Net production (t/ha/yr) Stem wood	8.6	4.6	5.3	3.3
Stem bark	0.8	0.2	0.3	0.2
Branches	0.5 + 1.2[c]	5.3 + 1.1[c]	6.1 + 0.8[c]	4.4 + 1.1[c]
Fruits etc.	0.2[c]	1.0[c]	0.7[c]	0.4[c]
Foliage	3.3[c]	3.6[d]	4.1[d]	2.7[d]
Root estimate	2.6	2.4	2.3	1.7

Eight or ten trees were sampled per stand and several root systems were excavated. Stand biomass values for the above 0.10 ha plots were derived from regressions on D^2H. Bark and roots were assumed to grow at the same relative rates as other woody parts. Nutrient contents were determined.

a. All-sided LAI was about 26.
b. Mid-summer values.
c. Litterfall, measured over 3 years.
d. Leaf biomass plus early summer leaf litterfall; total leaf litterfall in the *F. sylvatica* stands was 2.4 to 2.8 t/ha/yr.

Albrektson, A. (1980a). "Biomass of Scots Pine (*Pinus sylvestris* L.). Amount, Development, Methods of Mensuration." The University of Agricultural Sciences, Dept. of Silviculture. Report No.2. Umea, Sweden.

Albrektson, A. (1980b). Relations between tree biomass fractions and conventional silvicultural measurements. In: "Structure and Function of Northern Coniferous Forests - an Ecosystem Study" (T. Persson, ed.) pp.315-327. Ecol. Bull. 32, Swedish Natural Science Research Council.

Sweden.	62°10'N 14°50'E 295 m Ytterhogdal	60°55'N 14°25'E 300 m Siljansfors
Sandy till iron podzols.	*Pinus sylvestris*	*Picea abies* (51%)[a] *Pinus sylvestris*
Age (years)	84	145
Trees/ha	825	1222 + 215
Tree height (m)	20.4[b]	24.8[b]
Basal area (m²/ha)	25.3	17.2 + 16.5
Leaf area index		
Stem volume (m³/ha)		

Dry biomass (t/ha)

Stem wood	101.6	59.1 + 70.1
Stem bark	7.0	7.5 + 4.6
Branches	10.7	10.7 + 6.4
Fruits etc.	0.4	0.0 + 0.4
Foliage	7.8	6.5 + 2.0
Root estimate	16.9	23.5 + 20.6

CAI (m³/ha/yr)

Net production (t/ha/yr)

Stem wood		
Stem bark		
Branches		
Fruits etc.		
Foliage	1.13[c]	0.65[c] + 0.47[c]
Root estimate		

Stand biomass values were derived by proportional basal area allocation based on samples of at least 6 trees per stand. Root biomass values include stumps, which weighed 6.9 t/ha in the pure *P. sylvestris* stand. Dead tree parts weighed 2.2 and 6.9 t/ha in the left and right columns, respectively.
The right column gives values separately for *Picea abies* plus *Pinus sylvestris*.
a. Percentage of the total basal area.
b. Dominant tree height.
c. Biomass of new foliage.

Continued from p.225.

60°20'N   17°13'E   10 m   Sweden, Lisselbo.

Plantation.   *Pinus sylvestris*
Gravelly, sandy,
iron podzol.

	Unfertilized plot	Fertilized plots			
Age (years)	13	13	13	13	13
Trees/ha	1131	1194	1244	1281	1156
Tree height (m)	5.5	5.9	5.4	5.0	5.0
Basal area (m²/ha)	7.7	12.4	12.1	10.5	10.8
Leaf area index					
Stem volume (m³/ha)					
Dry biomass (t/ha)					
Stem wood	8.0	14.2	13.1	10.2	11.1
Stem bark	1.6	2.4	2.3	2.1	2.1
Branches	4.9	10.1	8.5	7.7	8.4
Fruits etc.	0.0	0.0	0.0	0.0	0.0
Foliage	3.6	7.3	7.2	6.2	6.6
Root estimate	0.9[a]	2.0[a]	1.9[a]	1.5[a]	1.7[a]
CAI (m³/ha/yr)					
Net production (t/ha/yr)					
Stem wood					
Stem bark					
Branches					
Fruits etc.					
Foliage	1.17[b]	2.28[b]	2.47[b]	2.39[b]	2.51[b]
Root estimate					

Sixty-six trees were sampled in all and stand biomass values for the above five
plots were derived from regressions on D. There was 0.6, 1.1, 1.2, 1.3 and 0.9 t/ha
of dead tree parts in columns left to right.
a. Stumps only.
b. Biomass of new foliage.

Continued from p.226.

60°48-51'N   16°25-31'E   170-205 m   Sweden, Jädraås.

Sandy, iron podzols.   *Pinus sylvestris* with *Picea abies et al.*

	$99\%^a$	$100\%^a$	$100\%^a$	$99\%^a$	$97\%^a$	$99\%^a$
Age (years)	9	12	14	14	26	27
Trees/ha	1421	1801	2538	2527	3459	3164
Tree height (m)	5.7^b	6.6^b	7.7^b	7.2^b	13.7^b	13.3^b
Basal area (m²/ha)	8.8	6.4	15.1	12.1	29.5	27.4
Leaf area index						
Stem volume (m³/ha)						
Dry biomass (t/ha)						
Stem wood	7.8	5.9	18.8	12.6	76.5	66.9
Stem bark	1.8	1.4	3.0	2.9	7.8	7.2
Branches	8.9	3.6	6.0	6.4	10.7	9.1
Fruits etc.	0.0	0.0	0.0	0.0	0.0	0.0
Foliage	9.1	3.3	4.3	6.1	10.3	4.9
Root estimate	2.5	1.4	3.5	2.8	10.9	10.1
CAI (m³/ha/yr)						
Net production (t/ha/yr)						
Stem wood						
Stem bark						
Branches						
Fruits etc.						
Foliage	3.19^c	0.90^c	1.17^c	2.36^c	1.92^c	1.57^c
Root estimate						

Six trees were sampled from each stand, and stand values were derived by proportional basal area allocation, except for the 14-year-old stand with 15.1 m²/ha basal area where 10 trees were sampled and stand biomass values were derived from regressions on D. Root biomass values include stumps which weighed 1.2, 0.8, 1.3, 1.2, 4.2 and 3.9 t/ha in columns left to right. There was 0.0, 0.6, 2.4, 1.3, 4.5 and 5.2 t/ha of dead branches in columns left to right.
a. Percentage of the total basal area accounted for by *P. sylvestris*.
b. Dominant tree height.
c. Biomass of new foliage.

Continued from p.227.

60°50-52'N   16°25-35'E   185-205 m   Sweden, Jädraås.

Sandy, iron   *Pinus sylvestris* with *Picea abies et al.*
podzols.

	$99\%^a$	$99\%^a$	$97\%^a$	$100\%^a$	$97\%^a$	$100\%^a$
Age (years)	28	29	34	50	77	100
Trees/ha	3102	1337	1116	1775	876	453
Tree height (m)	13.4^b	14.8^b	15.1^b	14.8^b	17.1^b	19.2^b
Basal area (m²/ha)	29.8	22.4	21.7	22.9	21.7	19.7
Leaf area index						
Stem volume (m³/ha)						

Dry biomass (t/ha)		$99\%^a$	$99\%^a$	$97\%^a$	$100\%^a$	$97\%^a$	$100\%^a$
	Stem wood	68.7	57.5	61.4	58.8	61.2	76.4
	Stem bark	7.2	5.7	5.3	6.6	5.7	4.9
	Branches	10.6	12.4	8.9	6.1	6.9	9.3
	Fruits etc.	0.0	0.2	0.1	0.0	0.2	0.2
	Foliage	6.6	5.7	5.4	4.4	5.4	3.5
	Root estimate	12.2	9.1	13.7	12.1	11.4	19.0

CAI (m³/ha/yr)

Net production (t/ha/yr)							
	Stem wood						
	Stem bark						
	Branches						
	Fruits etc.						
	Foliage	1.84^c	1.46^c	1.36^c	0.91^c	1.01^c	0.81^c
	Root estimate						

Six trees were sampled from each of the stands aged 29, 34 and 77 and stand values
were derived by proportional basal area allocation. Ten trees were sampled from
each of the other stands, and stand values were derived from regressions on D.
Root biomass values include stumps which weighed 4.7, 3.5, 5.0, 4.5, 4.8 and 6.5
t/ha in columns left to right. There was 6.9, 2.8, 2.8, 3.1, 2.4 and 2.3 t/ha of
dead branches in columns left to right.
a. Percentage of the total basal area accounted for by *P. sylvestris*.
b. Dominant tree height.
c. Biomass of new foliage.

Burger, H. (1939). Holz, Blattmenge und Zuwachs. IV Ein 80-jähriger Buchenbestand. *Mitt. schweiz. Anst. forstl. VersWes.* <u>21</u>, 307-348.

Burger, H. (1947). Holz, Blattmenge und Zuwachs. VIII Die Eiche. *Mitt. schweiz. Anst. forstl. VersWes.* <u>25</u>, 211-279.

Burger, H. (1949). Holz, Blattmenge und Zuwachs. X Die Buche. *Mitt. schweiz. Anst. forstl. VersWes.* <u>26</u>, 419-468.

Switzerland	47°03'N 7°17'E 480 m Aarburg	ca.47°N 8°E 600 m Sihlwald	47°30'N 8°45'E 505 m Winterthur
Plantations.	*Fagus sylvatica*		*Quercus* spp. (70%)[a] and *Fagus sylvatica*
	Thinned (Burger 1939)	Thinned (Burger 1949)	Thinned favouring *Quercus*. Heavy fluvio-glacial loam. (Burger 1947)
Age (years)	80	98	65
Trees/ha	572	460	376 + 732
Tree height (m)			25.7 12.0
Basal area (m²/ha)			19.5 + 8.3
Leaf area index	7.9^b	6.2^b	3.9^b+ 2.7^b
Stem volume (m³/ha)			227 + 66
Dry biomass (t/ha) — Stem wood			} 113.5 + 33.0
Dry biomass (t/ha) — Stem bark			
Dry biomass (t/ha) — Branches			16.5 + 10.4
Dry biomass (t/ha) — Fruits etc.			
Dry biomass (t/ha) — Foliage	3.2	3.0	2.1 + 0.9
Dry biomass (t/ha) — Root estimate			
CAI (m³/ha/yr)	9.8	9.4	6.8 + 1.7
Net production (t/ha/yr) — Stem wood	} 5.59		
Net production (t/ha/yr) — Stem bark			
Net production (t/ha/yr) — Branches			
Net production (t/ha/yr) — Fruits etc.			
Net production (t/ha/yr) — Foliage	3.18	2.95	2.1 + 0.9
Net production (t/ha/yr) — Root estimate			

Three or five trees were sampled in 3 or 4 crown classes per site, and many others were sampled at other sites. Stand biomass values were obtained by multiplying mean tree values by the numbers of trees per hectare in each crown class. Dry biomass values were derived from the author's data on fresh weights, water contents of branches and leaves, stem volumes and wood specific gravities.

The right column gives values separately for *Quercus* spp. plus *F. sylvatica*.

a. Percentage of the total basal area.

b. Projected leaf areas; the author gave 2-sided leaf areas.

Burger, H. (1942). Holz, Blattmenge und Zuwachs. VI Ein Plenterwald mittlerer Standortsgüte. Der bernische Staatswald Toppwald in Emmental. *Mitt. schweiz. Anst. forstl. VersWes.* 22, 377-445.

Burger, H. (1947). Holz, Blattmenge und Zuwachs. VIII Die Eiche. *Mitt. schweiz. Anst. forstl. VersWes.* 25, 211-279.

Burger, H. (1951). Holz, Blattmenge und Zuwachs. XI Die Tanne. *Mitt. schweiz. Anst. forstl. VersWes.* 27, 247-287.

Switzerland	46°55'N 7°35'E 950-1000 m Emmental *Abies* spp. *Picea abies*, and *Fagus sylvatica* Stoney clays pH 4.6-4.9 (Burger 1942)	47°23'N 8°11'E 400 m Staufen nr Lenzburg *Abies alba* Gravelly soil (Burger 1951)	ca.47°N 650 m Adlisberg *Quercus robur* Plantation Fertile loam (Burger 1947)
Age (years)	Mixed	56	13
Trees/ha	516^a	910	10300
Tree height (m)	5-36		7.0
Basal area (m²/ha)	44		30.1
Leaf area index	9.0^b	7.4^b	
Stem volume (m³/ha)	530		124
Dry biomass (t/ha)			
Stem wood			} 71.8
Stem bark			
Branches		21.3	16.8
Fruits etc.			
Foliage	13.1	12.6	5.3
Root estimate			
CAI (m³/ha/yr)	10	19.4	17.5
Net production (t/ha/yr)			
Stem wood			} 10.1
Stem bark			
Branches			
Fruits etc.			
Foliage			5.3
Root estimate			

Three or four trees of *Abies*, *Picea*, *Fagus* and *Quercus* were sampled in three or four crown classes per site, and stand biomass values were obtained by multiplying mean tree values by the numbers of trees per hectare in each crown class. Basal areas were estimated from mean tree diameters and numbers of trees per hectare. Dry biomass values were derived from the author's data on fresh weights, water contents of branches and leaves, stem volumes and wood specific gravities.

a. Trees over 8 cm D.

b. The author estimated all-sided leaf areas to be 20.8 and 17.0 in columns left and right, respectively.

Burger, H. (1941). Holz, Blattmenge und Zuwachs. V Fichten und Föhren verschiedener Herkunft auf verschiedenen Kulturorten. *Mitt. schweiz. Anst. forstl. VersWes.* 22, 10-62.

Switzerland	47°13'N 7°32'E 470 m Solothurn		46°38'N 9°45'E 1600 m Bergün	
Plantations.	*Picea abies*			
	a	*b*	*a*	*b*
Age (years)	40	40	40	40
Trees/ha	1800	2350	4800	4800
Tree height (m)	19.7	17.2	9.2	9.1
Basal area (m²/ha)	58.8	62.5	40.8	40.8
Leaf area index	12.0[c]	10.7[c]	7.7[c]	7.9[c]
Stem volume (m³/ha)				
Dry biomass (t/ha)				
Stem wood				
Stem bark				
Branches	91.8	73.6	58.6	60.0
Fruits etc.				
Foliage	19.8	16.2	14.9	14.4
Root estimate				
CAI (m³/ha/yr)	26.8	19.0	13.4	13.0
Net production (t/ha/yr)				
Stem wood	} 8.6	} 6.3	} 4.3	} 4.3
Stem bark				
Branches				
Fruits etc.				
Foliage				
Root estimate				

Trees in four crown classes were sampled in each plot, and stand biomass values were obtained by multiplying mean tree values by the numbers of trees per hectare in each crown class. Basal areas were estimated from mean tree diameters and numbers of trees per hectare. Needles were assumed to contain 55% water (percentage of fresh weight).
a. Seed origin at 500 m altitude in Switzerland.
b. Seed origin at 1850 m altitude in Switzerland.
c. All-sided LAI values given by the author have been divided by 2.3.

Burger, H. (1941). Holz, Blattmenge und Zuwachs. V Fichten und Föhren verschiedener Herkunft auf verschiedenen Kulturorten. *Mitt. schweiz. Anst. forstl. VersWes.* <u>22</u>, 10-62.

Burger, H. (1948). Holz, Blattmenge und Zuwachs. IX Die Föhre. *Mitt. schweiz. Anst. forstl. VersWes.* <u>25</u>, 435-493.

Switzerland	47°36'N 8°32'E 410 m Eglisau			ca.47°N 605 m Gurmels	46°52'N 9°32'E 660 m Chur
Plantations.	*Pinus sylvestris*				*P. sylvestris* (85%)[a]
	Three provenances (Burger 1941)			Sown after clearfelling.	and *Larix decidua* syn. *europaea*
				Thinned	Thinned
Age (years)	32	32	32	70	88
Trees/ha	1550	1550	1550	600	498 + 106
Tree height (m)	16.1	16.6	16.0	25	
Basal area (m²/ha)	28.5	33.5	28.1	34.4	37.2 + 6.5
Leaf area index	2.3[b]	2.6[b]	2.3[b]	2.8[b]	2.1[b]+ 1.0
Stem volume (m³/ha)				400	642
Dry biomass (t/ha) Stem wood				} 173.2	} 267.1
Stem bark					
Branches	35.2	42.5	43.7	20.7	24.9 + 2.6
Fruits etc.					
Foliage	4.8	5.4	5.1	5.2	5.5 + 0.4
Root estimate					
CAI (m³/ha/yr)	13.2	15.0	13.6	13.2	8.0 + 1.4
Net production (t/ha/yr) Stem wood	} 5.2	} 6.1	} 5.5	} 5.7	} 3.3 + 0.6
Stem bark					
Branches					
Fruits etc.					
Foliage					
Root estimate					

Three or four trees were sampled per stand, and over 200 trees were sampled in other stands. Stand biomass values were obtained by multiplying mean tree values by the numbers of trees per hectare in different crown classes. Dry weights were derived from the author's data on fresh weights, water contents of branches and needles, stem volumes and wood specific gravities. Basal areas were estimated from mean tree diameters and numbers of trees per hectare.
The right column gives values separately for *P. sylvestris* plus *Larix decidua*.
a. Percentage of the total basal area.
b. All-sided leaf areas can be obtained by multiplying by 2.8.

Christensen, B. (1978). Biomass and primary production of *Rhizophora apiculata* Bl. in a mangrove in southern Thailand. *Aquat. Bot.* 4, 43-52.

7°50'N 98°20'E   sea level   Thailand, Phuket Island, Ao Nam Bor.

Mangrove swamp.	*Rhizophora apiculata*, with a few *Ceriops tagal*, *Sonneratia alba*, *Rhizophora mucronata et al.*
	Regenerated after clearfelling.

Age (years)	14-15
Trees/ha	
Tree height (m)	11
Basal area (m²/ha)	
Leaf area index	3.7-4.2
Stem volume (m³/ha)	

Dry biomass (t/ha)

Stem wood	} 74.4 + 61.2^a
Stem bark	
Branches	15.8
Fruits etc.	0.3
Foliage	7.4
Root estimate	

CAI (m³/ha/yr)

Net production (t/ha/yr)

Stem wood	
Stem bark	} 20.0
Branches	
Fruits etc.	0.3
Foliage	6.7^b
Root estimate	

All trees were harvested in one 25 m² plot in February. Wood production was estimated from relationships between biomass and production for mangroves taken from an unpublished Ph.D. thesis (Aksornkoae 1975, Michigan State University).
a. Above-ground prop roots.
b. Estimated by observing leaf longevity on tagged branches for 16 months.

Ogawa, H., Yoda, K., Ogino, K. and Kira, T. (1965). Comparative ecological studies on three main types of forest vegetation in Thailand. II Plant biomass. *Nature and Life in S.E. Asia* 4, 49-80.

Ogawa, H., Yoda, K. and Kira, T. (1961). A preliminary survey on the vegetation of Thailand. *Nature and Life in S.E. Asia* 1, 21-157.

Thailand	18°30'N 98°40'E 300 m		17°00'N 99°30'E 300 m
	Mt Inthanon	Nr Tak	*Dipterocarpus alatus,*
Deep sandy soils.			*Mangifera caloneura,*
			Gymnosporia spp.
	Dipterocarp savanna forest[a]	Mixed savanna forest[b]	Evergreen gallery forest
Age (years)			
Trees/ha	1576	1340	16200
Tree height (m)	20–25[c]	18–25[c]	29[c]
Basal area (m²/ha)	15.3	19.1	82.2
Leaf area index	1.6	1.8	12.1
Stem volume (m³/ha)			
Dry biomass (t/ha) — Stem wood	} 58	} 74	} 467
Dry biomass (t/ha) — Stem bark			
Dry biomass (t/ha) — Branches	12	16	214
Dry biomass (t/ha) — Fruits etc.			
Dry biomass (t/ha) — Foliage	1.9	2.2	14.5
Dry biomass (t/ha) — Root estimate	10	11	54
CAI (m³/ha/yr)			
Net production (t/ha/yr) — Stem wood			
Net production (t/ha/yr) — Stem bark			
Net production (t/ha/yr) — Branches			
Net production (t/ha/yr) — Fruits etc.			
Net production (t/ha/yr) — Foliage			
Net production (t/ha/yr) — Root estimate			

Many trees were sampled, and biomass values for the above plots of 2500, 1000 and 100 m² (in columns left to right) were derived from regressions on D^2H. Values given above are from Ogawa *et al.* (1965) which updated those published earlier.
a. *Pentacme siamensis, Terminalia tomentosa, Terminalia mucronata et al.*
b. *Shorea robusta, Dipterocarpus tuberculatus, Parinarium anamense et al.*
c. Height of the top canopy.

Ogawa, H., Yoda, K., Ogino, K. and Kira, T. (1965). Comparative ecological studies on three main types of forest vegetation in Thailand. II Plant biomass. *Nature and Life in S.E. Asia* 4, 49-80.

Ogawa, H., Yoda, K., Kira, T., Ogino, K., Shidei, T., Ratanowongse, D. and Apasutaya, C. (1965). Comparative ecological study on three main types of forest vegetation in Thailand. I Structure and floristic composition. *Nature and Life in S.E. Asia* 4, 13-48.

19°30'N 99°00'E   ca.500 m   Thailand, Chiang Mai Province, Ping Kong.

Sandy soils.	*Dipterocarpus obtusifolius, Shorea obtusa, et al.*		*D. obtusifolius, Lagerstroemia calyculata et al.*
	Open savanna forest 58%[a]	Open savanna – monsoon forest 70%[a]	Dry monsoon forest 79%[a]
Age (years)			
Trees/ha	1488	906	713
Tree height (m)	10-19[b]	18-29[b]	17-26[b]
Basal area (m²/ha)	17.4	23.9	35.4
Leaf area index	1.8 + 0.9[c]	2.3 + 2.6[c]	3.9 + 1.8[c]
Stem volume (m³/ha)			

Dry biomass (t/ha)

Stem wood	} 55 } + 0.1[c]	} 112 } + 1.0[c]	} 209 } + 0.7[c]
Stem bark			
Branches	11 }	26 }	53 }
Fruits etc.			
Foliage	2.1 + 0.5[c]	3.0 + 1.4[c]	3.8 + 0.5[c]
Root estimate	10	16	25

CAI (m³/ha/yr)

Net production (t/ha/yr)

Stem wood			
Stem bark			
Branches			
Fruits etc.			
Foliage			
Root estimate			

Eight to sixteen trees were sampled per stand, representing 6 to 21 species, and roots were excavated. Stand values for the above 0.16 ha plots were derived from regressions on D²H. Only trees at least 4.5 cm D were included.
a. Percentage light interception.
b. Height of the top canopy.
c. Understorey shrubs.

Kira, T., Ogawa, H., Yoda, K. and Ogino, K. (1967). Comparative ecological studies on three main types of forest vegetation in Thailand. IV Dry matter production, with special reference to the Khao Chong rainforest. *Nature and Life in S.E. Asia* 5, 149-174.

Kira, T., Ogawa, H., Yoda, K. and Ogino, K. (1964). Primary production of a tropical rainforest of southern Thailand. *Bot. Mag., Tokyo* 77, 428-429.

Ogawa, H., Yoda, K., Ogino, K. and Kira, T. (1965). (See 1967 reference above) II Plant biomass. *Nature and Life in S.E. Asia* 4, 49-80.

7°35'N 99°45'E   ca.100 m   Thailand, Trang Province, Khao Chong Reserve.

Deep sandy soil.

Padbruggea pubescens, Eugenia clarkeana
and 83 other species.

Evergreen tropical rainforest.

	Age (years)	Mature	Mature
	Trees/ha	1175	1338
	Tree height (m)	26-36^a	26-36^a
	Basal area (m²/ha)	39.9	34.5
	Leaf area index	10.7 (or 10.7)^b	11.4
	Stem volume (m³/ha)		
Dry biomass (t/ha)	Stem wood	} 254 (or 292)^b	} 206
	Stem bark		
	Branches	106^c (or 104)^b	80^c
	Fruits etc.		
	Foliage	7.7^c (or 7.8)^b	8.2^c
	Root estimate	32	30

	CAI (m³/ha/yr)		
Net production (t/ha/yr)	Stem wood		} 4.18 + 0.78^d
	Stem bark		
	Branches		} 1.69 + 0.27^d + 11.39^e
	Fruits etc.		
	Foliage		12.0^e
	Root estimate		0.7

One hundred and nineteen trees representing 80 species were sampled, and roots of three trees were excavated. Stand values for the above two 0.16 ha plots were derived from regressions on D²H. Only trees at least 4.5 cm D were included. Production values given above are from Kira *et al.* (1967) which updated those given by Kira *et al.* (1964).

a. Height of the top canopy.

b. Alternative values obtained by completely clearfelling a 400 m² plot.

c. Including climbers, which consisted of about 18 t/ha wood and 2.4 t/ha leaves.

d. Mortality.

e. Litterfall, measured over 42 days and expressed as annual values.

Hughes, M.K. (1971). Tree biocontent, net production and litterfall in a deciduous woodland. *Oikos* 22, 62-73.

Hughes, M.K. (1969). "Investigations of the Ecosystem Energetics of an English Woodland." Unpublished Ph.D. thesis, University of Durham, England.

54°40'N 1°20'E   67 m   U.K., England, Durham.

Alnus glutinosa (55%)[a], *Betula pendula* (44%)[a],

and *Acer pseudoplatanus*.

Variously coppiced and logged in last 100 years, but untouched for previous 15 years.

Age (years)		5 to 80
Trees/ha		1600
Tree height (m)		12
Basal area (m²/ha)		ca.21
Leaf area index		3.6
Stem volume (m³/ha)		

Dry biomass (t/ha)

Stem wood	
Stem bark	106.9
Branches	
Fruits etc.	0.4
Foliage	1.7
Root estimate	

CAI (m³/ha/yr)

Net production (t/ha/yr)

Stem wood	
Stem bark	$4.26 + 0.34^b$
Branches	
Fruits etc.	0.34^b
Foliage	1.78^b
Root estimate	

Twelve *A. glutinosa* and 13 *B. pendula* trees were sampled, and stand biomass values for 8 plots totalling 0.32 ha were derived from regressions on D.
a. Percentage of the total stem number.
b. Litterfall, measured over 2 years.

Ovington, J.D. and Madgwick, H.A.I. (1959a). The growth and composition of natural stands of birch. I Dry matter production. *Pl. Soil* <u>10</u>, 271-283.

Ovington, J.D. and Madgwick, H.A.I. (1959b). The growth and composition of natural stands of birch. II The uptake of nutrients. *Pl. Soil* 10, 389-400.

52°29'N 0°15'W   ca.50 m   U.K., England, near Peterborough, Holme Fen Nature
                              Reserve.

Deep peat.

Betula verrucosa syn. *pendula* (77-96%)[a]

and *Betula pubescens*.

Age (years)	6	24	27	32	38	42	46	53	55
Trees/ha	10450	4990	2480	4210	1500	1340	740	1020	880
Tree height (m)	2.1	9.1	12.8	9.5	12.3	13.0	18.8	18.5	17.6
Basal area (m²/ha)		14.1	19.5	16.2	17.0	20.6	25.6	26.0	25.0
Leaf area index	0.4	6.2	3.5	4.1	1.7	2.8	4.1	5.9	6.5

Stem volume (m³/ha)

Dry biomass (t/ha)

	6	24	27	32	38	42	46	53	55
Stem wood	} 0.9	} 48.0	} 68.2	} 52.8	} 58.9	} 58.5	} 102.3	} 139.8	} 134.5
Stem bark									
Branches	0.4	11.8	8.8	9.4	10.9	9.1	16.9	30.6	27.0
Fruits etc.									
Foliage	0.1	2.4	1.3	1.6	0.7	1.1	1.6	2.3	2.5
Root estimate	0.3	16.9				25.8			49.8

CAI (m³/ha/yr)

Net production (t/ha/yr)

	6	24	27	32	38	42	46	53	55
Stem wood	} 2.6	} 3.2	} 4.0	} 4.0	} 4.2	} 4.4	} 4.6	} 4.8	} 5.4
Stem bark									
Branches	1.6	1.9	2.8	2.8	2.2	2.0	1.7	1.5	1.3
Fruits etc.									
Foliage	1.2	1.2	1.5	1.4	1.4	1.3	1.4	1.5	1.6
Root estimate									

Two trees of average size were sampled from each stand in August, and two root systems were excavated. Stand biomass values for the above 0.1 ha plots were obtained by multiplying mean tree values by the numbers of trees per hectare. There was 0.1 to 4.5 t/ha of dead branches in each stand. Nutrient contents were determined.
The production values refer to ages 10-15, 15-20, 20-25, 25-30, 30-35, 35-40, 40-45, 45-50 and 50-55 in columns left to right, derived from smoothed curves of estimated cumulative dry biomass with age, including estimated mortality and litterfall.
a. Percentage of the total tree number.

Ford, E.D. and Newbould, P.J. (1970). Stand structure and dry weight production through the sweet chestnut (*Castanea sativa* Mill.) coppice cycle. *J. Ecol.* 58, 275-296.

Ford, E.D. and Newbould, P.J. (1971). The leaf canopy of a coppiced deciduous woodland. *J. Ecol.* 59, 843-862.

51°05'N 0°55'E 10 m U.K., England, near Ashford, Ham Street Woods.

Fertile
loam.

Castanea sativa

Managed coppice.

Age (years)	1	2	5	9
Trees/ha	76190[a]	40000[a]	18000[a]	14000[a]
Tree height (m)	1.0	2.5	5.0	7.0
Basal area (m²/ha)		2.1	16.0	36.8
Leaf area index	1.1	2.1	3.6	5.6
Stem volume (m³/ha)				

Dry biomass (t/ha)		1	2	5	9
	Stem wood				
	Stem bark	0.8	2.5	10.5	28.5
	Branches				
	Fruits etc.	0.0	0.0	0.0	0.0
	Foliage	0.4[b]	0.7[b]	1.9[b]	3.6[b]
	Root estimate				

CAI (m³/ha/yr)

Net production (t/ha/yr)		1	2	5	9
	Stem wood				
	Stem bark	0.6[c]	1.8[c]	6.5[c]	7.3[c]
	Branches				
	Fruits etc.	0.0	0.0	0.0	0.0
	Foliage	0.4[b]	0.7[b]	1.9[b]	3.6[b]
	Root estimate				

Ten to fifty stems were sampled per age, and stand biomass values for the above 45 m² plots were derived from regressions on H and D². Wood increments were estimated by repeated sampling during one year.
a. Number of stems per hectare.
b. Leaf biomass at the end of the growing season.
c. Including estimated shoot mortality.

Satchell, J.E. (1981). "The Ecology of Meathop Wood." Institute of Terrestrial Ecology, Merlewood Research Station, Grange-over-Sands, England. Unpublished manuscript.

Sykes, J.M. and Bunce, R.G.H. (1970). Fluctuations in litterfall in a mixed decidu-ous woodland over a three-year period 1966-1968. *Oikos* 21, 326-329.

Satchell, J.E. (1971). Feasibility study of an energy budget for Meathop Wood. In: "Productivity of Forest Ecosystems" (P. Duvigneaud, ed.) pp.619-630. UNESCO, Paris.

54°12'N 2°54'W   45 m   U.K., England, Lake District, near Grange-over-Sands, Meathop.

Brown earth with humus, pH 4.1-7.3

Quercus robur (43%)[a], *Fraxinus excelsior* (29%)[a], *Betula* spp. (13%)[a], *Acer pseudoplatanus* (3%)[a], with understorey of *Corylus avellana* (12%)[a]

	1962	1967	1972
Age (years)	to 70-81	to 70-81	to 70-81
Trees/ha		519[b]	537[b]
Tree height (m)			14.6
Basal area (m²/ha)		$18.2 + 4.8^c$	$20.7 + 4.7^c$
Leaf area index		$4.5 + 0.8^c$	$4.5 + 0.8^c$
Stem volume (m³/ha)			

Dry biomass (t/ha)

	1962	1967	1972	
Stem wood			65.7	
Stem bark	$79.6 + 8.2^c$	$93.3 + 14.4^c$	10.2	$+ 15.1^c$
Branches			33.6	
Fruits etc.				
Foliage		$3.7 + 0.4^c$	$3.7 + 0.4^c$	
Root estimate			74.4	

Net production / CAI (m³/ha/yr)

	1962	1967	1972	
Stem wood			1.87	
Stem bark		$2.73^d + 1.23^{cd}$	0.38	$+ 0.15^c$
Branches			0.95	$+ 1.60^e$
Fruits etc.			0.60^e	
Foliage		$3.7 + 0.4^c$	$3.7 + 0.4^c$	
Root estimate				

Seventy-two trees and 30 *C. avellana* were sampled in winter, and roots were exca-vated. Stand biomass values for a 1.0 ha plot were derived from regressions on stem girths. Increments for 1962-67 were derived by core sampling, and for 1967-72 by remeasurement. In 1972 there was 5.1 t/ha of overstorey dead wood, and 2.4 t/ha of understorey dead wood. Leaf biomass values are for June. These data update those given by Satchell (1971).
a. Percentage of the total biomass.
b. There were 752 to 759 stems per hectare.
c. Understorey shrubs.
d. Excluding woody litterfall.
e. Litterfall, measured over 4 to 5 years; mean leaf litterfall was 3.2 t/ha/yr.

Peterken, G.F. and Newbould, P.J. (1966). Dry matter production by *Ilex aquifolium* L. in the New Forest. *J. Ecol.* <u>54</u>, 143-150.

50°50'N 1°35'W   ca.100 m   U.K., England, New Forest.

Well-drained podzols over gravel subsoils.	*Ilex aquifolium* Beneath an overstorey of broadleaved trees.			Unshaded	
Age (years)	80	94	92	100	82
Trees/ha	3100	5510	5700	3130	5400
Tree height (m)	9	9	5	8	8
Basal area (m²/ha)					
Leaf area index	2.4^a	2.1^a	1.4	5.6	5.6
Stem volume (m³/ha)					

Dry biomass (t/ha)

Stem wood	} 39.6	} 52.6	} 25.2	} 85.3	} 152.3
Stem bark					
Branches	15.7	15.4	10.0	36.2	37.6
Fruits etc.					
Foliage	4.1	3.7	2.5	9.3	18.0
Root estimate	22.6				

CAI (m³/ha/yr)

Net production (t/ha/yr)

Stem wood					
Stem bark					
Branches	} 3.6^b	} 3.8^b	} 2.2^b	} 9.7^b	} 15.4^b
Fruits etc.					
Foliage					
Root estimate					

One tree in each of three size-classes was sampled, and one root system was excavated. Stand biomass values for the above 100-170 m² plots were obtained by multiplying mean tree values by the numbers of trees per hectare in each size class.
a. The overstorey canopies had summer LAI values of about 2.8.
b. Total production, including woody litterfall (assumed to be equal to the number of new branch scars multiplied by mean branch weight) and new foliage production (derived from the average foliage longevity).

242   UNITED KINGDOM   *Picea*

Ford, D.E. (1981). A high rate of dry matter production in an early pole stage plantation of Sitka spruce and its relation to canopy structure and development. *Forestry* (in press).

Deans, J.D. (1979). Fluctuations of the soil environment and fine root growth in a young Sitka spruce plantation. *Pl. Soil* 52, 195-208.

Deans, J.D. (1981). Dynamics of coarse root production in a young plantation of *Picea sitchensis*. *Forestry* 54, 139-155.

55°19'N 3°33'W   355 m   U.K., Scotland, near Moffat, Greskine Forest.

Plantation.   *Picea sitchensis*
Peaty gley
soil.

Fertilizers were applied at planting.

Age (years)	17
Trees/ha	3817
Tree height (m)	8
Basal area (m²/ha)	26.6
Leaf area index	9.1[a]
Stem volume (m³/ha)	ca.110

Dry biomass (t/ha):

Stem wood	} 56.3
Stem bark	
Branches	25.0
Fruits etc.	0.0
Foliage	26.6
Root estimate	$(4.9 + 20.1)^b$

CAI (m³/ha/yr)   ca.20

Net production (t/ha/yr):

Stem wood	} 16.43^c
Stem bark	
Branches	4.30^c
Fruits etc.	
Foliage	6.01^d
Root estimate	$(5.28 + 3.15)^b$

Seven trees were sampled at age 16, and 9 trees were sampled at age 18. Seven root systems were excavated, and fine roots were extensively sampled by soil coring. Stand biomass values for a 990 m² plot were derived from regressions on basal area per tree.
a. Projected LAI was 10.6 at age 16, and 7.5 at age 18.
b. Fine plus thick roots.
c. Woody litterfall was negligible, and there was no tree mortality.
d. New foliage biomass.

Wright, T.W. and Will, G.M. (1958). The nutrient content of Scots and Corsican pines growing on sand dunes. *Forestry* <u>31</u>, 13-25.

57°39'N 3°40'W 10 m U.K., Scotland, Morayshire, Culbin Forest.

Plantations.
Poor aolian
sand dunes.

	Pinus nigra var. *calabrica* syn. var. *maritima*			*Pinus sylvestris*		
Age (years)	18	28	48	18	28	64
Trees/ha	5189	3608	1112	5189	2125	815
Tree height (m)	4.6	8.5	13.6	5.5	11.9	16.5
Basal area (m²/ha)						
Leaf area index						
Stem volume (m³/ha)	25	65	147	50	138	157
Dry biomass (t/ha) Stem wood	11.3	39.8	74.0	28.4	67.0	83.2
Stem bark	4.9	12.9	21.3	7.3	8.4	14.2
Branches	6.4	10.9	11.2	13.1	14.0	16.7
Fruits etc.						
Foliage	3.1	4.8	5.6	6.2	4.7	4.7
Root estimate						
CAI (m³/ha/yr)						
Net production (t/ha/yr) Stem wood						
Stem bark						
Branches						
Fruits etc.						
Foliage						
Root estimate						

Three trees of average size were sampled per stand at various times of year. Stand biomass values for plots of 400 or 800 m² were obtained by multiplying mean tree values by the numbers of trees per hectare. Nutrient contents were determined.

Ovington, J.D. (1957). Dry matter production of *Pinus sylvestris* L. *Ann. Bot.* 21, 287-316.

Ovington, J.D. (1959). Mineral content of plantations of *Pinus sylvestris* L. *Ann. Bot.* 23, 75-88.

Ovington, J.D. (1961). Some aspects of energy flow in plantations of *Pinus sylvestris* L. *Ann. Bot.* 25, 12-20.

52°30'N 1°53'E   50 m   U.K., England, Thetford Chase.

Sandy soils.   *Pinus sylvestris*

	Plan-tation	Naturally regenerated		Plantations					
Age (years)	7	11	14	17	20	23	31	35	55
Trees/ha	4810	4230	5190	5640	5400	3640	2370	1890	760
Tree height (m)	1.4	2.9	3.6	4.9	5.8	8.2	12.6	14.2	16.0
Basal area (m²/ha)	0.1	4.1	7.2	14.4	19.6	25.3	36.0	32.5	30.8
Leaf area index	0.8^a	2.1^a	2.5^a	3.3^a	3.9^a	1.9^a	3.1^a	3.6^a	2.7^a

Stem volume (m³/ha)

Dry biomass (t/ha)

Stem wood	}1.0	}5.2	}8.4	}16.2	}27.1	}44.3	}81.7	}98.8	}96.7
Stem bark									
Branches	1.0	4.3	7.8	9.4	12.0	13.8	9.4	9.6	12.3
Fruits etc.	0.0	0.0	0.0	0.1	0.5	0.5	0.6	0.6	1.0
Foliage	2.1	5.8	6.7	9.0	10.5	5.1	8.3	9.8	7.2
Root estimate	3.4	10.6	10.4	12.8	14.0	28.1	27.7	44.0	34.1

CAI (m³/ha/yr)

Net production (t/ha/yr)

Stem wood	}1.1	}1.1	}2.6	}3.6	}11.0	}8.5	}9.3	}5.0	
Stem bark									
Branches	1.1	1.4	3.8	2.3	3.0	3.0	3.8	0.8	
Fruits etc.	0.0	0.0	0.7	1.3	1.0	0.4	0.8	0.8	
Foliage	1.7	3.9	4.3	3.6	3.7	3.6	3.5	3.3	
Root estimate									

One to three trees of average size were sampled per stand in August, and their roots were excavated. Stand biomass values were obtained by multiplying mean tree values by the numbers of trees per hectare. The production values refer to ages 7-11, 11-14, 14-17, 17-20, 20-23, 23-31, 31-35 and 35-55 in columns left to right, derived from smoothed curves of estimated cumulative dry biomass with age, including estimated mortality, thinnings and litterfall. Nutrient contents were determined.
a. All-sided LAI values can be obtained by multiplying by 2.8.

Ovington, J.D. and Madgwick, H.A.I. (1959). Distribution of organic matter and plant nutrients in a plantation of Scots pine. *Forest Sci.* 5, 344-355.

Attiwell, P.M. and Ovington, J.D. (1968). Determination of forest biomass. *Forest Sci.* 14, 13-15.

55°30'N 2°40'W   168 m   U.K., Scotland, near Jedburgh.

Plantation.
Moderately drained   *Pinus sylvestris*
heavy clay.

Age (years)	33
Trees/ha	4260
Tree height (m)	15
Basal area (m²/ha)	ca.43
Leaf area index	
Stem volume (m³/ha)	

Dry biomass (t/ha)	Stem wood	} 118.8 (or 120.6)^{a}
	Stem bark	
	Branches	14.0 (or 14.4)^{a}
	Fruits etc.	
	Foliage	7.3 (or 7.6)^{a}
	Root estimate	36.1

	CAI (m³/ha/yr)	
Net production (t/ha/yr)	Stem wood	
	Stem bark	
	Branches	
	Fruits etc.	
	Foliage	
	Root estimate	

Four trees were sampled within each of five diameter classes, and the roots of 17 trees were excavated. Stand biomass values were obtained by multiplying mean tree values by the numbers of trees per hectare in each diameter class. There was 9.7 t/ha of dead branches and 9.6 t/ha of standing dead trees. Nutrient contents were determined.
a. Alternative values derived from regressions on tree girth.

Miller, H.G., Cooper, J.M. and Miller, J.D. (1976). Effect of nitrogen supply on nutrients in litterfall and crown leaching in a stand of Corsican pine. *J. appl. Ecol.* <u>13</u>, 233-248.

Miller, H.G. and Miller, J.D. (1976). Effect of nitrogen supply on net primary production in Corsican pine. *J. appl. Ecol.* <u>13</u>, 249-256.

57°39'N   3°40'W   ca.50 m   U.K., Scotland, Morayshire, Culbin Forest.

Plantations.
Blown sand.

Pinus nigra var. *maritima*

All plots received 21 kg/ha P and 120 kg/ha K at age 36, plus the following amounts of N at ages 36, 37 and 38.

		Nil	84 kg/ha	168 kg/ha	336 kg/ha	504 kg/ha
Age (years)		39	39	39	39	39
Trees/ha		2110	2110	2110	2110	2110
Tree height (m)		12.9	13.3	13.3	13.2	13.1
Basal area (m²/ha)		39.4	41.1	41.7	42.0	42.4
Leaf area index		2.8^a	3.8^a	4.1^a	4.8^a	5.0^a
Stem volume (m³/ha)						
Dry biomass (t/ha)	Stem wood	78.4	80.4	86.2	86.8	92.5
	Stem bark	13.3	14.8	15.2	15.0	16.7
	Branches	14.8	18.0	16.9	16.7	16.7
	Fruits etc.					
	Foliage	7.4	9.8	11.8	13.4	14.8
	Root estimate	28.1	36.1	33.8	31.7	37.0
CAI (m³/ha/yr)						
Net production (t/ha/yr)	Stem wood	3.58	4.90	6.88	7.48	5.70
	Stem bark	0.78	1.01	1.12	0.94	1.14
	Branches	$1.79 + 0.33^b$	$2.20 + 0.28^b$	$2.05 + 0.24^b$	$1.87 + 0.20^b$	$1.90 + 0.36^b$
	Fruits etc.	0.16^b	0.13^b	0.13^b	0.16^b	0.17^b
	Foliage	3.18^c	4.01^c	4.97^c	5.37^c	5.46^c
	Root estimate	1.46	4.39	3.16	2.81	4.37

Thirty trees were sampled at age 36, and 15 trees were sampled per treatment at age 39. Roots were excavated. Stand values for three 400 m² plots per treatment were derived from regressions on basal area per tree.
a. All-sided LAI values can be obtained by multiplying by 2.85.
b. Litterfall, measured over 6 years.
c. Foliage increment plus 85% of the total litterfall; mean foliage litterfall was 2.13, 2.72, 3.29, 3.48 and 3.44 t/ha/yr in columns left to right.

Miller, H.G., Miller, J.D. and Cooper, J.M. (1980). Biomass and nutrient accumulation at different growth rates in thinned plantations of Corsican pine. *Forestry* <u>53</u>, 23-39.

54-58°N 2-6°W -- U.K. Scotland.

Plantations.   *Pinus nigra* var. *maritima*

Stands with a maximum mean annual increment of 12 m³/ha/yr.[a]

Age (years)	20	30	40	50	60	70	80
Trees/ha	4046	1678	908	614	474	403	358[b]
Tree height (m)	6.4	11.2	15.4	18.8	21.6	23.8	25.4
Basal area (m²/ha)	24.8	26.8	28.3	30.4	32.2	34.3	35.5
Leaf area index							
Stem volume (m³/ha)	122	215	294	359	413	464	500

	Stem wood	39.9	70.0	95.6	116.6	134.1	150.6	162.3
Dry biomass (t/ha)	Stem bark	7.6	12.2	16.0	19.2	21.9	24.4	26.2
	Branches	5.7	12.2	17.8	22.4	26.2	29.8	32.4
	Fruits etc.							
	Foliage	9.0	9.8	9.6	8.9	8.0	7.1	6.3
	Root estimate	23.4	29.7	34.5	38.2	41.0	43.6	45.3

CAI (m³/ha/yr)		14.2	16.4	15.8	14.0	11.8	9.4	7.2
	Stem wood							
Net production (t/ha/yr)	Stem bark							
	Branches							
	Fruits etc.							
	Foliage							
	Root estimate							

Regressions relating biomass to stem volumes, based on data collected by Miller *et al.* (1976) (see p.246) were applied to stand volume tables. Values given above refer to stands prior to each thinning. Root values include the stumps. There was 2.4, 4.4, 6.2, 7.6, 8.8, 9.9 and 10.7 t/ha of dead branches in columns left to right. Nutrient contents were determined.

a. Tables for other yield classes can be obtained from the Macaulay Institute for Soil Research, Craigiebuckler, Aberdeen, Scotland.

b. An error in the authors' Table 6 has been corrected.

Whittaker, R.H. (1966). Forest dimensions and production in the Great Smoky Mountains. *Ecology* 47, 103–121.

Whittaker, R.H. and Woodwell, G.M. (1971). Measurement of net primary production of forests. In: "Productivity of Forest Ecosystems" (P. Duvigneaud, ed.) pp.159–175. UNESCO, Paris.

35°28-47'N ca.84°W   (alt. given below)   U.S.A. Great Smoky Mountains.

	Surrey Fork Flats *Acer saccharum* (43%)[a] *Tsuga canadensis* (35%)[a] *Fagus grandifolia* (17%)[a] et al. 870 m	Porter Creek Flats *A. saccharum* (29%)[a] *Aesculus octandra* (19%)[a] *Halesia carolina* (19%)[a] et al. 730 m
Age (years)	Mature	Mature
Trees/ha	1210[b]	1110[b]
Tree height (m)	33[c]	32[c]
Basal area (m²/ha)	64.0	53.2
Leaf area index		
Stem volume (m³/ha)	890[d]	764[d]
Dry biomass (t/ha) Stem wood		
Stem bark		
Branches	} 606.5	} 496.2
Fruits etc.		
Foliage	3.5	3.8
Root estimate		
CAI (m³/ha/yr)	5.9[d]	5.3[d]
Net production (t/ha/yr) Stem wood		
Stem bark		
Branches	} 8.0[e]	} 11.5[e]
Fruits etc.		
Foliage	3.5	3.8
Root estimate		

Stand biomass values for 0.1 ha plots were derived from published regressions on D, stem volumes and from branch/stem biomass ratios and other relationships. The foliage production of *T. canadensis* was taken as its foliage biomass divided by its mean foliage age.

a. Percentage of the total volume increment; *A. octandra* syn. *flava*.
b. Stems over 1.9 cm D.
c. Canopy height.
d. Parabolic volumes.
e. Excluding woody litterfall and any mortality.

Bickelhaupt, D.H., Leaf, A.L. and Richards, N.A. (1973). Effect of branching habit on above-ground dry weight estimates of *Acer saccharum* stands. In: "IUFRO Biomass Studies", pp. 219-230. College of Life Sciences and Agriculture, University of Maine, Orono, U.S.A.

Young, H.E. (1972). Biomass sampling methods for puckerbrush studies. In: "Forest Biomass Studies", pp. 179-190. College of Life Sciences and Agriculture, University of Maine, Orono, U.S.A.

U.S.A.	ca.42°30'N 76°W 500-600 m New York, Alleghany Uplands Heiberg Forest *Acer saccharum* Silt loams with fragipan. (Bickelhaupt *et al.*, 1973)	45-47°N 68-70°W -- Maine *A. saccharum* with *Fagus grandifolia* *et al.* (Young 1972)	
Age (years)	40-45	18	49
Trees/ha	1350		
Tree height (m)	20-22^a	6.1	
Basal area (m²/ha)	22.8		
Leaf area index			
Stem volume (m³/ha)			
Dry biomass (t/ha)			
Stem wood			
Stem bark	}29.2^b	}43.4	}124.4
Branches			
Fruits etc.			
Foliage	2.7	3.2	2.9
Root estimate			
CAI (m³/ha/yr)			
Net production (t/ha/yr)			
Stem wood			
Stem bark			
Branches			
Fruits etc.			
Foliage			
Root estimate			

Bickelhaupt *et al.* sampled 10 trees in September and derived stand biomass values from regressions on D.
Young (1972) harvested, and determined the fresh weights of all trees within one plot per stand; the plots were at least as long and as wide as the trees were tall; dry weights were estimated from the water contents of subsamples.
a. Height of the dominant trees.
b. Including 15.3 t/ha of stems over 3.8 cm D.

Whittaker, R.H. (1966). Forest dimensions and production in the Great Smoky Mountains. *Ecology* <u>47</u>, 103-121.

Whittaker, R.H. and Woodwell, G.M. (1971). Measurement of net primary production of forests. In: "Productivity of Forest Ecosystems" (P. Duvigneaud, ed.) pp.159-175. UNESCO, Paris.

35°28-47'N ca.84°W   1310 m   U.S.A., Great Smoky Mountains, Trillium Gap.

Aesculus octandra (40%)[a]

Tilia heterophylla (26%)[a]

Prunus serotina (16%)[a]

et al.

Age (years)		222[b]
Trees/ha		1450[c]
Tree height (m)		34.0[b]
Basal area (m²/ha)		54.2
Leaf area index		6.2
Stem volume (m³/ha)		720[d]
Dry biomass (t/ha)	Stem wood	387.3
	Stem bark	34.7
	Branches	74.0
	Fruits etc.	0.2
	Foliage	4.0
	Root estimate	78.0
	CAI (m³/ha/yr)	5.5[d]
Net production (t/ha/yr)	Stem wood	4.3[e]
	Stem bark	0.5[e]
	Branches	2.2[e]
	Fruits etc.	0.3
	Foliage	4.0
	Root estimate	1.8

Stand biomass values for a 0.2 ha plot were derived from published regressions of biomass on D and from stem volumes, branch/stem biomass ratios and other relationships.

a. Percentage of the total volume increment; *A. octandra* syn. *flava*.
b. Weighted mean age and height.
c. Stems over 1.9 cm D.
d. Parabolic volumes.
e. Excluding woody litterfall and any mortality.

Cleve, K. van, Viereck, L.A. and Schlentner, R.L. (1971). Accumulation of nitrogen in alder (*Alnus*) ecosystems near Fairbanks, Alaska. *Arctic Alpine Res.* <u>3</u>, 101–114.

---

ca.64°45'N 148°15'W　　--　U.S.A., Alaska, near Fairbanks, Tanana River floodplain.

Alluvial sand
and silt.　　　　*Alnus incana* subsp. *tenufolia* (83-89%)[a] with *Salix* spp.

Age (years)	5	15	20
Trees/ha	49699[b]	4563[b]	7142[b]
Tree height (m)	3	5-8	6-7
Basal area (m² /ha)	15.4	19.6	28.9
Leaf area index			
Stem volume (m³ /ha)			

Dry biomass (t/ha)

Stem wood	} 5.43	} 20.23	} 31.41
Stem bark			
Branches	1.43	5.81	9.16
Fruits etc.	0.00	0.14	0.03
Foliage	1.89	1.63	2.14
Root estimate	5.90	15.20	24.73

CAI (m³ /ha/yr)

Net production (t/ha/yr)

Stem wood	} 2.24
Stem bark	} + 5.93[c]
Branches	0.67 }
Fruits etc.	ca.0.10
Foliage	1.89[d]
Root estimate	

All trees were sampled within each of the above 100 m² plots, and roots were excavated in one 9 m² area at each site. There was 10.5 and 16.4 t/ha of dead stems and branches in the 15- and 20-year-old stands, respectively. Nitrogen contents were determined.

a. Percentage of the above-ground biomass.
b. Number of *A. incana* stems; there were also 54768, 1091 and 99 *Salix* stems per hectare in columns left to right.
c. Mortality and branch death.
d. Mean foliage biomass.

Zavitkovski, J. and Stevens, R.D. (1972). Primary productivity of red alder ecosystems. *Ecology* <u>53</u>, 235-242.

Zavitkovski, J., Isebrands, J.G. and Crow, T.R. (1976). "Application of Growth
Analysis in Forest Biomass Studies." Proc. 3rd N. Am. For. Biol. Workshop (C.P.P.
Reid and C.H. Fechner, eds) pp.196-226. Colorado State University, Fort Collins,
Colorado, U.S.A.

44-46°N ca.124°W   50-300 m   U.S.A., Oregon, coastal range.

Latosolic
and alluvial
soils.   *Alnus rubra*

Age (years)	5	10	20	30	40	50
Trees/ha	22490^a	6503^a	1881^a	910^a	544^a	365^a
Tree height (m)	5	11	20	25	27	29
Basal area (m²/ha)	ca.12	ca.26	ca.36	ca.40	ca.35	ca.30
Leaf area index		6.5				
Stem volume (m³/ha)						

Dry biomass (t/ha)

	5	10	20	30	40	50
Stem wood						
Stem bark	} 20.0	} 50.0	} 140.0	} 180.0	} 200.0	} 205.0
Branches						
Fruits etc.						
Foliage	5.4	5.5	5.0	4.2	3.5	3.0
Root estimate			32.0^b	40.0^b	45.0^b	50.0^b

CAI (m³/ha/yr)

Net production (t/ha/yr)

	5	10	20	30	40	50
Stem wood						
Stem bark	} 10.0^c	} $10.6+3.3^d+2.8^e$	} 10.0^c	} 7.0^c	} 5.0^c	} 4.0^c
Branches						
Fruits etc.						
Foliage	5.4	5.5	5.0	4.2	3.5	3.0
Root estimate		2.9				

In all, 119 trees were sampled in 22 stands, and 28 root systems were excavated.
Stand biomass values for plots of 1 to 809 m² in 50 stands were derived from regressions on D²H. The production values refer to years 5-10, 10-15, 15-25, 25-35,
35-45 and 45-50 in columns left to right, estimated from graphs of cumulative biomass plus mortality with age. The values for ages 10-15 (second column from the
left) were taken from Zavitkovski and Stevens (1972, their Table 2).
a. Estimated from a power curve relating tree numbers per hectare to stand age.
b. Extractable roots plus 10%.
c. Including estimated mortality and litterfall.
d. Mortality.
e. Woody litterfall.

Turner, J., Cole, D.W. and Gessel, S.P. (1976). Mineral nutrient accumulation and cycling in a stand of red alder (*Alnus rubra*). *J. Ecol.* <u>64</u>, 965-974.

47°24'N 122°15'W 210 m U.S.A., Washington, Cedar River, Thompson Research Center.

Compacted coarse till soils. *Alnus rubra*

Age (years)	31-37
Trees/ha	716
Tree height (m)	22.4[a]
Basal area (m²/ha)	23.3
Leaf area index	
Stem volume (m³/ha)	

Dry biomass (t/ha)

Stem wood	128.0	⎫
Stem bark	23.6	⎪
Branches	19.4	⎬ + 9.5[b]
Fruits etc.		⎪
Foliage	4.1	⎭
Root estimate	35.2	

CAI (m³/ha/yr)

Net production (t/ha/yr)

Stem wood	⎫ 7.6 + 8.1[c]
Stem bark	⎭
Branches	2.1 + 4.4[d]
Fruits etc.	
Foliage	4.1[e]
Root estimate	

Above-ground biomass values were estimated by harvesting all trees within a 45 m² plot in July. Root biomass was estimated using regressions calculated by Zavitkovski and Stevens (1972) (see p.252). Nutrient contents were determined.

a. Height of the dominant trees.
b. Understorey shrubs.
c. Mortality.
d. Woody litterfall.
e. New foliage biomass; foliage litterfall was 3.4 t/ha/yr.

Parker, G.R. and Schneider, G. (1975). Biomass and productivity of an alder swamp in northern Michigan. *Can. J. For. Res.* 5, 403-409.

Parker, G.R. and Schneider, G. (1974). Structure and edaphic factors of an alder swamp in northern Michigan. *Can. J. For. Res.* 4, 499-508.

Voigt, G.K. and Steucek, G.L. (1969). Nitrogen distribution and accretion in an alder ecosystem. *Proc. Soil Sci. Soc. Am.* 33, 946-949.

U.S.A.	46°31'N 84°21'W 200-400 m Michigan, nr Sault Ste Marie, Dunbar Experimental Forest		42°00'N 72°09'W ca.250 m Connecticut, nr Union
	Alnus rugosa, Fraxinus nigra, Populus balsamifera et al.		*A. rugosa*
	Impeded drainage (21% 36% 20%)[a]	Sandy loam (60% 25% 9%)[a]	Old mill pond. (Voigt and Steucek 1969)
Age (years)			7-22
Trees/ha			
Tree height (m)			
Basal area (m²/ha)			
Leaf area index			
Stem volume (m³/ha)			

		Dry biomass (t/ha)		
	Stem wood	34.2 ⎫	18.0 ⎫	⎫
	Stem bark	2.8 ⎬ + 0.6[b]	2.2 ⎬ + 1.1[b]	⎬ 16.9
	Branches	10.9 ⎭	5.9 ⎭	⎭
	Fruits etc.			
	Foliage	$3.5^c + 0.2^{bc}$	$2.9^c + 0.2^{bc}$	0.7
	Root estimate			4.7

CAI (m³/ha/yr)

		Net production (t/ha/yr)	
	Stem wood	⎫ 1.33^d	⎫ 1.25^d
	Stem bark	⎭	⎭
	Branches	0.58^d	0.48^d
	Fruits etc.		
	Foliage	3.49^c	2.86^c
	Root estimate		

At Dunbar, Parker and Schneider (1974, 1975) sampled 39 *A. rugosa* and 38 *F. nigra* in late summer, and derived biomass values for ten 16 m² plots per stand from regressions on D.

Voigt and Steucek (1969) sampled 6 'clumps' of *A. rugosa* in the autumn, including the roots. Nitrogen contents were determined.

a. Percentage of the total biomass accounted for by *A. rugosa, F. nigra* and *P. balsamifera* (left to right within the brackets).

b. Understorey shrubs.

c. Including the current year's twigs.

d. Excluding woody litterfall.

Young, H.E. (1972). Biomass sampling methods for puckerbrush studies. In: "Forest
   Biomass Studies", pp.179-190. College of Life Sciences and Agriculture,
   University of Maine, Orono, U.S.A.

45-47°N 68-70°W   --   U.S.A., Maine.

Alnus serrulata et al.

'Puckerbrush' stands.

Age (years)	4	5	6	20
Trees/ha				
Tree height (m)	2	2.7	3	5.5
Basal area (m²/ha)				
Leaf area index				
Stem volume (m³/ha)				

Dry biomass (t/ha)	Stem wood				
	Stem bark	} 7.0	} 19.0	} 17.3	} 31.3
	Branches				
	Fruits etc.				
	Foliage	3.8	4.3	6.3	2.3
	Root estimate	3.2		9.0	

CAI (m³/ha/yr)					
Net production (t/ha/yr)	Stem wood				
	Stem bark				
	Branches				
	Fruits etc.				
	Foliage				
	Root estimate				

All trees were harvested within one plot per stand and their fresh weights were
measured. Dry weights were estimated from the water contents of subsamples. The
plots were at least as long and as wide as the trees were tall. Roots were excavated
in the 4- and 6-year-old stands.

Weaver, G.T. and DeSelm, H.R. (1973). Biomass distributional patterns in adjacent coniferous and deciduous forest ecosystems. In: "IUFRO Biomass Studies", pp. 415-427. College of Life Sciences and Agriculture, University of Maine, Orono, U.S.A.

35°15-25'N 82°55' to 83°03'W   1524-1954 m   U.S.A., North Carolina, Balsam Mountains.

Sandy acid loams,
high in organic matter,   *Betula lutea* syn. *alleghaniensis*
low in exchangeable   with understorey shrubs.
cations.

		40-60	ca.110
Age (years)		40-60	ca.110
Trees/ha			
Tree height (m)			
Basal area (m²/ha)			
Leaf area index			
Stem volume (m³/ha)			

		40-60	ca.110
Dry biomass (t/ha)	Stem wood	} 75[a]	} 113[a]
	Stem bark		
	Branches	25[a]	42[a]
	Fruits etc.		
	Foliage	4[a]	3[a]
	Root estimate		

CAI (m³/ha/yr)			
Net production (t/ha/yr)	Stem wood		
	Stem bark		
	Branches		
	Fruits etc.		
	Foliage		
	Root estimate		

Over 50 trees were sampled from four 40-60 year-old and two 110-year-old stands, and biomass values for several 400 m² plots within each stand were derived from regressions on D.
a. Including the understorey shrubs.

Young, H.E. (1972). Biomass sampling methods for puckerbrush studies. In: "Forest Biomass Studies", pp.179-190. College of Life Sciences and Agriculture, University of Maine, Orono, U.S.A.

Young, H.E. (1973). Biomass variation in apparently homogeneous puckerbrush stands. In: "IUFRO Biomass Studies", pp.197-206. College of Life Sciences and Agriculture, University of Maine, Orono, U.S.A.

45-47°N 68-70°W   --   U.S.A., Maine.

'Puckerbrush' stands.	*Betula alleghaniensis* syn. *lutea et al.*	*Betula papyrifera et al.*		*Betula populifolia et al.*		
					Eddington	Old Town
Age (years)	25	22	24	12	27	40
Trees/ha						
Tree height (m)	8.8	6.7	10.4	3 to 4	7.6	
Basal area (m²/ha)						
Leaf area index						
Stem volume (m³/ha)						
Dry biomass (t/ha) — Stem wood						
Stem bark	59.5	30.0	52.0	28.8	47.3	36.3
Branches					7.6	2.3
Fruits etc.						
Foliage	2.2	1.9	3.5	6.3	2.6	1.7
Root estimate				9.0		
CAI (m³/ha/yr)						
Net production (t/ha/yr) — Stem wood						
Stem bark						
Branches						
Fruits etc.						
Foliage						
Root estimate						

All trees were harvested within one plot per stand and their fresh weights were measured. Dry weights were estimated from the water contents of subsamples. The plots were at least as long and as wide as the trees were tall. Roots were excavated in the 12-year-old stand.

Whittaker, R.H. and Niering, W.A. (1975). Vegetation of the Santa Catalina mountains, Arizona. V Biomass, production and diversity along the elevation gradient.
Ecology 56, 771–790.

Whittaker, R.H. and Niering, W.A. (1968). Vegetation of the Santa Catalina mountains, Arizona. IV Limestone and acid soils. *J. Ecol.* 56, 523–544.

ca.32°20'N 110°50'W   (alt. given below) U.S.A., Arizona, Santa Catalina Mtns.

Small trees and shrubs.

	Carnegiea gigantea, Cercidium microphyllum, Fouquieria splendens, et al. 1020 m	*C. microphyllum Franseria deltoides F. splendens, et al.* 870 m	*Cercocarpus breviflorus et al.* 1810 m
Age (years)			
Trees/ha	3870	1460	3930
Tree height (m)	3.4^a	3.1^a	3.7^a
Basal area (m²/ha)	10.9	3.0	2.7
Leaf area index	0.9	0.6	1.4
Stem volume (m³/ha)	18.5^b	4.5^b	5.0^b
Dry biomass (t/ha)			
Stem wood	} 8.0 + 0.1^c	} 2.4 + 0.2^c	} 4.9
Stem bark			
Branches	4.0 + 0.6^c	1.4 + 0.9^c	1.8 + 0.1^c
Fruits etc.			
Foliage	0.2 + 0.1^c	0.1 + 0.1^c	0.5 + 0.3^c
Root estimate			
CAI (m³/ha/yr)			
Net production (t/ha/yr)			
Stem wood	0.25^d }	0.12^d }	0.32^d }
Stem bark	0.04^d } + 0.13^{cd}	0.04^d } + 0.26^{cd}	0.06^d } + 0.02^{cd}
Branches	0.30^d }	0.20^d }	0.37^d }
Fruits etc.	0.02 + 0.03^c	0.02 + 0.05^c	0.04 + 0.03^c
Foliage	0.23 + 0.17^c	0.14 + 0.22^c	0.46 + 0.10^c
Root estimate			

Small trees and shrubs were sampled in one 0.1 ha plot per stand, and stand biomass
values were derived from regressions on D, wood volumes and surface areas, and from
other relationships. All trees and shrubs over 1 cm D were included.
a. Weighted mean heights.
b. Parabolic volumes.
c. Understorey shrubs.
d. Excluding woody litterfall and mortality.

Whittaker, R.H., Bormann, F.H., Likens, G.E. and Siccama, T.G. (1974). The Hubbard
  Brook ecosystem study: forest biomass and production. *Ecol. Monogr.* <u>44</u>, 233-252.
Bormann, F.H., Siccama, T.G., Likens, G.E. and Whittaker, R.H. (1970). The Hubbard
  Brook ecosystem study: composition and dynamics of the tree stratum. *Ecol.
  Monogr.* <u>40</u>, 373-388.
Gosz, J.R., Likens, G.E. and Bormann, F.H. (1972). Nutrient content of litterfall in
  the Hubbard Brook Experimental Forest, New Hampshire. *Ecology* <u>53</u>, 769-784.
Whittaker, R.H. and 4 others. (1979). *Ecology* <u>60</u>, 203-220.

43°55'N 71°40'W   (alt. given below)   U.S.A., New Hampshire, Hubbard Brook.

Bouldery, glacial till podzolic soils.	*Fagus grandifolia, Acer saccharum, Betula lutea* syn. *alleghaniensis, et al.*		
	$(46\% \ 32\% \ 20\%)^a$	$(26\% \ 43\% \ 29\%)^a$	$(27\% \ 26\% \ 25\%)^a$
	500-630 m	630-710 m	710-785 m
Age (years)	83^b	95^b	124^b
Trees/ha	2420	1290	1290
Tree height (m)	10.8^b	16.7^b	16.9^b
Basal area (m²/ha)	22.0	23.7	26.3
Leaf area index	5.5^c	5.7^c	6.2^c
Stem volume (m³/ha)	121^d	172^d	194^d
Dry biomass (t/ha) Stem wood	60.1	92.8	104.2
Stem bark	6.9	9.4	10.9
Branches	26.9	41.9	40.3
Fruits etc.	0.2	0.1	0.2
Foliage	3.5	2.8	3.0
Root estimate	23.5	20.6	30.6
CAI (m³/ha/yr)	2.2^d	3.7^d	3.8^d
Net production (t/ha/yr) Stem wood	2.02 ⎫	2.94 ⎫	3.15 ⎫
Stem bark	0.24 ⎬ + ca.2.1^e	0.32 ⎬ + ca.2.1^e	0.35 ⎬ + ca.2.1^e
Branches	2.23 ⎭	3.16 ⎭	3.49 ⎭
Fruits etc.	0.18	0.16	0.21
Foliage	2.85	3.53	3.74
Root estimate	1.39^f	2.02^f	2.23^f

Ninety-three trees were sampled, 81 roots were excavated, and stand biomass values
were derived from regressions on D, wood volumes, and surface areas, and from other
relationships. There was about 2.7 t/ha of dead branches in each stand. Nutrient
contents were determined.
a. Percentage of the total biomass accounted for by *F. grandifolia, A. saccharum*
   and *B. lutea* (written left to right within the brackets).
b. Weighted mean ages and heights. c. Including all-sided leaf areas of a few
   conifers. d. Parabolic volumes. e. Estimated woody litterfall and mortality;
   measured foliage litterfall was 2.55, 2.82 and 2.88 t/ha/yr in columns left to
   right, or 2.65, 3.06 and 3.22 t/ha/yr including consumption losses.
f. Assuming that roots grew at the same relative rates as above-ground woody parts.

Whittaker, R.H. (1966). Forest dimensions and production in the Great Smoky Mountains. *Ecology* 47, 103-121.
Whittaker, R.H. and Woodwell, G.M. (1971). Measurement of net primary production of forests. In: "Productivity of Forest Ecosystems" (P. Duvigneaud, ed.) pp.159-175. UNESCO, Paris.
Young, H.E. (1972). Biomass sampling methods for puckerbush studies. In: "Forest Biomass Studies", pp.179-190. College of Life Sciences and Agriculture, University of Maine, Orono, U.S.A.

U.S.A.	35°28-47'N ca.84°W 1580 m Great Smoky Mountains, Newfoundland Gap.		45-47°N 68-70°W -- Maine
	Fagus grandifolia (73%)[a] *Picea rubens* (24%)[a] *et al.*	*F. grandifolia* (80%)[a] *Acer saccharum* (6%)[a] *et al.*	*F. grandifolia* *et al.* (Young 1972)
Age (years)	135[b]	84[b]	26
Trees/ha	2140[c]	2170[c]	
Tree height (m)	15.6[b]	13.4[b]	7.9
Basal area (m²/ha)	27.7	22.0	
Leaf area index	5.2[d]	4.4	
Stem volume (m³/ha)	185[e]	120[e]	
Dry biomass (t/ha) Stem wood	118	87	} 78.5
Stem bark	8	6	
Branches	40	35	
Fruits etc.			
Foliage	5	3	1.8
Root estimate			
CAI (m³/ha/yr)	2.8[e]	1.6[e]	
Net production (t/ha/yr) Stem wood	2.6[f]	} 3.3[f]	
Stem bark	0.2[f]		
Branches	1.8[f]		
Fruits etc.			
Foliage	4.2	2.7	
Root estimate			

Whittaker (1966) derived stand biomass values for plots of at least 0.1 ha from published regressions on D and from stem volumes, branch/stem biomass ratios and other relationships.
Young (1972) harvested all trees within one plot of at least 64 m², measured their fresh weights and estimated dry weights from the water contents of subsamples.
a. Percentage of the total volume increment.
b. Weighted mean ages and heights.
c. Stems over 1.9 cm D.
d. All-sided LAI was 14.8.
e. Parabolic volumes.
f. Excluding woody litterfall and mortality.

Whittaker, R.H. and Niering, W.A. (1975). Vegetation of the Santa Catalina mountains, Arizona. V Biomass, production and diversity along the elevation gradient.
Ecology <u>56</u>, 771-790.

Whittaker, R.H. and Niering, W.A. (1968). Vegetation of the Santa Catalina mountains, Arizona. IV Limestone and acid soils. *J. Ecol.* <u>56</u>, 523-544.

---

ca.32°20'N   110°50'W   1220 m   U.S.A., Arizona, Santa Catalina Mountains, near
Tucson.

Fouquieria splendens, *Prosopis juliflora* syn. *glandulosa*
syn. *velutina*, *et al.* with understorey shrubs.

Age (years)	
Trees/ha	$20 + 640^c$
Tree height (m)	2.5^a
Basal area (m²/ha)	0.2
Leaf area index	1.6
Stem volume (m³/ha)	0.3^b

Dry biomass (t/ha)

Stem wood	$\}0.56 + 0.22^c$
Stem bark	
Branches	$0.04 + 0.87^c$
Fruits etc.	
Foliage	$0.04 + 0.42^c$
Root estimate	

CAI (m³/ha/yr)

Net production (t/ha/yr)

Stem wood	0.04^d
Stem bark	$0.04^d \} + 0.25^{cd}$
Branches	0.01^d
Fruits etc.	$0.01 + 0.04^c$
Foliage	$0.06 + 0.33^c$
Root estimate	

Ten to 15 trees were sampled of each of the major species and stand biomass values
for a 0.1 ha plot were derived from regressions on D, wood volumes and surface
areas, and from other relationships. All trees and shrubs over 1 cm D were included.
a. Weighted mean height.
b. Parabolic volume.
c. Understorey shrubs.
d. Excluding woody litterfall and mortality.

Chew, R.M. and Chew, A.E. (1965). The primary productivity of a desert-shrub
   (*Larrea tridentata*) community. *Ecol. Monogr.* <u>35</u>, 355-375.

Whittaker, R.H. and Niering, W.A. (1975). Vegetation of the Santa Catalina mountains, Arizona. V Biomass, production and diversity along the elevation gradient.
   Ecology <u>56</u>, 771-790.

U.S.A.	31°55'N 109°09'W 1370 m Arizona, San Simon valley. *Larrea tridentata* (72%)[a] *Flourensia cernua* (9%)[a] Shallow soil over gravel alluvium. (Chew and Chew 1965)	ca.32°20'N 110°50'W 760 m Arizona, Santa Catalina Mtns. *Larrea divaricata et al.* (Whittaker and Niering 1975)
Age (years)	17 (5 to 65)	
Trees/ha	6140	49500
Tree height (m)		1.7[b]
Basal area (m² /ha)		4.2
Leaf area index	0.9	0.6
Stem volume (m³ /ha)		3.5[c]
Dry biomass (t/ha) Stem wood	} 4.6	} 2.5
Dry biomass (t/ha) Stem bark		
Dry biomass (t/ha) Branches		1.3
Dry biomass (t/ha) Fruits etc.		
Dry biomass (t/ha) Foliage		0.4
Dry biomass (t/ha) Root estimate		
CAI (m³ /ha/yr)		
Net production (t/ha/yr) Stem wood	} 1.4[d]	0.17[e]
Net production (t/ha/yr) Stem bark		0.02[e]
Net production (t/ha/yr) Branches		0.19[e]
Net production (t/ha/yr) Fruits etc.		0.00
Net production (t/ha/yr) Foliage		0.54
Net production (t/ha/yr) Root estimate		

Chew and Chew (1965) sampled 61 shrubs in summer, excavated 17 root systems, and
derived stand biomass values from weight-volume and volume-age relationships; stand
production was estimated from age-cumulative biomass relationships and age-frequency
distributions.
Whittaker and Niering (1975) sampled shrubs within a 0.1 ha plot and derived stand
biomass values from regressions on D, wood volumes and surface areas, and other
relationships; all shrubs over 1 cm D were included.
a. Percentage of the vegetation area covered.   *b.* Weighted mean height.
c. Parabolic volume.   *d.* Including litterfall; about 65%, i.e. 0.9 t/ha/yr, of this
   total production, was leaves or leaved stems.
e. Excluding woody litterfall and mortality.

Whittaker, R.H. (1966). Forest dimensions and production in the Great Smoky Mountains. *Ecology* <u>47</u>, 103-121.

---

ca.35°30'N 84°W   700 m   U.S.A., Great Smoky Mountains.

Liriodendron tulipifera (77%)[a], *Acer rubrum* (8%)[a], *Robinia pseudoacacia, et al.*

Vigorous successional stand on abandoned farmland.

Age (years)	29[b]
Trees/ha	1820[c]
Tree height (m)	22.4[b]
Basal area (m²/ha)	34.2
Leaf area index	7.4
Stem volume (m³/ha)	310[d]

Dry biomass (t/ha)		
	Stem wood	150
	Stem bark	17
	Branches	49
	Fruits etc.	
	Foliage	4.7
	Root estimate	

CAI (m³/ha/yr)		14.4[d]
Net production (t/ha/yr)	Stem wood	
	Stem bark	
	Branches	19.9[e]
	Fruits etc.	
	Foliage	4.1
	Root estimate	

Stand biomass values for a 0.1 ha plot were derived from published regressions on D and from stem volumes, branch/stem biomass ratios and other relationships.
a. Percentage of the total volume increment.
b. Weighted mean age and height.
c. Stems over 1.9 cm diameter.
d. Parabolic volume.
e. Excluding woody litterfall and mortality.

Sollins, P., Reichle, D.E. and Olson, J.S. (1973). "Organic Matter Budget and Model for a Southern Appalachian *Liriodendron* Forest." EDFP-IBP-73/2. Oak Ridge National Laboratory, Tennessee, U.S.A.

Harris, W.F., Kinerson, R.S. and Edwards, N.T. (1977). Comparison of below-ground biomass of natural deciduous forest and loblolly pine plantations. *Pedobiologia* 17, 369-381.   Edwards, N.T. and Harris, W.F. (1977). *Ecology* 58, 431-437.

Reichle, D.E., Edwards, N.T., Harris, W.F. and Sollins, P. (1981). In: "Dynamic Properties of Forest Ecosystems" (D.E. Reichle, ed.), p.657. Cambridge Univ. Press.

35°55'N 84°17'W   290 m   U.S.A., Tennessee, Oak Ridge Reservation.

Colluvial cherty silt loams, high in organic matter, pH 5.8.

Liriodendron tulipifera (78%)[a] with *Quercus* spp., *Pinus echinata*, *Carya tomentosa* and understorey trees.

Age (years)	40[b]	48[b]
Trees/ha		
Tree height (m)		
Basal area (m² /ha)	19.2[c]	22.1[c]
Leaf area index	$5.6 + 1.4^d$	$6.0 + 1.1^d$
Stem volume (m³ /ha)		

Dry biomass (t/ha)

Stem wood	$\left.\right\}76.2 + 5.4^d$	$\left.\right\}94.4 + 5.9^d$
Stem bark		
Branches	$19.6 + 2.0^d$	$27.1 + 2.1^d$
Fruits etc.		0.2
Foliage	$2.9 + 0.5^d$	$3.2 + 0.5^d$
Root estimate	$31.5 + 3.6^d$	$38.9 + 4.0^d$

CAI (m³ /ha/yr)

Net production (t/ha/yr)

Stem wood	$\left.\right\}1.11 + 0.39^e + 0.59^d + 0.52^{de}$
Stem bark	
Branches	$0.57 + 0.46^e + 0.02^d + 0.04^{de}$
Fruits etc.	$0.20 + 0.02^d$
Foliage	$3.09 + 0.50^d + 0.10^f$
Root estimate	$2.55 + 0.37^d$ (or ca.9.0)^g

About 250 trees were sampled and stand biomass values for several 500 m² plots were derived from regressions on D. Root values include the stumps. Values given above were taken from Sollins *et al.* (1973); Reichle *et al.* (1981) gave overstorey and understorey stem increment values of 2.25 and 0.06 t/ha/yr, respectively.

a. Percentage of the total above-ground biomass.
b. Mean age of the dominant and co-dominant trees.
c. Trees at least 2.54 cm D.
d. Understorey shrubs and saplings.
e. Woody litterfall and mortality.
f. Consumption.
g. Updated value from Harris *et al.* (1977).

Harris, W.F., Goldstein, R.A. and Henderson, G.S. (1973). Analysis of forest biomass pools, annual primary production and turnover of biomass for a mixed deciduous forest watershed. In: "IUFRO Biomass Studies", pp.43-64. College of Life Sciences and Agriculture, University of Maine, Orono, U.S.A.

Harris, W.F. and Henderson, G.S. (1981). In: "Dynamic Properties of Forest Ecosystems" (D.E. Reichle, ed.) pp.658-661. Cambridge University Press, Cambridge, London, New York, Melbourne.

35°58'N 84°17'W   265-360 m   U.S.A., Tennessee, Walker Branch Site.

Red-yellow podzols, pH 4.0-6.5. Ultisols.

Liriodendron tulipifera with *Carya* spp., *Quercus alba, Quercus rubra, et al.*

Age (years)	30-80
Trees/ha	
Tree height (m)	12-25
Basal area (m²/ha)	19.0
Leaf area index	
Stem volume (m³/ha)	

Dry biomass (t/ha)

Stem wood	} 83.5
Stem bark	
Branches	21.2
Fruits etc.	
Foliage	3.9
Root estimate	30.6[a]

CAI (m³/ha/yr)

Net production (t/ha/yr)

Stem wood	
Stem bark	} 3.55 + 1.18[b] + 0.63[c]
Branches	
Fruits etc.	
Foliage	3.90[d]
Root estimate	2.0

Many trees were sampled (about 150 at all four Walker Branch sites), roots were excavated in 2-3 pits, and stand biomass values for 50-100 circular plots were derived from regressions on D.
a. Including stumps, which represented about half the 'root' biomass.
b. Mortality.
c. Woody litterfall.
d. Leaf production; leaf litterfall was 3.70 t/ha/yr.

Carter, M.C. and White, E.H. (1971). "Dry Weight and Nutrient Accumulation in Young Stands of Cottonwood (*Populus deltoides* Bartr.)." Agr. Exp. Stn, Auburn University, Circular 190. Alabama, U.S.A.

---

ca.31°10'N 88°00'W   20-100 m   U.S.A., Alabama, near Mobile.

Alluvial soils
of various   *Populus deltoides*
quality.

Age (years)	6	7	7	7	8	8	8	9
Trees/ha	17347	964	2446	8056	1161	2693	2199	1507
Tree height (m)	9.9	21.9	17.7	14.9	20.0	18.3	17.1	18.5
Basal area (m²/ha)	18.8	23.9	23.4	20.2	20.4	23.6	18.4	18.4
Leaf area index								
Stem volume (m³/ha)	61	180	148	131	142	170	111	126

Dry biomass (t/ha)									
	Stem wood	26.1	56.5	39.8	47.0	56.0	56.3	36.9	45.2
	Stem bark	7.2	9.5	7.9	8.6	8.6	7.9	8.3	8.3
	Branches	3.4	12.2	14.6	7.2	10.8	14.9	7.7	10.6
	Fruits etc.								
	Foliage	2.5	3.8	2.7	3.6	3.6	2.7	2.5	2.0
	Root estimate								

CAI (m³/ha/yr)

Net production (t/ha/yr)	
	Stem wood
	Stem bark
	Branches
	Fruits etc.
	Foliage
	Root estimate

One tree was sampled in each of three size classes from each stand in August. Biomass values for two 200 m² plots per stand were derived from regressions on D. Nutrient contents were determined.

Blackmon, B.G., Baker, J.B. and Cooper, D.T. (1979). Nutrient use by three geographic sources of eastern cottonwood. *Can. J. For. Res.* <u>9</u>, 532-534.

33°35'N   89-90°W   50-150 m   U.S.A., Mississippi, near Stoneville.

Plantations.
Fine, silty
loam.

Populus deltoides

All stands thinned at age 3.

	S. Illinois provenance	Mississippi provenance	Louisiana provenance
Age (years)	11	11	11
Trees/ha	452	476	487
Tree height (m)	22	22	22
Basal area (m²/ha)			
Leaf area index			
Stem volume (m³/ha)			
Dry biomass (t/ha)			
Stem wood	} 63.0	} 68.8	} 70.8
Stem bark			
Branches	5.4	4.9	8.7
Fruits etc.			
Foliage	1.1	1.3	2.0
Root estimate			
CAI (m³/ha/yr)			
Net production (t/ha/yr)			
Stem wood			
Stem bark			
Branches			
Fruits etc.			
Foliage	1.1	1.3	2.0
Root estimate			

One average-sized tree was sampled of each provenance in each of 4 blocks, and stand values were obtained by multiplying mean tree values by the numbers of trees per hectare. Nutrient contents were determined.

Koerper, G.J. and Richardson, C.J. (1980). Biomass and net annual primary production regressions for *Populus grandidentata* on three sites in northern lower
Michigan. *Can. J. For. Res.* <u>10</u>, 92-101.

45°34'N 84°30'W   230-270 m   U.S.A., Michigan, Cheboygan County.

Populus grandidentata

		82%[a]	79%[a]	48%[a]
		Good loamy sand	Intermediate quality loamy sand	Infertile Rubicon sand
Age (years)		52	52	60
Trees/ha				
Tree height (m)		26.5	23.1	15.6
Basal area (m²/ha)		30.5	27.3	11.4
Leaf area index				
Stem volume (m³/ha)				
Dry biomass (t/ha)	Stem wood	126.5	95.6	27.2
	Stem bark	24.6	21.1	6.3
	Branches	18.0	10.3	4.0
	Fruits etc.			
	Foliage	2.4	1.8	1.0
	Root estimate			
CAI (m³/ha/yr)				
Net production (t/ha/yr)	Stem wood	4.90[b]	3.09[b]	1.08[b]
	Stem bark	0.97[b]	0.73[b]	0.28[b]
	Branches	2.75[b]	1.68[b]	0.61[b]
	Fruits etc.			
	Foliage	2.41	1.76	0.96
	Root estimate			

Ten or eleven trees were sampled per site and stand biomass values for fifteen
100 m² plots per site were derived from regressions on D. There was 5.5, 3.6 and
1.3 t/ha of dead branches in columns left to right. Values given above are for
P. grandidentata only; other species (*Acer*, *Quercus* etc.) were ignored.
a. Percentage of total basal area accounted for by *P. grandidentata*.
b. Excluding woody litterfall and any mortality.

Gosz, J.R. (1980). Biomass distribution and production budget for a non-aggrading forest ecosystem. *Ecology* <u>61</u>, 507-514.

Bray, J.R. and Dudkiewicz, L.A. (1963). The composition, biomass and productivity of two *Populus* forests. *Bull. Torrey Bot. Club* <u>90</u>, 298-308.

U.S.A.	35°50'N 105°50'W 3109-3231 m New Mexico Sangre de Cristo Range, Tesuque watershed. *Populus tremuloides* Deep gravelly, stoney loams pH 6.0-6.4 (Gosz 1980)	ca.47°N 95°W 200-400 m Minnesota, Itasca *P. tremuloides* Fertile silty, sandy uplands (Bray and Dudkiewicz 1963)
Age (years)	ca.80	41
Trees/ha	2270	1600
Tree height (m)		16-17
Basal area (m²/ha)	36.4	30.9
Leaf area index		6.6
Stem volume (m³/ha)		
Dry biomass (t/ha) — Stem wood	116.2	181.9
Dry biomass (t/ha) — Stem bark	20.5	(181.9)
Dry biomass (t/ha) — Branches	7.1	20.9
Dry biomass (t/ha) — Fruits etc.		
Dry biomass (t/ha) — Foliage	2.0	3.7
Dry biomass (t/ha) — Root estimate		
CAI (m³/ha/yr)		
Net production (t/ha/yr) — Stem wood	$0.40 + 0.85^a + 0.62^b$	$>4.4^d$
Net production (t/ha/yr) — Stem bark		
Net production (t/ha/yr) — Branches		$>0.8^e$
Net production (t/ha/yr) — Fruits etc.		
Net production (t/ha/yr) — Foliage	$1.77^a + 0.42^c$	3.8^f
Net production (t/ha/yr) — Root estimate		

Gosz (1980) sampled 14 trees and derived stand values for fifty-five 100 m² plots from regressions on D; there were 1740 dead trees per hectare (7.0 m²/ha basal area) weighing 13.8 t/ha, plus 3.1 t/ha of dead branches, and 1.2 t/ha of root sprouts, none of which is included above.
Bray and Dudkiewicz (1963) sampled 6 trees during July-September, and derived stand values by assigning biomass in proportion to the 'effective canopy area' per tree for 4 trees at each of 40 random points.
a. Litterfall, measured over 3 years.   *b*. Mortality.
c. Consumption, decay and other losses. *d*. Mean (not current) annual increment.
e. New twigs, plus old wood divided by its age (41) and excluding any woody litter-fall.   *f*. Foliage biomass plus consumption.

Whittaker, R.H. and Niering, W.A. (1975). Vegetation of the Santa Catalina mountains, Arizona. V Biomass, production and diversity along the elevation gradient. *Ecology* 56, 771-790.

Alban, D.H., Perala, D.A. and Schlaegel, B.E. (1978). Biomass and nutrient distribution in aspen, pine and spruce stands on the same soil type in Minnesota. *Can. J. For. Res.* 8, 290-299.

U.S.A.	ca.32°20'N 110°50'W 2250 m Arizona Santa Catalina Mtns *Populus tremuloides* (72%)[a] *Robinia neomexicana* (7%)[a] *et al.* (Whittaker and Niering 1975)	47°20'N 94°30'W 400 m Minnesota Pike Bay Expt. Forest *P. tremuloides*, *Populus grandidentata* with *Acer saccharum*, *Acer rubrum et al.* (Alban *et al.* 1978)
Age (years)	34 [b]	40
Trees/ha	2350 710[c]	1334 + 1655[f]
Tree height (m)	16.1[b]	20.3 11.0[f]
Basal area (m²/ha)	31.6 + 0.4[c]	34.7 + 7.0[f]
Leaf area index	6.4	
Stem volume (m³/ha)	218[d]	286[g]+ 39[fg]
Dry biomass (t/ha) — Stem wood	} 103 + 0.7[c]	119.4
Dry biomass (t/ha) — Stem bark		28.3
Dry biomass (t/ha) — Branches	16 + 0.2[c]	17.0
Dry biomass (t/ha) — Fruits etc.		
Dry biomass (t/ha) — Foliage	5.2 + 0.0[c]	3.8
Dry biomass (t/ha) — Root estimate		38
CAI (m³/ha/yr)	5.4[d]	
Net production (t/ha/yr) — Stem wood	3.25[e] } + 0.06[ce]	
Net production (t/ha/yr) — Stem bark	0.68[e]	
Net production (t/ha/yr) — Branches	2.17[e] + 0.06[ce]	
Net production (t/ha/yr) — Fruits etc.	0.45	
Net production (t/ha/yr) — Foliage	3.75 + 0.07[c]	
Net production (t/ha/yr) — Root estimate		

Whittaker and Niering (1975) sampled 10-15 trees of each species and derived stand biomass values for a 0.1 ha plot from regressions on D, wood volumes and surface areas, and from other relationships; all trees and shrubs over 1 cm D were included. Alban *et al.* (1978) sampled 10 trees in spring, and felled two trees in August to estimate foliage biomass and to extract roots; stand biomass values for ten 80 m² plots were derived from regressions on D²H; nutrient contents were determined.
a. Percentage of the total stem volume.   b. Weighted mean age and height.
c. Understorey shrubs.   d. Parabolic volumes.
e. Excluding woody litterfall and any mortality.
f. Values for trees other than *Populus* spp. (all species are included in the biomass values).   g. Volumes inside bark.

Young, H.E. (1972). Biomass sampling methods for puckerbrush studies. In: "Forest Biomass Studies", pp. 179-190. College of Life Sciences and Agriculture, University of Maine, Orono, U.S.A.

Young, H.E. (1973). Biomass variation in apparently homogeneous puckerbrush stands. In: "IUFRO Biomass Studies", pp. 197-206. College of Life Sciences and Agriculture, University of Maine, Orono, U.S.A.

45-47°N 68-70°W   --   U.S.A., Maine.

Populus tremuloides, Populus grandidentata, with

Betula populifolia, Acer rubrum et al.

'Puckerbrush' stands

nr Old Town

Age (years)	6	7	20	40	41
Trees/ha					
Tree height (m)	4.9	5.0	10.7		13.7
Basal area (m²/ha)					
Leaf area index					
Stem volume (m³/ha)					

Dry biomass (t/ha)

Stem wood	}	}	}	} 49.8	}
Stem bark	} 21.6	} 23.8	} 74.5	}	} 117.9
Branches	}	}	}	3.3	}
Fruits etc.					
Foliage	3.7	3.9	1.9	2.7	2.1
Root estimate		5.7			

CAI (m³/ha/yr)

Net production (t/ha/yr)

Stem wood					
Stem bark					
Branches					
Fruits etc.					
Foliage					
Root estimate					

All trees were harvested within one plot per stand and their fresh weights were measured. Dry weights were estimated from the water contents of subsamples. The plots were at least as long and as wide as the trees were tall. Roots were excavated in the 7-year-old stand.

Cooper, A.W. (1980). Above-ground biomass accumulation and net primary production during the first 70 years of succession in *Populus grandidentata* Michx. stands on poor sites in northern lower Michigan. In:"Proceedings of Workshop on Forest Succession", Mountain Lake, Va, U.S.A.

45°35'N 84°45'W   235 m   U.S.A., Michigan, Cheboygan County, Douglas Lake.

Infertile, freely-drained, weakly podzolized, Rubicon sands.

Populus tremuloides, Populus grandidentata, with *Acer rubrum, Quercus rubra, Pinus strobus et al.*

Regeneration after fire.

Age (years)	20	30	40	50	60	70
Trees/ha						
Tree height (m)						
Basal area (m²/ha)				15.7		
Leaf area index						
Stem volume (m³/ha)						

Dry biomass (t/ha)

	20	30	40	50	60	70
Stem wood / Stem bark	}8 + 3	}20 + 6	}32 + 11	}43 + 16	}52 + 24	}56 + 36
Branches	1.4+0.5	2.6+1.1	3.5+1.9	4.0+2.6	4.3+3.3	4.3+4.2
Fruits etc.						
Foliage	0.8+0.2	1.0+0.3	1.2+0.5	1.2+0.6	1.1+0.8	0.9+1.1
Root estimate						

CAI (m³/ha/yr)

Net production (t/ha/yr)

	20	30	40	50	60	70
Stem wood / Stem bark	}(0.6+0.3)[a]	}(1.1+0.3)[a]	}(1.4+0.4)[a]	}(1.5+0.4)[a]	}(1.1+0.5)[a]	}(0.2+0.8)[a]
Branches / Fruits etc. / Foliage / Root estimate	}0.8+0.3	}1.3+0.4	}1.5+0.5	}1.4+0.6	}1.4+0.7	}1.2+0.9

Fifty-two *Populus*, and 79 trees of the other species, were sampled in July-August. Stand biomass values for plots of 200-400 m² were derived from regressions on D, and equations were calculated relating biomass and production to age. Values are given above for *Populus* spp. plus other species (left and right, respectively, in each column). There was about 0.4, 0.6, 0.9, 1.1, 1.2 and 1.4 t/ha of dead branches in columns left to right.

a. Including woody litterfall, but excluding any mortality.

Pastor, J. and Bockheim, J.G. (1981). Biomass and production of an aspen-mixed hardwood-spodosol ecosystem in northern Wisconsin. *Can. J. For. Res.* <u>11</u>, 132–138.

45°50'N 89°40'W   ca.300 m   U.S.A., Wisconsin.

Acid, glacial-outwash, sandy loams.	*Populus tremuloides* (58%)[a], *Acer saccharum* (19%)[a], *Populus grandidentata* (10%)[a], *Acer rubrum, et al.*

Age (years)		39–63
Trees/ha		1341
Tree height (m)		31.6
Basal area (m²/ha)		
Leaf area index		
Stem volume (m³/ha)		
Dry biomass (t/ha)	Stem wood	124.0
	Stem bark	24.0
	Branches	23.3
	Fruits etc.	2.4
	Foliage	
	Root estimate	20.2
CAI (m³/ha/yr)		
Net production (t/ha/yr)	Stem wood	3.80[b]
	Stem bark	0.80[b]
	Branches	3.32[b]
	Fruits etc.	
	Foliage	2.36
	Root estimate	1.20

Nine trees were sampled of *P. tremuloides* and 9 of *A. saccharum*. Stand biomass values for three 400 m² plots were derived from regressions on D. There was 3.5 t/ha of dead branches. Roots were assumed to grow at the same relative rates as above-ground parts.
a. Percentage of the total basal area.
b. Excluding woody litterfall and any mortality.

Crow, T.R. (1978). Biomass and production in three contiguous forests in northwest Wisconsin. *Ecology* <u>59</u>, 265-273.

45°30'N 89°20'W   600 m   U.S.A., Wisconsin, 30 km SE of Rhinelander.

Well-drained acidic sandy loams.	*Populus tremuloides, Acer rubrum, Betula papyrifera, Acer saccharum* with understorey shrubs.		
	(53% 13% 11% 3%)[a]	(33% 21% 17% 6%)[a]	(12% 26% 18% 11%)[a]
Age (years)	ca.50	ca.50	ca.50
Trees/ha	2842	2080	1868
Tree height (m)	14.4	14.4	14.7
Basal area (m²/ha)	18.0	16.5	18.8
Leaf area index	5.0	4.9	5.5

Stem volume (m³/ha)

Dry biomass (t/ha)

Stem wood	66.1	65.6	78.8
Stem bark	11.3	10.3	11.1
Branches	13.5 }+ 1.7[c]	16.8 }+ 0.8[c]	25.7 }+ 0.4[c]
Fruits etc.			
Foliage	2.4[b]	2.8[b]	3.7[b]
Root estimate			

CAI (m³/ha/yr)

Net production (t/ha/yr)

Stem wood	4.37	3.35	2.61
Stem bark	0.72	0.49	0.32
Branches	0.82 }+ 0.42[c]	0.83 }+ 0.21[c]	0.90 }+ 0.08[c]
Fruits etc.			
Foliage	2.41[b]	2.77[b]	3.73[b]
Root estimate			

Trees in several diameter classes of all the main species were sampled. Biomass values for about eighty 4 m² plots per stand were derived by proportional basal area allocation.

a. Percentage of the total basal area accounted for by *P. tremuloides, A. rubrum, B. papyrifera* and *A. saccharum* (written left to right within the brackets).

b. Including current year's twigs.

c. Understorey shrubs.

Safford, L.O. and Filip, S.M. (1974). Biomass and nutrient content of 4-year-old fertilized and unfertilized northern hardwood stands. *Can. J. For. Res.* <u>4</u>, 549-554.

ca.44°00'N 71°20'W	-- U.S.A., New Hampshire, Barlett Experimental Forest.	
	Prunus pensylvanica (53%)[a] *Betula alleghaniensis* syn. *lutea* (11%)[a] *Betula papyrifera, et al.* Clearfelled and scarified	*P. pensylvanica* (94%)[a] *Betula* spp. *et al.* Clearfelled, scarified and fertilized (4.5 t/ha limestone, 1165 kg/ha NPK)
Age (years)	4	4
Trees/ha		
Tree height (m)	1.2-2.2	1.2-2.9
Basal area (m²/ha)		
Leaf area index		
Stem volume (m³/ha)		
Dry biomass (t/ha) Stem wood	} 3.9	} 18.9
Stem bark		
Branches	0.6 } + 1.6[b]	2.5 } + 1.1[b]
Fruits etc.		
Foliage	1.0	1.8
Root estimate		
CAI (m³/ha/yr)		
Net production (t/ha/yr) Stem wood		
Stem bark		
Branches		
Fruits etc.		
Foliage		
Root estimate		

All woody vegetation was sampled in nine 1 m² plots in late September at each of the two sites. Nutrient contents were determined.
a. Percentage of the total woody plant biomass.
b. Understorey shrubs.

Lawson, G.J., Cottam, G. and Loucks, O.L. (1981). In: "Dynamic Properties of Forest Ecosystems" (D.E. Reichle, ed.), pp. 663-664. Cambridge University Press, Cambridge, London, New York, Melbourne.

43°02'N 89°24'W 274 m	U.S.A., Wisconsin.	
	Noe Woods	Nakoma
	Quercus alba, *Quercus velutina,* *Prunus serotina*	*Q. alba* and *Q. velutina*
	Silt loam, pH 5.8.	Urban oak remnant.
Age (years)	130	130
Trees/ha	422	
Tree height (m)	23.7	27.4
Basal area (m²/ha)	33.2	
Leaf area index	4.4	2.9
Stem volume (m³/ha)		
Dry biomass (t/ha) Stem wood	} 209.2	} 99.0
Stem bark		
Branches	50.8	26.0
Fruits etc.	0.1	
Foliage	3.9	2.6
Root estimate	66.0	
CAI (m³/ha/yr)		
Net production (t/ha/yr) Stem wood	} $2.71 + 3.49^{a} + 1.49^{b}$	} 2.23^{c}
Stem bark		
Branches	$0.73 + 0.90^{b}$	0.59^{c}
Fruits etc.	0.08^{b}	
Foliage	4.29^{b}	2.59
Root estimate	6.60	

There was 20.6 t/ha of standing dead wood at Noe Woods.
a. Mortality.
b. Litterfall.
c. Excluding woody litterfall and any mortality.

Rochow, J.J. (1974a). Estimates of above-ground biomass and primary productivity in a Missouri forest. *J. Ecol.* 62, 567–577.
Rochow, J.J. (1974b). Litterfall relations in a Missouri forest. *Oikos* 25, 80–85.
Rochow, J.J. (1975). Mineral nutrient pool and cycling in a Missouri forest. *J. Ecol.* 63, 985–994.
Whittaker, R.H. (1966). Forest dimensions and production in the Great Smoky Mountains. *Ecology* 47, 103–121.

U.S.A.	38°48'N 92°12'W 175–245 m Missouri, SE Boone County. *Quercus alba* (30%)[a] other *Quercus* spp. (28%)[a] *Acer saccharum* (11%)[a] et al. Loess covered uplands (Rochow 1974 a, b, 1975)	35°28–47'N ca.84°W 300 m Great Smoky Mountains, Oak Ridge. *Q. alba* (57%)[d] *Quercus velutina* (28%)[d] et al. (Whittaker 1966)
Age (years)	35–92	Mature
Trees/ha	522	660[e]
Tree height (m)		29[f]
Basal area (m²/ha)		31.4
Leaf area index		
Stem volume (m³/ha)		350[g]
Dry biomass (t/ha) Stem wood		
Stem bark		
Branches	$94.8 + 3.3^{b}$	366.5
Fruits etc.		
Foliage		3.5
Root estimate		
CAI (m³/ha/yr)		4.6[g]
Net production (t/ha/yr) Stem wood		
Stem bark	$1.7 + 1.1^{c}$	12.0[h]
Branches	$+ 0.5^{b}$	
Fruits etc.		
Foliage	3.5^{c}	3.5
Root estimate		

Rochow (1974a) estimated the biomass of trees over 4 cm D in eighteen 800 m² plots from their parabolic wood volumes, wood specific gravities, and litterfall; nutrient contents were determined.
Whittaker (1966) derived stand biomass values for a 0.1 ha plot from published regressions on D, from stem volumes, branch/stem biomass ratios, and other relationships.
a. Percentage of the total basal area.   *b*. Saplings.
c. Litterfall, measured over 3 years.
d. Percentage of the total volume increment.
e. Stems over 1.9 cm D.   *f*. Canopy height.
g. Parabolic volume.   *h*. Excluding woody litterfall and any mortality.

Ovington, J.D., Heitkamp, D. and Lawrence, D.B. (1963). Plant biomass and productivity of prairie, savanna, oakwood and maize field ecosystems in central Minnesota. *Ecology* <u>44</u>, 52–63.

Whittaker, R.H. (1963). Net production of heath balds and forest heaths in the Great Smoky Mountains. *Ecology* <u>46</u>, 176–182.

Whittaker, R.H. (1966). Forest dimensions and production in the Great Smoky Mountains. *Ecology* 47, 103–121.

U.S.A.	45°24'N 93°10'W 400 m Minnesota, 50 km N of Minneapolis *Quercus borealis,* with *Pinus banksiana,* *Pinus strobus,* *Quercus* spp. *et al.* Infertile sands and peats	35°28–47'N ca.84°W Great Smoky Mountains 1450 m Gregory Bald Q. *borealis* (80%)[b] *Magnolia acuminata* *et al.*	1390 m Parson Bald Q. *borealis* (51%)[b] *Quercus alba* (29%)[b] *et al.* (Whittaker 1963, 1966)
Age (years)	up to 90	Mature	146[c]
Trees/ha	over 800[a]	2660[d]	2600[d]
Tree height (m)	16	14[e]	7.5[c]
Basal area (m²/ha)	ca.25	24.6	22.0
Leaf area index			3.5
Stem volume (m³/ha)		137[f]	65[f]
Dry biomass (t/ha) Stem wood	} 111	} 132.2	46
Stem bark			6
Branches	49		31
Fruits etc.			
Foliage	4	2.8	3
Root estimate	16		
CAI (m³/ha/yr)		2.3[f]	1.2[f]
Net production (t/ha/yr) Stem wood		} 4.7[g]	1.0[g]
Stem bark			0.2[g]
Branches			1.5[g]
Fruits etc.			
Foliage		2.8	2.3
Root estimate			

Ovington *et al.* (1963) sampled 3 trees and took soil core samples of the roots; stand biomass values for a 900 m² plot were obtained by multiplying mean tree values by the numbers of trees per hectare.

Whittaker (1963, 1966) estimated stand values for plots of at least 0.1 ha from the weight of clippings of current year's twigs, from published regressions, from stem volumes, branch/stem biomass ratios and other relationships.

a. Number of *Q. borealis.*

b. Percentage of the total volume increment.

c. Weighted mean age and height.   *d.* Stems over 1.9 cm D.

e. Canopy height.   *f.* Parabolic volumes.

g. Excluding woody litterfall and mortality.

Whittaker, R.H. and Woodwell, G.M. (1968). Dimension and production relations of trees and shrubs in the Brookhaven forest, New York. *J. Ecol.* <u>56</u>, 1–25.

Whittaker, R.H. and Woodwell, G.M. (1969). Structure, production and diversity of the oak-pine forest at Brookhaven, New York. *J. Ecol.* <u>57</u>, 155–174.

Woodwell, G.M., Whittaker, R.H. and Houghton, R.A. (1975). Nutrient concentrations in plants in the Brookhaven oak-pine forest. *Ecology* <u>56</u>, 318–332.

40°40'N 73°10'W   25 m   U.S.A., New York, Long Island, Brookhaven.

Sandy, well-drained, podzolic soils	*Quercus coccinea* (31%)[a], *Quercus alba* (18%)[a], *Pinus rigida* (18%)[a], *Quercus velutina* (3%)[a], et al. with understorey shrubs.

Age (years)	43^b
Trees/ha	1854
Tree height (m)	7.6^b
Basal area (m²/ha)	15.6
Leaf area index	$3.4^c + 0.4^d$
Stem volume (m³/ha)	69.9

Dry biomass (t/ha)

Stem wood	$35.1 + 0.5^d$
Stem bark	$8.1 + 0.2^d$
Branches	$14.7 + 0.4^d$
Fruits etc.	$0.2 + 0.0^d$
Foliage	$4.1 + 0.3^d$
Root estimate	$33.3 + 3.0^d$

CAI (m³/ha/yr)	1.59

Net production (t/ha/yr)

Stem wood	$1.49 + 0.05^d$ ⎫
Stem bark	$0.27 + 0.01^d$ ⎬ $+ 0.60^e$
Branches	$2.47 + 0.21^d$ ⎭
Fruits etc.	$(0.22 + 0.02^d)^f$
Foliage	$(3.51 + 0.31^d)^g$
Root estimate	$2.60 + 0.73^d$

Fifteen trees were sampled of each major species and roots were excavated. Stand values for several 0.1 ha plots were derived from regressions on D, stem volumes and surface areas, and from other relationships. All trees and shrubs over 1 cm D were included. Roots were assumed to grow at the same relative rates as the above-ground parts. Nutrient contents were determined.

a. Percentage of the total stem volume.
b. Weighted mean age and height (*Quercus* spp. were 38–45 years old, *P. rigida* was 84–100 years). c. Including all-sided LAI of *P. rigida*. d. Understorey shrubs.
e. Mean woody litterfall measured over 3 years; the increment values include estimated mortality. f. Including litterfall of 0.16 t/ha/yr of seeds etc. and miscellaneous matter. g. Including foliage litterfall of 2.61 t/ha/yr.

Reiners, W.A. (1972). Structure and energetics of three Minnesota forests. *Ecol. Monogr.* 42, 71-94.

Reiners, W.A. and Reiners, N.M. (1970). Energy and nutrient dynamics of forest floors in three Minnesota forests. *J. Ecol.* 58, 497-519.

Whittaker, R.H. and Niering, W.A. (1975). Vegetation of the Santa Catalina Mountains, Arizona. V Biomass, production and diversity along the elevation gradient. *Ecology* 56, 771-790.

U.S.A.	45°30'N 193°20'W 400 m Minnesota, N of Minneapolis	ca.32°20'N 110°50'W 1310 m Arizona, Santa Catalina Mtns
	Quercus ellipsoidalis (75%)[a] *Acer rubrum* (8%)[a] *et al.*	*Quercus oblongifolia* (56%)[b] *Quercus emoryi* (44%)[b] with understorey shrubs.
	(Reiners 1972)	(Whittaker and Niering 1975)
Age (years)	45-50	117[c]
Trees/ha	1788	190
Tree height (m)	ca.15	5.3[c]
Basal area (m²/ha)	26.5	4.0
Leaf area index		1.8
Stem volume (m³/ha)		10.7[d]

Dry biomass (t/ha)		
Stem wood	92.4	$\Big\}$ 6.1
Stem bark	12.2	
Branches	16.2	2.9
Fruits etc.	0.0	
Foliage	3.4	0.4 + 1.5[e]
Root estimate		

CAI (m³/ha/yr)		6.3[d]
Net production (t/ha/yr)		
Stem wood	3.56	0.14[g] $\Big\}$ + 0.01[eg]
Stem bark	0.48	0.03[g]
Branches	1.22 + 1.12[f]	0.22 + 0.02[e]
Fruits etc.	0.01	0.05 + 0.12[e]
Foliage	3.44	0.28 + 0.33[e]
Root estimate		

Reiners (1972) derived stand biomass values for sixteen 100 m² plots from regressions on D and from the stem volumes of sample trees; increment values refer to the previous year's growth.

Whittaker and Niering (1975) sampled 10-15 trees per species and derived stand biomass values for a 0.1 ha plot from regressions on D, wood volumes and surface areas, and from other relationships; all trees over 1 cm D were included.

a. Percentage of the total biomass.   b. Percentage of the total stem volume.
c. Weighted mean age and height.   d. Parabolic volumes.
e. Understorey shrubs.
f. Woody litterfall; total litterfall was 4.57 t/ha/yr (Reiners and Reiners 1970).
g. Excluding woody litterfall and any mortality.

Whittaker, R.H. (1963). Net production of heath balds and forest heaths in the Great Smoky Mountains. *Ecology* 46, 176-182.

Whittaker, R.H. (1966). Forest dimensions and production in the Great Smoky Mountains. *Ecology* 47, 103-121.

35°28-47'N ca.84°W	(alt. given below)	U.S.A., Great Smoky Mountains.	
	970 m	820 m	820 m
	Mt LeConte	Greenbrier Cove	Cherokee Orchard
	Quercus prinus (19%)[a] *Nyssa sylvatica* (15%)[a] *et al.* with understorey shrubs	*Q. prinus* (56%)[a] *Quercus borealis* (12%)[a] *Acer rubrum* (18%)[a] *et al.*	*Q. prinus* (17%)[a] *Q. borealis* (16%)[a] *Liriodendron tulipifera* (18%)[a] *et al.*
Age (years)		Mature	Mature
Trees/ha	1410[b] + 17290[c]	2130[b]	2240[b]
Tree height (m)	10 3.5[c]	30[d]	21[d]
Basal area (m²/ha)	10.3 + 9.6[c]	35.6	26.9
Leaf area index	3.0[c]	6.3	
Stem volume (m³/ha)	44.1[e] + 15.9[ce]	402[e]	203[e]

Dry biomass (t/ha)			
Stem wood			
Stem bark			
Branches	} 40.0 + 24.3[c]	} 415.8	} 166.4
Fruits etc.			
Foliage		4.2	3.6
Root estimate			

CAI (m³/ha/yr)	0.6[e] + 0.6[ce]	6.2[e]	9.0[e]

Net production (t/ha/yr)			
Stem wood			
Stem bark			
Branches	} 2.2[f] + 3.2[cf]	} 14.0[f]	} 19.0[f]
Fruits etc.			
Foliage		4.2	3.6
Root estimate			

Stand values for plots of at least 0.1 ha were estimated from the weights of clippings of current year's twigs, from published regressions, from stem volumes, branch/stem biomass ratios and other relationships.
a. Percentage of the total volume increment.
b. Stems over 1.9 cm D.
c. Understorey shrubs.
d. Canopy height.
e. Parabolic volumes.
f. Excluding woody litterfall and mortality; foliage production of trees and shrubs at Mt LeConte (left column) was 2.3 t/ha/yr.

Harris, W.F., Goldstein, R.A. and Henderson, G.S. (1973). Analysis of forest biomass pools, annual primary production and turnover of biomass for a mixed deciduous forest watershed. In: "IUFRO Biomass Studies", p.43-64. College of Life
Sciences and Agriculture, University of Maine, Orono, U.S.A.

Harris, W.F. and Henderson, G.S. (1981). In: "Dynamic Properties of Forest Ecosystems", (D.E. Reichle, ed.), pp. 658-661. Cambridge University Press,
Cambridge, London, New York and Melbourne.

35°58'N 84°17'W   265-360 m   U.S.A., Tennessee, Walker Branch Sites.

Red-yellow podzols pH 4.0-6.5. Ultisols.	*Quercus prinus,* with *Carya* spp., *Quercus alba, Quercus rubra, Quercus velutina, et al.*	*Carya* spp., *Q. alba,* *Q. prinus,* *Q. velutina, et al.*
Age (years)	30-80	30-80
Trees/ha		
Tree height (m)	12-25	25
Basal area (m²/ha)	25.8	20
Leaf area index		
Stem volume (m³/ha)		

Dry biomass (t/ha)

Stem wood	}102.9	}90.5
Stem bark		
Branches	30.3	26.9
Fruits etc.		
Foliage	4.7	4.2
Root estimate	38.0[a]	32.2[a]

CAI (m³/ha/yr)

Net production (t/ha/yr)

Stem wood	}5.40 + 0.55[b] + 0.55[c]	}4.83 + 1.18[b] + 0.70[c]
Stem bark		
Branches		
Fruits etc.		
Foliage	4.70[d]	4.20[d]
Root estimate		

Many trees were sampled per site (about 150 at all four Walker Branch sites), roots
were excavated in 2 or 3 pits per site, and stand values for 50-100 circular plots
per site were derived from regressions on D. There was 1.2 t/ha of standing dead
wood in the right column.

a. Including stumps which represented about half the 'root' biomass.

b. Mortality.

c. Woody litterfall.

d. Leaf biomass; leaf litterfall was 3.9 and 4.1 t/ha/yr in columns left and right,
respectively.

Day, F.P. and Monk, C.D. (1977a). Net primary production and phenology on a southern Appalachian watershed. *Am. J. Bot.* <u>64</u>, 1117-1125.

Day, F.P. and Monk, C.D. (1974). Vegetation patterns on a southern Appalachian watershed. *Ecology* <u>55</u>, 1064-1074.

Day, F.P. and Monk, C.D. (1977b). Seasonal nutrient dynamics in the vegetation on a southern Appalachian watershed. *Am. J. Bot.* <u>64</u>, 1126-1139.

35°03'N 83°26'W   726-993 m   U.S.A., North Carolina, Nantahala Mountains.

Porter loams, pH 5.2 on 53° slope.

Quercus prinus (21%)[a], *Acer rubrum* (9%)[a], *Quercus coccinea* (8%)[a], *et al.*, with understorey shrubs including *Rhododendron* spp.

Age (years)		60-200
Trees/ha		3044
Tree height (m)		25
Basal area (m²/ha)		25.6
Leaf area index		
Stem volume (m³/ha)		

Dry biomass (t/ha)	Stem wood	$\}$ 97.1 + 11.0[b]
	Stem bark	
	Branches	22.1 + 4.2[b]
	Fruits etc.	0.4
	Foliage	3.5 + 2.1[b]
	Root estimate	30.7

CAI (m³/ha/yr)

Net production (t/ha/yr)	Stem wood	$\}$ 3.77 + 1.34[c]
	Stem bark	
	Branches	
	Fruits etc.	0.42[c]
	Foliage	4.19[d]
	Root estimate	

Stand biomass values for twenty-five 0.125 ha plots were derived from published regressions on D. There was 9.6 t/ha of standing dead trees. Nutrient contents were determined.

a. Percentage of the total basal area (*Quercus* species accounted for 43%).
b. Understorey shrubs; shrubs are included in the production values.
c. Litterfall.
d. New foliage biomass; leaf litterfall was 3.20 t/ha/yr.

Johnson, F.L. and Risser, P.G. (1974). Biomass, annual net primary production, and dynamics of six mineral elements in a Post oak - Blackjack oak forest. *Ecology* 55, 1246-1258.

35°15'N 97°20'W   ca.250 m   U.S.A., Oklahoma, 17 km E of Norman, Lake Thunderbird.

Sandy,   *Quercus stellata* and *Quercus marilandica* (98.6%)[a],
red-yellow
podzols   with understorey shrubs.

Age (years)	to 80
Trees/ha	2600[b]
Tree height (m)	
Basal area (m²/ha)	18.3
Leaf area index	4.8[c]
Stem volume (m³/ha)	

Dry biomass (t/ha)		
Stem wood	} 109.5	
Stem bark		
Branches	64.9	} + 1.4[d]
Fruits etc.		
Foliage	4.8	
Root estimate	39.0	

CAI (m³/ha/yr)

Net production (t/ha/yr)		
Stem wood	} 3.69	
Stem bark		
Branches	3.64 + 0.53[e]	} + 0.30[d]
Fruits etc.		
Foliage	4.76	
Root estimate	2.24	

Five trees of each of the two *Quercus* species were sampled in June and roots were core sampled. Stand values for two 100 m² plots were derived from regressions on D. Roots were assumed to grow at the same relative rates as above-ground parts. There was 14.3 t/ha of dead branches. Nutrient contents were determined.
a. Percentage of the total basal area accounted for by the two *Quercus* species.
b. Including 1420 *Quercus* trees.
c. Peak value, attained in June.
d. Understorey shrubs.
e. Woody litterfall measured over one year.

Monk, C.D., Child, G.I. and Nicholson, S.A. (1970). Biomass, litter and leaf surface area estimates of an oak - hickory forest. *Oikos* 21, 138-141.

Rolfe, G.L., Akhtar, M.A. and Arnold, L.E. (1978). Nutrient distribution and flux in a mature oak - hickory forest. *Forest Sci.* 24, 122-130.

U.S.A.	33°57'N 83°24'W ca.300 m Georgia, Athens. *Quercus* spp., *Carya pallida*, et al. (Monk *et al.* 1970)	37°30'N 88-89°W 200-400 m Illinois, Shawnee Hills, Pope County. *Quercus* spp. (85%)[d], *Acer saccharum*, *Carya glabra*, et al. Silty loams	
		Mesic site	Xeric site
Age (years)	>50	150[e]	150[e]
Trees/ha		2130[f]	1905[f]
Tree height (m)			
Basal area (m²/ha)			
Leaf area index	ca.4		
Stem volume (m³/ha)			
Dry biomass (t/ha) — Stem wood	$137.2 + 0.2^a$	130.6	127.3
Dry biomass (t/ha) — Stem bark			
Dry biomass (t/ha) — Branches		53.6	51.9
Dry biomass (t/ha) — Fruits etc.			
Dry biomass (t/ha) — Foliage	$4.4 + 0.03^a$	5.6	5.2
Dry biomass (t/ha) — Root estimate		48.8	46.8
CAI (m³/ha/yr)			
Net production (t/ha/yr) — Stem wood	$>0.8^b$	7.1[g]	6.2[g]
Net production (t/ha/yr) — Stem bark			
Net production (t/ha/yr) — Branches			
Net production (t/ha/yr) — Fruits etc.			
Net production (t/ha/yr) — Foliage	4.6^c		
Net production (t/ha/yr) — Root estimate			

Monk *et al.* (1970) sampled 23 trees and estimated stand values for the tree layer in twenty 10 m² plots from regressions on D.
Rolfe *et al.* (1978) sampled 11 trees (in Pope County), excavated 4 root systems, and derived stand values for three 400 m² plots per site from regressions on D; nutrient contents were determined.
a. Understorey shrubs.
b. Stem increment plus new twigs, excluding the increment of large branches and any woody litterfall.
c. New foliage biomass plus estimated consumption.
d. Percentage of the total biomass.   *e.* Average age of *Quercus* dominants.
f. Trees over 3.8 cm D.   *g.* Total litterfall only, measured over 2½ years.

Whittaker, R.H. (1963). Net production of heath balds and forest heaths in the Great Smoky Mountains. *Ecology* 44, 176–182.

Whittaker, R.H. (1962). Net production relations of shrubs in the Great Smoky Mountains. *Ecology* 43, 357–377.

ca.35°40'N 83°30'W   (alt. given below) U.S.A., Tennessee, Great Smoky Mountains, Mount LeConte.

Leached acid podzols with a low nutrient status, on 15-40° slopes.	*Rhododendron carolinianum, Rhododendron catawbiense, Kalmia latifolia, Vaccinium constablaei, Pyrus melanocarpa, et al.*						
	2010 m^a	2010 m^b	1500 m	1430 m	1500 m	1560 m	1490 m
Age (years)							Mature
Trees/ha							42700
Tree height (m)	1.3 ^{c}	2.5^c	0.7^c	1.5^c	1.8^c	1.7^c	3.8^d
Basal area (m²/ha)							35.0
Leaf area index	2.0	2.8	2.0	2.6	4.5	4.8	6.1
Stem volume (m³/ha)							60.1^e
Dry biomass (t/ha): Stem wood, Stem bark, Branches, Fruits etc., Foliage, Root estimate	17.6	25.0	8.3	19.8	27.7	37.2	107.8
CAI (m³/ha/yr)							2.3^e
Net production (t/ha/yr): Stem wood, Stem bark, Branches, Fruits etc., Foliage, Root estimate	4.1^f	4.9^f	2.3^f	3.8^f	5.9^f	6.3^f	9.8^f

Stand values were estimated for ten 1 m² plots per stand from the weight of clippings of current year's twigs, from published regressions, from stem volumes, branch/stem biomass ratios, and other relationships.
a. Values in this column refer to a plot of almost pure *R. carolinianum*.
b. Values in this column refer to a plot of almost pure *R. catawbiense*.
c. Canopy heights.
d. Weighted mean height.
e. Parabolic volume.
f. Excluding woody litterfall and mortality.

Schlesinger, W.H. (1978). Community structure, dynamic and nutrient cycling in the Okefenokee cypress swamp forest. *Ecol. Monogr.* 48, 43-65.

30°31'-31°08'N 82°08-38'W   32-40 m   U.S.A., Georgia, Okefenokee.

Peat bog.

Taxodium distichum var. *nutans* (81%)[a],

Ilex cassine, *Nyssa sylvatica* var. *biflora*,

with understorey shrubs.

Age (years)	
Trees/ha	$1430 + 500^{b}$
Tree height (m)	13.9
Basal area (m²/ha)	$51.9 + 12.3^{b}$
Leaf area index	$1.1^{c} + 0.3^{cb}$
Stem volume (m³/ha)	

Dry biomass (t/ha)

Stem wood	253.3
Stem bark	36.7
Branches	8.2 $\Big\} + 5.3^{b}$
Fruits etc.	0.1
Foliage	2.3^{d}
Root estimate	

CAI (m³/ha/yr)

Net production (t/ha/yr)

Stem wood	2.32
Stem bark	0.35 $\Big\} + 0.43^{e}$
Branches	0.63 $\Big\} + 1.06^{b} + 0.10^{be}$
Fruits etc.	$0.07 + 0.34^{e}$
Foliage	2.33^{de}
Root estimate	

Twenty-three trees were sampled in July and stand biomass values for two 0.1 ha plots were derived from regressions on D. There was 49 t/ha of standing dead trees, 2.7 t/ha of dead branches and 0.7 t/ha of *Tillandsia usneoides* (Spanish moss), none of which is included above. Nutrient contents were determined.

a. Percentage of the total basal area.
b. Understorey shrubs.
c. All-sided LAI values were 2.4 for trees and 0.5 for shrubs.
d. Including current year's twigs.
e. Litterfall, measured over one year; there was also 0.08 t/ha/yr of litterfall from *T. usneoides*.

Connor, W.H. and Day, J.W. (1976). Productivity and composition of a baldcypress water tulepo site and a bottomland hardwood site in a Louisiana swamp. *Am. J. Bot.* <u>63</u>, 1354-1364.

29°45'N   90°30'W   1 to 2 m   U.S.A., Louisiana, near New Orleans, Barataria Bay.

Swamp.	*Taxodium distichum* (46%)[a], *Nyssa aquatica* (48%)[a], et al.	*N. aquatica* (35%)[a], *Populus heterophylla* (14%)[a], *Acer rubrum* (10%)[a], *T. distichum, et al.*
	Soft clay and peat.	**Organic matter over grey clay.**
Age (years)	50-95	< 30
Trees/ha	1730	2970
Tree height (m)		
Basal area (m²/ha)		
Leaf area index		
Stem volume (m³/ha)		

Dry biomass (t/ha)			
Stem wood			
Stem bark		} 372.0	} 165.3
Branches			
Fruits etc.			
Foliage			
Root estimate			

CAI (m³/ha/yr)

Net production (t/ha/yr)			
Stem wood			
Stem bark		} 5.00	} 8.00
Branches			
Fruits etc.		} 6.20[b]	} 5.74[b]
Foliage			
Root estimate			

Stand biomass values for the above two 0.1 ha plots were derived from regressions on D published by Monk *et al.* (1970) (see p.285).
a. Percentage of the total basal area.
b. Total litterfall measured over one year.

Grier, C.C., Vogt, K.A., Keyes, M.R. and Edmonds, R.L. (1981). Biomass distribution
and above- and below-ground production in young and mature *Abies amabilis* zone
ecosystems of the Washington Cascades. *Can. J. For. Res.* 11, 155-167.
Grier, C.C. and Milne, W.A. (1981). Regression equations for calculating component
biomass of young *Abies amabilis* (Dougl.) Forbes. *Can. J. For. Res.* 11, 184-187.
Gholz, H.L., Grier, C.C., Campbell, A.G. and Brown, A.T. (1979). "Equations for
Estimating Biomass and Leaf Area of Plants in the Pacific Northwest." Forest
Research Laboratory, Oregon State University, Corvallis, USA. Research Paper 41.

47°19'N 121°35'W   1140 m   U.S.A., Washington, Findley Lake.

Sandy to gravelly clay loams overlying morainal deposits.

Abies amabilis with some *Abies procera*, *Tsuga heterophylla*, *Tsuga mertensiana* and understorey shrubs.

Age (years)	23	180
Trees/ha	110500	510
Tree height (m)	1.4 (0.3 to 2.6)	22.1 (3.1 to 39.0)
Basal area (m² /ha)	45.7	74.3
Leaf area index		
Stem volume (m³ /ha)		

Dry biomass (t/ha)

Stem wood	25.0 ⎫	293.9
Stem bark	2.7 ⎬ + 2.2^a	62.2
Branches	7.8 ⎭	67.8
Fruits etc.		
Foliage	13.6 + 0.3^a	21.7
Root estimate	24.7	137.7

CAI (m³ /ha/yr)

Net production (t/ha/yr)

Stem wood	2.85 ⎫	1.58
Stem bark	0.29 ⎬ + 0.30^b + 0.06^a	0.36
Branches	0.90 ⎭ + 0.47^c	0.38 + 1.15^c
Fruits etc.		
Foliage	0.22 + 1.04^c	0.00 + 1.03^c
Root estimate	1.78 + 10.04^d	0.70 + 11.53^d

Fifty-six trees were sampled from the 23-year-old stand in November, and stand
biomass values were derived from regressions on D. Regressions published by Gholz
et al. (1979) were used to estimate the biomass of the older stand and the biomass
of the thick roots. Roots over 5 mm diameter weighed 15.5 and 124.9 t/ha in columns
left and right, respectively. There was 2.0 and 7.9 t/ha of dead branches, and 60.5
and 157.0 t/ha of standing dead trees in columns left and right, respectively.
a. Understorey shrubs.
b. Mortality, estimated only in this 23-year-old stand.
c. Litterfall.
d. Estimated root turnover of trees and shrubs.

Turner, J. and Singer, M.J. (1976). Nutrient distribution and cycling in a sub-alpine coniferous forest ecosystem. *J. appl. Ecol.* <u>13</u>, 295-301.

Fujimori, T., Kawanabe, S., Saito, H., Grier, C.C. and Shidei, T. (1976). Biomass and primary production in forests of three major vegetation zones of the north-western United States. *J. Jap. For. Soc.* <u>58</u>, 360-373.

U.S.A.	47°52'N 123°00'W 1200 m Washington, Findley Lake. *Abies amabilis* (93%)[a], *Tsuga mertensiana* with understorey shrubs. Unthinned. Coarse sandy loam. (Turner and Singer 1976)	44°00'N 122°40'W 1300 m Oregon, Wildcat Mtn Reserve. *Abies procera* (56%)[a], *Pseudotsuga menziesii* (40%)[a], *Abies amabilis* (4%)[a]. Brown podzolic sandy loam. (Fujimori *et al.* 1976)
Age (years)	up to 170	100–130
Trees/ha	620	350
Tree height (m)		49.9[b]
Basal area (m²/ha)	96.7	98.1
Leaf area index		
Stem volume (m³/ha)		1989
Dry biomass (t/ha) — Stem wood	265.0 ⎫	683.1
Stem bark	38.7 ⎪	111.9
Branches	17.7 ⎬ + 17.7[c]	67.8
Fruits etc.	⎪	
Foliage	15.7 ⎭	17.5
Root estimate		
CAI (m³/ha/yr)		19.3
Net production (t/ha/yr) — Stem wood		6.7[d]
Stem bark		1.0[d]
Branches	⎫ 0.94[f]	2.0[d] (or 5.6)[de]
Fruits etc.	⎭	
Foliage	2.08[f]	3.2
Root estimate		

Turner and Singer (1976) derived stand values for a 450 m² plot from regressions calculated by Dice (1970) (see p.329), and from a sample of 6 trees; there were 490 dead trees per hectare weighing 127.7 t/ha; nutrient contents were determined. Fujimori *et al.* (1976) derived stand values for a 0.34 ha plot from regressions on D²H for the main species, and by proportional basal area allocation for the minor species; foliage litterfall was similar to the foliage increment given above.
a. Percentage of the total basal area.
b. Mean height of the dominant trees.   *c.* Understorey shrubs.
d. Excluding woody litterfall and mortality.
e. Alternative value derived by branch ring analysis.
f. Litterfall only.

Whittaker, R.H. (1966). Forest dimensions and production in the Great Smoky Mountains. *Ecology* <u>47</u>, 103-121.

35°28-47'N ca.84°W   (alt. given below)   U.S.A., Great Smoky Mountains.

Abies fraseri, Picea rubens, et al.

	$(69\% \ 28\%)^a$	$(67\% \ 31\%)^a$	$(96\% \ 3\%)^a$	$(77\% \ 18\%)^a$
	1620 m	1620 m	1920 m	1900 m
Age (years)	Mature	45-55	Mature	Mature
Trees/ha	1410^b	1470^b	710^b	5580^b
Tree height (m)	26^c	21^c	17^c	10^c
Basal area (m²/ha)	50.2	59.7	40.0	56.3
Leaf area index			5.3^d	
Stem volume (m³/ha)	488^e	472^e	276^e	209^e

Dry biomass (t/ha)				
Stem wood				
Stem bark				
Branches	} 310	} 300	} 210	} 200
Fruits etc.				
Foliage				
Root estimate				

	5.3^e	8.5^e	2.4^e	2.7^e
CAI (m³/ha/yr)				

Net production (t/ha/yr)				
Stem wood				
Stem bark	} 6.4^f	} 10.6^f	} 2.9^f	} 4.1^f
Branches				
Fruits etc.				
Foliage	2.8	3.4	1.8	2.4
Root estimate				

Stand biomass values for plots of at least 0.1 ha were derived from published regressions on D, from stem volumes, branch/stem biomass ratios, and other relationships.

a. Percentage of the total volume increment accounted for by *A. fraseri* and *P. rubens*, written left and right, respectively, within the brackets.
b. Stems over 1.9 cm D.
c. Weighted mean heights.
d. All-sided LAI was 12.3.
e. Parabolic volumes.
f. Excluding woody litterfall and mortality.

Whittaker, R.H. and Niering, W.A. (1975). Vegetation of the Santa Catalina Mountains, Arizona. V Biomass, production and diversity along the elevation gradient. *Ecology* 56, 771–790.

Whittaker, R.H. and Niering, W.A. (1968). Vegetation of the Santa Catalina Mountains, Arizona. IV Limestone and acid soils. *J. Ecol.* 56, 523–544.

ca.32°20'N 110°50'W   (alt. given below)   U.S.A., Arizona, Santa Catalina Mountains, near Tucson.

	N facing slope of Mt Lemmon. *Abies lasiocarpa* (85%)[a], *Pseudotsuga menziesii* (10%)[a], et al. with understorey of *Jamesia americana*, et al.	Marshall Gulch. *Abies concolor* (55%)[a], *P. menziesii* (20%)[a], *Pinus strobiformis*, et al.
	2720 m	2340 m
Age (years)	106[b]	124[b] (50–145)
Trees/ha	590 160[c]	1510
Tree height (m)	33.5[b]	25.5[b]
Basal area (m²/ha)	57.8 + 0.3[c]	58.6
Leaf area index	6.4[d]	6.7[d]
Stem volume (m³/ha)	837[e]	746[e]

Dry biomass (t/ha)		
Stem wood	} 290 + 0.40[c]	} 270
Stem bark		
Branches	50 + 0.20[c]	74
Fruits etc.		
Foliage	16.2 + 0.04[c]	16.7
Root estimate		

CAI (m³/ha/yr)	5.7[e]	5.6[e]

Net production (t/ha/yr)		
Stem wood	2.75[f] } + 0.02[cf]	3.95[f]
Stem bark	0.48[f] }	0.85[f]
Branches	1.30[f] + 0.02[cf]	1.64[f]
Fruits etc.	0.42 + 0.00[c]	0.54
Foliage	3.65 + 0.04[c]	4.12
Root estimate		

Ten to fifteen trees were sampled of each of the major species, and stand values for the above two 0.1 ha plots were derived from regressions on D, wood volumes and surface areas, and from other relationships. All trees and shrubs over 1 cm D were included.

a. Percentage of the total stem volume.
b. Weighted mean ages and heights.
c. Understorey shrubs.
d. All-sided LAI values were 14.7 and 15.5 in columns left and right, respectively.
e. Parabolic volumes.
f. Excluding woody litterfall and mortality.

Westman, W.E. and Whittaker, R.H. (1975). The pygmy forest region of northern California: studies on biomass and primary productivity. *J. Ecol.* <u>63</u>, 493-520.

Jenny, H., Arkley, R.J. and Schultz, A.M. (1969). The pygmy forest podsol ecosystem and its dune associates of the Mendocino coast. *Madroño* <u>20</u>, 60-74.

---

ca.39°20'N 123°45'W   146 m   U.S.A., California, Mendocino, near Fort Bragg.

Podzols.   *Cupressus pygmaea* (68%)[a], *Vaccinium ovatum* (17%)[a], *Pinus contorta* subsp. *bolanderi* (9%)[a].

Age (years)		up to 136
Trees/ha		>150000
Tree height (m)		1.2^b 2.6^b
Basal area (m²/ha)		1.7^c
Leaf area index		2.1^d
Stem volume (m³/ha)		2.0^e
	Stem wood	12.1
Dry biomass (t/ha)	Stem bark	7.0
	Branches	2.8
	Fruits etc.	1.4
	Foliage	3.3^f
	Root estimate	8.3
CAI (m³/ha/yr)		0.07^e
	Stem wood	0.22^g
Net production (t/ha/yr)	Stem bark	0.11^g
	Branches	1.02 ⎫
	Fruits etc.	0.53 ⎬ *h*
	Foliage	1.19^f ⎭
	Root estimate	0.96

Extensive sampling was done and roots were excavated. Stand values were derived from regressions on various dimensions following Whittaker *et al.* (1974) (see p.259). Values given above are the means of five 100 m² plots.
a. Percentage of the total biomass; *C. pygmaea* is also known as *C. goveniana* var. *pigmaea.*
b. Weighted mean heights of *C. pygmaea* (1.2 m) and *P. contorta* (2.6 m).
c. 'Basal area' of *P. contorta* only, measured 10 cm from the ground.
d. Including all-sided foliage area of *P. contorta.*
e. Parabolic volume.   *f.* Including the current year's twigs.
g. Excluding mortality.
h. Including total litterfall of 0.60 t/ha/yr.

294   U.S.A.   *Juniperus*

Gholz, H.L. (1981). Environmental limits on aboveground net primary production, leaf
   area and biomass in vegetation zones of the Pacific Northwest. *Ecology* (in press).
Gholz, H.L., Grier, C.C., Campbell, A.G. and Brown, A.T. (1979). "Equations for
   Estimating Biomass and Leaf Area of Plants in the Pacific Northwest." Forest Research Laboratory, Oregon State University, Corvallis, USA. Research Paper No.41.
Gholz, H.L., Fitz, F. and Waring, R.H. (1976). Leaf area differences associated with
   old-growth forest communities in the western Oregon Cascades. *Can. J. For. Res.*
   6, 49-57.

44-45°N   ca.121°W   1356 m   U.S.A., Oregon, Cascade Mountains.

High lava plain.

Juniperus occidentalis

Age (years)	up to 350
Trees/ha	199[a]
Tree height (m)	1
Basal area (m²/ha)	27.8
Leaf area index	0.9[b]
Stem volume (m³/ha)	

Dry biomass (t/ha)

Stem wood	} 9
Stem bark	
Branches	7
Fruits etc.	
Foliage	4
Root estimate	

CAI (m³/ha/yr)

Net production (t/ha/yr)

Stem wood	} 0.2[c]
Stem bark	
Branches	ca.0.1[c]
Fruits etc.	
Foliage	1.0
Root estimate	

Stand biomass values for a plot of over 0.25 ha were derived from regressions on D.
a. Trees over 10 cm D; there were 470 trees/ha less than 10 cm D.
b. All-sided LAI was 2.0.
c. Excluding woody litterfall and mortality.

Landis, T.D. and Mogren, E.W. (1975). Tree strata biomass of subalpine spruce-fir stands in southwestern Colorado. *Forest Sci.* <u>21</u>, 9-12.

---

ca.37°30'N   107°W   3100-3500 m   U.S.A., Colorado, San Juan National Forest.

Picea engelmannii (85%)[a] and *Abies lasiocarpa*.

	Near Durango		Wolf Creek Pass	
Age (years)	15-250	15-250	15-250	15-250
Trees/ha				
Tree height (m)				
Basal area (m²/ha)				
Leaf area index				
Stem volume (m³/ha)				
Dry biomass (t/ha) Stem wood	120	135	95	120
Stem bark	30	30	25	25
Branches	28	30	20	28
Fruits etc.				
Foliage	18	18	15	17
Root estimate				
CAI (m³/ha/yr)				
Net production (t/ha/yr) Stem wood				
Stem bark				
Branches				
Fruits etc.				
Foliage				
Root estimate				

A total of 29 *P. engelmanii* were sampled over two years, and stand values for the above four 800 m² plots were derived from regressions on D.
a. Approximate percentage of the total biomass.

Singer, F.P. and Hutnik, R.J. (1966). Accumulation of organic matter in red pine and Norway spruce plantations of various spacings. *Penn. State Univ. Res. Briefs* 1, 22-28.

Alban, D.H., Perala, D.A. and Schlaegel, B.E. (1978). Biomass and nutrient distribution in aspen, pine and spruce stands on the same soil type in Minnesota. *Can. J. For. Res.* 8, 290-299.

U.S.A. Plantations.	ca.41°N 78°W -- Central Pennsylvania *Picea abies* (Singer and Hutnik 1966)			47°20'N 94°30'W 400 m Minnesota Pike Bay Expt. Forest *Picea glauca* Fine sandy loams. (Alban *et al.* 1978)
Age (years)	42	42	42	40
Trees/ha	1076	2242	2990	2187
Tree height (m)				14.4
Basal area (m²/ha)				41.1
Leaf area index				
Stem volume (m³/ha)				256[a]
Dry biomass (t/ha) Stem wood	}106.4	}139.2	}135.2	88.6 ⎫
Stem bark				11.1
Branches	36.9	39.3	27.9	35.1 ⎬ + 0.1[b]
Fruits etc.				
Foliage	15.5	18.4	15.8	17.9 ⎭
Root estimate				34
CAI (m³/ha/yr)				
Net production (t/ha/yr) Stem wood				
Stem bark				
Branches				
Fruits etc.				
Foliage				
Root estimate				

Singer and Hutnik (1966) sampled 9 trees and derived stand values for the above three 400 m² plots from regressions on D and H; there was 0.0, 6.2 and 3.7 t/ha of dead branches in columns left to right.
Alban *et al.* (1978) sampled 10 trees in spring and excavated the roots of two further trees in August; stand values for ten 80 m² plots were derived from regressions on D²H; nutrient contents were determined.
a. Volume inside bark.
b. Understorey shrubs.

Cleve, K. van (1981). In: "Dynamic Properties of Forest Ecosystems" (D.E. Reichle, ed.), pp. 648-650. Cambridge University Press, Cambridge, London, New York, and Melbourne.

Barney, R.J. and Cleve, K. van (1973). Black spruce fuel weights and biomass in two interior Alaska stands. *Can. J. For. Res.* 3, 304-311.

64°45'N   148°15'W   (alt. given below)   U.S.A., Alaska, near Fairbanks, Bonanza Creek Experimental Forest.

Picea mariana with understorey shrubs.

	-- 'Moss' site, pH 5.4-6.4.	167 m Poorly drained alluvial muskeg with permafrost, pH 5.3.	470 m Silt loam muskeg, pH 4.6-5.6.
Age (years)	130	51	55
Trees/ha	5000	27335	14820
Tree height (m)	13.7	2.9	3.1
Basal area (m²/ha)	34.7	18.2	22.0
Leaf area index			
Stem volume (m³/ha)			

Dry biomass (t/ha)

Stem wood	$\Big\} 86.1$	$\Big\} 8.00$	$\Big\} 14.02$
Stem bark			
Branches	$13.0 \Big\} + 4.7^a$	$3.59 \Big\} + 0.71^a$	$4.69 \Big\} + 0.95^a$
Fruits etc.	5.3	1.25	0.72
Foliage	$8.9 + 0.4^a$	$3.76 + 0.45^a$	$4.59 + 0.54^a$
Root estimate	51.7	12.5	10.4

CAI (m³/ha/yr)

Net production (t/ha/yr)

Stem wood	$\Big\} 1.18 + 0.07^b$	$\Big\} 0.29$	$\Big\} 0.70$
Stem bark			
Branches	$0.26 + 0.17^b$	0.11	0.22
Fruits etc.	$0.07 + 0.07^b$	$0.02 \Big\}^c$	$0.02 \Big\}^c$
Foliage	$0.07 + 0.23^b$	0.09	0.17
Root estimate			

The fresh weights of all trees were measured within three 40 m² plots in each stand, and dry weights were estimated from the water contents of subsamples. There was 31.1, 4.3 and 2.6 t/ha of standing dead wood in columns left to right, according to van Cleve (1981) but only 2.2 and 0.6 t/ha in the centre and right columns, respectively, according to Barney and van Cleve (1973).
a. Understorey shrubs.
b. Litterfall.
c. Excluding woody, foliage and all other litterfall.

Whittaker, R.H. (1963). Net production of heath balds and forest heaths in the Great Smoky Mountains. *Ecology* <u>46</u>, 176-182.

Whittaker, R.H. (1966). Forest dimensions and production in the Great Smoky Mountains. *Ecology* <u>47</u>, 103-121.

Weaver, G.T. and DeSelm, H.R. (1973). Biomass distribution patterns in adjacent coniferous and deciduous forest ecosystems. In: "IUFRO Biomass Studies", pp.415-427. College of Life Sciences and Agriculture, University of Maine, Orono, U.S.A.

U.S.A.	ca.35°40'N 83°30'W 1740 m Tennessee, Great Smoky Mtns. *Picea rubens* (83%)[a] *et al.* with understorey of *Rhododendron catawbiense* (11%)[a], *et al.* (Whittaker 1963, 1966)	ca.35°20'N 83°00'W 1524-1954 m North Carolina, Balsam Mtns. *P. rubens*, with *Abies fraseri*, and understorey shrubs. Sandy acid loams. (Weaver and DeSelm 1973)
Age (years)		40-60
Trees/ha	$640 + 3720^b$	
Tree height (m)	$23 \quad 4.5^b \ (20.2)^c$	
Basal area (m²/ha)	$50.5 + 5.4^b$	
Leaf area index	2.1^b	
Stem volume (m³/ha)	$514^d + 10.5^{bd}$	
Dry biomass (t/ha) Stem wood		} 138^f
Stem bark		
Branches	} $300.0 + 21.0^b$	25^f
Fruits etc.		
Foliage		16^f
Root estimate		
CAI (m³/ha/yr)	$3.4^d + 0.5^{bd}$	
Net production (t/ha/yr) Stem wood		
Stem bark		
Branches	} $6.1^e + 2.0^{be}$	
Fruits etc.		
Foliage		
Root estimate		

Whittaker (1963, 1966) estimated stand values for a plot of at least 0.1 ha from the weight of clippings of current year's twigs, from published regressions, from stem volumes, branch/stem biomass ratios, and other relationships.

Weaver and DeSelm (1973) sampled over 50 trees and derived stand values for fourteen 400 m² plots from regressions on D.

a. Percentage of the total volume increment.
b. Understorey shrubs.   c. Weighted mean height.
d. Parabolic volumes.
e. Excluding woody litterfall and mortality; total foliage production of trees and shrubs was 2.6 t/ha/yr.
f. Mean of all 14 plots, and including shrubs.

Adams, W.R. (1928). Studies in tolerance of New England forest trees. VIII Effect of spacing in a jack pine plantation. *Vermont Agric. Exp. Stn Bull.* <u>282</u>, 49 pp.

Alban, D.H., Perala, D.A. and Schlaegel, B.E. (1978). Biomass and nutrient distribution in aspen, pine and spruce stands on the same soil type in Minnesota. *Can. J. For. Res.* <u>8</u>, 290-299.

U.S.A. Plantations.	ca.44°28'N 73°12'W 50-100 m Vermont, near Burlington. *Pinus banksiana* Spacing experiment on deep glacial sands. (Adams 1928)				47°20'N 94°30'W 400 m Minnesota Pike Bay Expt. Forest *P. banksiana* with understorey shrubs. Fine sandy loams. (Alban *et al.* 1978)
Age (years)	10	10	10	10	40
Trees/ha	25057	6726	2990	1680	1580
Tree height (m)	3.7	4.2	4.1	4.2	18.4
Basal area (m²/ha)	29.3	18.3	11.7	8.1	35.1
Leaf area index					
Stem volume (m³/ha)					263[a]
Dry biomass (t/ha) Stem wood	} 21.6	} 12.9	} 3.6	} 4.7	106.6
Stem bark					12.5
Branches	6.9	8.7	5.3	8.2	23.6 } + 2.9[b]
Fruits etc.					
Foliage	5.7	5.8	4.5	4.2	5.6
Root estimate	7.1	3.9	4.1	3.1	28
CAI (m³/ha/yr)					
Net production (t/ha/yr) Stem wood					
Stem bark					
Branches					
Fruits etc.					
Foliage					
Root estimate					

Adams (1928) sampled one average-sized tree, including the roots, in each spacing treatment, and obtained stand values for one 550 m² plot per treatment by multiplying mean tree values by the numbers of trees per hectare; nutrient contents were determined.
Alban *et al.* (1978) sampled 10 trees in spring, and the roots of 2 further trees were excavated in August; stand values for ten 80 m² plots were derived from regressions on D²H; nutrient contents were determined.
a. Volume inside bark.
b. Understorey shrubs.

Ohmann, L.F. and Grigal, D.F. (1979). Early revegetation and nutrient dynamics
    following the 1971 Little Sioux forest fire in northeastern Minnesota. *Forest
    Sci. Monogr.* 21, 80 pp.

ca.48°N 92°W   390–470 m   U.S.A., Minnesota, Little Sioux.

Loamy glacial sands and gravels.

Pinus banksiana (40%)[a], *Populus tremuloides* (33%)[a],
Betula papyrifera, Populus grandidentata et al.
with understorey shrubs.

Regeneration after fire in May 1971.

	1971	1972	1973	1974	1975
Age (years)	1	2	3	4	5
Trees/ha	98414	71757	56700	42500	39843
Tree height (m)					
Basal area (m²/ha)					
Leaf area index					
Stem volume (m³/ha)					

Dry biomass (t/ha)

	1971	1972	1973	1974	1975
Stem wood					
Stem bark					
Branches	$0.4 + 0.2^b$	$1.7 + 0.8^b$	$2.1 + 1.0^b$	$2.6 + 1.7^b$	$8.0 + 2.8^b$
Fruits etc.					
Foliage					
Root estimate					

CAI (m³/ha/yr)

Net production (t/ha/yr)

	1971	1972	1973	1974	1975
Stem wood					
Stem bark	0.30	1.23	0.95	0.47	5.19
Branches					
Fruits etc.					
Foliage	0.38^c	0.89^c	1.11^c	1.28^c	2.51^c
Root estimate					

All vegetation was sampled in August in ten 0.6 m² plots at seven sites in each
year. Wood production values were derived from regressions of cumulative biomass
against time. Nutrient contents were determined.
a. Mean percentage of the total tree number at seven sites over the five years.
b. Understorey shrubs; the production values include the shrubs.
c. Biomass of deciduous leaves, plus conifer needle litterfall assuming that the
   needles were retained for three years.

Green, D.C. and Grigal, D.F. (1979). Jack pine biomass accretion on shallow and deep soils in Minnesota. *Proc. Soil Sci. Soc. Am.* 43, 1233-1237.

Green, D.C. and Grigal, D.F. (1978). "Generalized Biomass Estimation Equations for Jack Pine (*Pinus banksiana* Lamb.)." Minn. For. Res. Notes 268.

Green, D.C. and Grigal, D.F. (1980). Nutrient accumulations in Jack pine st·nds on deep and shallow soils over bedrock. *Forest Sci.* 26, 325-333.

47°50'N 91°50'W   410-550 m   U.S.A., Minnesota, near Ely.

Sandy loams, pH 3.9-4.5 derived from four rock types, namely: *Pinus banksiana* (over 80%)[a] *et al.*

	till	gabbro	granite	greenstone
Age (years)	>50	>50	>50	>50
Trees/ha	>400	>400	>400	>400
Tree height (m)	14.3	9.3	10.2	11.2
Basal area (m²/ha)				
Leaf area index				
Stem volume (m³/ha)				

Dry biomass (t/ha)

	till	gabbro	granite	greenstone
Stem wood	106.5	61.7	56.0	59.5
Stem bark	11.5	7.1	6.6	6.8
Branches	19.3 + 3.0[b]	10.9	9.9	10.7
Fruits etc.				
Foliage	7.4	4.8	4.5	4.8
Root estimate				

CAI (m³/ha/yr)

Net production (t/ha/yr)

Stem wood	
Stem bark	
Branches	
Fruits etc.	
Foliage	
Root estimate	

Stand biomass values for *P. banksiana* were estimated from regressions on D, and biomass values for other species were estimated using published regressions. Nutrient contents were determined.
a. Percentage of the total basal area.
b. Tall shrubs; the other three stands had less than 1 t/ha of shrubs.

302 U.S.A.   *Pinus*

Zavitkovski, J., Jeffers, R.M., Nienstaedt, H. and Strong, T.F. (1981). Biomass
production of several Jack pine provenances at three Lake States locations. *Can.
J. For. Res.* <u>11</u>, 441-447.

U.S.A. Plantations.	45°38'N 87°59'W ca.300 m Pembine, Wisconsin *Pinus banksiana* Four provenances				48°18'N 89°10'W ca.300 m Watersmeet, Michigan. *b*	47°36'N 91°21'W ca.300 m Isabella, Minnesota. *b*
	46°48'[a]	45°48'[a]	46°12'[a]	44°12'[a]		
Age (years)	24	24	24	24	25	25
Trees/ha	3660	3703	3574	3617	2852	3069
Tree height (m)	10-12	10-12	10-12	10-12	10-12	8-11
Basal area (m²/ha)						
Leaf area index						
Stem volume (m³/ha)						
Dry biomass (t/ha) — Stem wood / Stem bark	87.0	90.6	86.8	100.2	78.3	53.0
Dry biomass (t/ha) — Branches / Fruits etc. / Foliage	21.8	22.7	22.0	26.7	21.3	19.2
Dry biomass (t/ha) — Root estimate						
CAI (m³/ha/yr)						
Net production (t/ha/yr) — Stem wood						
Net production (t/ha/yr) — Stem bark						
Net production (t/ha/yr) — Branches						
Net production (t/ha/yr) — Fruits etc.						
Net production (t/ha/yr) — Foliage						
Net production (t/ha/yr) — Root estimate						

Thirty-four trees were sampled, and stand biomass values for four 20 m² plots per
provenance were derived from regressions on D and D²H.
a. Latitude of the seed origin in the U.S.A.
b. Values in these columns are the means of the four provenances; the authors gave
individual provenance values.

Whittaker, R.H. and Niering, W.A. (1975). Vegetation of the Santa Catalina Mountains, Arizona. V Biomass, production and diversity along the elevation gradient. *Ecology* 56, 771-790.

Harris, W.F., Goldstein, R.A. and Henderson, G.S. (1973). Analysis of forest biomass pools, annual primary production and turnover of biomass for a mixed deciduous forest watershed. In: "IUFRO Biomass Studies", pp.43-64. Univ. of Maine, U.S.A.

Harris, W.F. and Henderson, G.S. (1981). In: "Dynamic Properties of Forest Ecosystems" (D.E. Reichle, ed.), pp.658-661. Cambridge University Press, Cambridge.

U.S.A.	ca.32°20'N 110°50'W 2040 m Arizona, Santa Catalina Mtns *Pinus cembroides* (69%)[a] *Juniperus deppeana* (16%)[a] et al. (Whittaker and Niering 1975)	35°58'N 84°17'W 265-360 m Tennessee, Walker Branch Site *Pinus echinata*, with *Liriodendron tulipifera* et al. Red-yellow podzols, pH 4.0-6.5 (Harris *et al.* 1973, 1981)
Age (years)	115[b]	30
Trees/ha	570 + 160[c]	
Tree height (m)	2.7[b]	12-25
Basal area (m²/ha)	4.3 + 0.6[c]	21.3
Leaf area index	0.7[d]	
Stem volume (m³/ha)	6.6[e]	

Dry biomass (t/ha)		
Stem wood	} 9.1 + 1.0[c]	} 89.6
Stem bark		
Branches	4.8 + 0.9[c]	27.6
Fruits etc.		
Foliage	1.4 + 1.5[c]	4.6
Root estimate		34.0[g]

CAI (m³/ha/yr)	5.5[e]	

Net production (t/ha/yr)		
Stem wood	0.07[f] } + 0.12[cf]	} 4.16 + 1.90[h] + 0.73[i]
Stem bark	0.02[f]	
Branches	0.13[f] + 0.20[cf]	
Fruits etc.	0.04 + 0.05[c]	
Foliage	0.40 + 0.80[c]	4.60[j]
Root estimate		

Whittaker and Niering (1975) sampled 10-15 trees per species and derived stand biomass values for a 0.1 ha plot from regressions on D, wood volumes and surface areas, and from other relationships; all trees and shrubs over 1 cm D were included. Harris *et al.* (1973, 1981) sampled many trees (about 150 at all 4 Walker Branch sites), excavated roots in 3 pits, and derived stand biomass values for 50-100 circular plots from regressions on D.
a. Percentage of the total stem volume.   *b*. Weighted mean age and height.
c. Understorey shrubs.   *d*. All-sided LAI was 2.0.   *e*. Parabolic volumes.
f. Excluding woody litterfall and any mortality.
g. Including stumps, which represented about half the 'root' biomass.   *h*. Mortality.
i. Woody litterfall.   *j*. Foliage production; foliage litterfall was 3.4 t/ha/yr.

McKee, W.H. and Shoulders, E. (1974). Slash pine biomass response to site preparation and soil properties. *Proc. Soil Sci. Soc. Am.* <u>38</u>, 144-148.

---

ca.31°N   92-93°W   50 m   U.S.A., Louisiana

Plantations.   *Pinus elliottii*
Fine silty loam soils,
saturated in winter,
unless drained.

	Untreated	Disked	Ploughed with drainage furrows
Age (years)	7	7	7
Trees/ha	ca.2000	ca.2000	ca.2000
Tree height (m)	7.6[a]	7.4[a]	8.3[a]
Basal area (m²/ha)	16.6	15.8	19.8
Leaf area index			
Stem volume (m³/ha)			
Dry biomass (t/ha) Stem wood	16.6	13.7	21.8
Stem bark	6.2	5.5	8.6
Branches	6.9	7.5	10.1
Fruits etc.	0.0	0.0	0.0
Foliage	8.9	7.5	10.1
Root estimate			
CAI (m³/ha/yr)			
Net production (t/ha/yr) Stem wood			
Stem bark			
Branches			
Fruits etc.			
Foliage			
Root estimate			

Four trees were sampled per treatment in September, and stand biomass values for plots of 0.14 ha per treatment were derived by proportional basal area allocation. Nutrient contents were determined.
a. Mean heights of sample trees.

Hanley, D.P. (1976). "Tree Biomass and Productivity Estimated for Three Habit Types of Northern Idaho." Bulletin 14. College of Forestry, Wildlife and Range Sciences, University of Idaho, Moscow, U.S.A.

48°20'N 116°50'W   over 500 m   U.S.A., Idaho, Priest River (and see below).

Pinus monticola, Larix occidentalis, Thuja plicata,   47°45'N 116°30'W

Tsuga heterophylla, Pseudotsuga menziesii, Abies grandis.   Deception Creek

	$\{$24% 17%$\}^a$ 54% 1% 3% 1%	$\{$28% 12%$\}^a$ 51% 1% 4% 1%	$\{$33% 2%$\}^a$ 58% 1% 5% 0%	$\{$30% 5%$\}^a$ 13% 39% 2% 11%	$\{$38% 1%$\}^a$ 0% 57% 1% 5%
Age (years)	105	110	100	over 250	20
Trees/ha	1056	584	756	105	2728
Tree height (m)		25.0	23.2	32.9	4.0
Basal area (m² /ha)	62.9	51.4	56.2	49.8	11.2
Leaf area index					
Stem volume (m³ /ha)					

Dry biomass (t/ha)

Stem wood	235	205	203	240	8.7
Stem bark	37	31	23	32	0.8
Branches	41	31	26	32	0.7
Fruits etc.					
Foliage	16	10	13	12	3.8
Root estimate	82	69	58	84	11.2

CAI (m³ /ha/yr)

Net production (t/ha/yr)

Stem wood	3.0^b	2.2^b	3.3^b	2.15^b	2.45^b
Stem bark	0.4^b	0.3^b	0.4^b	0.26^b	0.25^b
Branches	0.4^b	0.3^b	0.4^b	0.26^b	0.57^b
Fruits etc.					
Foliage	6.2^c	4.7^c	4.4^c	2.86^c	1.45^c
Root estimate	0.8	0.7	0.7	0.98	1.58

Stand biomass values for 'growth and yield' plots were derived from published regressions on D and H. In most instances, biomass was estimated at the beginning and end of a period of 10 years; the initial biomass values are given above. Increment values refer to the potential production of fully stocked stands as represented by the selected sample plots.

a. Percentage of the total basal area occupied by *P. monticola, L. occidentalis, T. plicata, T. heterophylla, P. menziesii* and *A. grandis*, e.g. 24%, 17%, 54%, 1%, 3% and 1%, respectively, in the far left column.

b. Including estimated mortality and new regeneration.

c. Derived from the foliage biomass and mean leaf longevity of each species.

Continued from p.305.

46°35'N   115°37'W   over 500 m   U.S.A., Idaho, Clearwater National Forest (and see below).

Pinus monticola, Abies grandis,   47°47'N   116°30'W
Pseudotsuga menziesii and *Larix occidentalis.*   Fernan District

	$\{94\%\ \ 5\%\}^a$ $\{1\%\ \ 0\%\}$	$\{84\%\ \ 11\%\}^a$ $\{3\%\ \ 0\%\}$	$\{78\%\ \ 16\%\}^a$ $\{6\%\ \ 0\%\}$	$\{15\%\ \ 43\%\}^a$ $\{19\%\ \ 21\%\}$
Age (years)	103	103	103	105
Trees/ha	803	628	699	1127
Tree height (m)	32.3	33.5	34.4	
Basal area (m²/ha)	67.7	61.3	62.2	53.5
Leaf area index				
Stem volume (m³/ha)				
Dry biomass (t/ha)				
Stem wood	344	350	381	200
Stem bark	30	35	44	28
Branches	34	35	38	41
Fruits etc.				
Foliage	15	15	16	21
Root estimate	69	71	65	55
CAI (m³/ha/yr)				
Net production (t/ha/yr)				
Stem wood	8.1[b]	7.1[b]	8.2[b]	2.9[b]
Stem bark	0.8[b]	0.7[b]	0.8[b]	0.3[b]
Branches	0.8[b]	0.7[b]	0.8[b]	0.6[b]
Fruits etc.				
Foliage	3.8[c]	3.8[c]	4.6[c]	5.4[c]
Root estimate	1.8	1.7	1.6	0.7

Same as p.305, except:

a. Percentage of the total basal area occupied by *P. monticola, A. grandis, P. menziesii* and *L. occidentalis*, e.g. 94%, 5%, 1% and 0%, respectively, in the far left column.

Continued from p.306.

46°35'N 115°37'W   over 500 m   U.S.A., Idaho, Clearwater National Forest.

Pinus monticola, Abies grandis,
Pseudotsuga menziesii and *Larix occidentalis.*

		{77% 20%}[a] {1% 0%}	{62% 22%}[a] {3% 3%}	{60% 33%}[a] {6% 1%}	{82% 15%}[a] {1% 1%}	{71% 22%}[a] {5% 2%}
Age (years)		103	103	103	103	103
Trees/ha		309	390	242	295	332
Tree height (m)		27.4	29.0	36.6	31.4	31.4
Basal area (m²/ha)		56.9	88.4	63.8	60.6	62.2
Leaf area index						
Stem volume (m³/ha)						
Dry biomass (t/ha)	Stem wood	337	539	441	350	365
	Stem bark	29	56	41	30	37
	Branches	34	56	43	35	37
	Fruits etc.					
	Foliage	15	24	19	15	16
	Root estimate	73	119	74	70	74
CAI (m³/ha/yr)						
Net production (t/ha/yr)	Stem wood	6.6[b]	9.8[b]	7.5[b]	8.2[b]	7.2[b]
	Stem bark	0.7[b]	1.0[b]	0.8[b]	0.8[b]	0.7[b]
	Branches	0.7[b]	1.0[b]	0.8[b]	0.8[b]	0.7[b]
	Fruits etc.					
	Foliage	3.4[c]	5.8[c]	4.8[c]	4.2[c]	3.5[c]
	Root estimate	1.7	2.4	1.7	1.7	1.5

Same as p.306.

Westman, W.E. and Whittaker, R.H. (1975). The pygmy forest region of northern California: studies on biomass and primary productivity. *J. Ecol.* 63, 493-520.

39°20'N 123°45'W   (alt. given below)   U.S.A., California, Mendocino, near Fort Bragg.

Noyo and Noyo-Blacklock podzols

Pinus muricata (89%)[a], *Pseudotsuga menziesii,*
Sequoia sempervirens, Myrica californica.

	80 m	90 m	100 m	170 m	75 m
Age (years)	84^d	94^d	89^d	94^d	62^d
Trees/ha	520^b+5440^c	500^b+3520^c	480^b+7000^c	400^b+6800^c	740^b+10720^c
Tree height (m)	11^d	18^d	24^d	28^d	24^d
Basal area (m²/ha)	53.4	53.9	64.3	102.3	99.3
Leaf area index					
Stem volume (m³/ha)	349^e	485^e	786^e	1409^e	1187^e

Dry biomass (t/ha)

Stem wood / Stem bark / Branches / Fruits etc. / Foliage / Root estimate	167	232	382	692	575

CAI (m³/ha/yr)	0.8^e	2.1^e	4.6^e	6.8^e	8.6^e

Net production (t/ha/yr)

Stem wood / Stem bark / Branches / Fruits etc. / Foliage / Root estimate	2.7^f	5.4^f	10.5^f	13.7^f	16.1^f

---

Stand biomass values for the above 250 to 500 m² plots were derived from regressions on various dimensions following extensive sampling and using the methods of Whittaker *et al.* (1974) (see p.259).
a. Mean percentage of the total basal area.
b. Number of *P. muricata* trees.
c. Number of trees of other species.
d. Weighted mean ages and heights.
e. Parabolic volumes.
f. Including foliage production, but excluding woody litterfall and any mortality; total litterfall was estimated to be up to 4.3 t/ha/yr.

Wiegert, R.G. and Monk, C.D. (1972). Litter production and energy accumulation in three plantations of longleaf pine (*Pinus palustris* Mill.). *Ecology* <u>53</u>, 949-953.

---

ca.33°35'N 81°45'W   150 m   U.S.A., South Carolina, near Aiken.

Plantations.

Pinus palustris

Age (years)	7	11	13
Trees/ha	924	1412	1680
Tree height (m)			
Basal area (m²/ha)			
Leaf area index			
Stem volume (m³/ha)			

Dry biomass (t/ha)				
	Stem wood	} 3.6	} 23.9	} 57.8
	Stem bark			
	Branches	0.9	6.2	16.7
	Fruits etc.			
	Foliage	2.1	4.0	9.2
	Root estimate			

CAI (m³/ha/yr)				
Net production (t/ha/yr)	Stem wood			
	Stem bark			
	Branches	} 0.02[a]	} 0.08[a]	} 0.05[a]
	Fruits etc.			
	Foliage	0.60[a]	3.40[a]	5.20[a]
	Root estimate			

Thirteen trees were sampled from the 7-year-old stand, and 5 trees were sampled from each of the other two stands, all in March-April. Stand biomass values for the above three 0.25 ha plots were obtained by multiplying mean tree values by the numbers of trees per hectare. There was 0.0, 2.6 and 6.2 t/ha of dead branches in columns left to right.
a. Litterfall only, measured over 10-12 months.

Klemmedson, J.O. (1975). Nitrogen and carbon regimes in an ecosystem of young dense ponderosa pine in Arizona. *Forest Sci.* 21, 163-168.

Gholz, H.L. (1981). Environmental limits on above-ground net primary production, leaf area, and biomass in vegetation zones of the Pacific Northwest. *Ecology* (in press).

Gholz, H.L., Grier, C.C., Campbell, A.G. and Brown, A.T. (1979). "Equations for Estimating Biomass and Leaf Area of Plants in the Pacific Northwest." Forest Research Laboratory, Oregon State University, Corvallis, USA. Research Paper no.41.

U.S.A.	ca.35°20'N 111°40'W 2500 m Arizona, Flagstaff, Wing Mountain.	44-45°N 121-122°W 870 m Oregon, Cascade Mountains
	Pinus ponderosa	*P. ponderosa*
	Unthinned natural stand on loamy soils derived from basalt. (Klemmedson 1975)	Typic vitrandept (Gholz 1981)
Age (years)	49	Mature
Trees/ha	22500	215[c]
Tree height (m)		8
Basal area (m²/ha)	67	26.1
Leaf area index		2.5[d]
Stem volume (m³/ha)		
Dry biomass (t/ha)		
Stem wood	}121.0	}99
Stem bark		
Branches	20.2	30
Fruits etc.		
Foliage	10.6	7
Root estimate	38.0[a]	
CAI (m³/ha/yr)		
Net production (t/ha/yr)		
Stem wood	}3.88[b]	}1.0[b]
Stem bark		
Branches		0.2
Fruits etc.		
Foliage		1.8[e]
Root estimate		

Klemmedson (1975) sampled all trees within ten 2 m² plots within several even-aged stands, and expressed values per hectare; nitrogen contents were determined.

Gholz (1981) derived stand biomass values for a plot at least 0.25 ha from regressions published by Gholz *et al.* (1979) based on trees sampled in northern Arizona.

a. Assumed to be 20% of the total biomass.
b. Excluding woody litterfall and mortality.
c. Trees over 10 cm D; there were 490 trees/ha less than 10 cm D.
d. All-sided LAI was 7.0.
e. Assumed to be 25% of the foliage biomass.

Whittaker, R.H. and Niering, W.A. (1975). Vegetation of the Santa Catalina Mountains, Arizona. V Biomass, production and diversity along the elevation gradient.
Ecology 56, 771-790.

Whittaker, R.H. and Niering, W.A. (1968). Vegetation of the Santa Catalina Mountains, Arizona. IV Limestone and acid soils. *J. Ecol.* 56, 523-544.

ca.32°20'N 110°50'W   (alt. given below)   U.S.A., Arizona, Santa Catalina Mountains, near Tucson.

Pinus ponderosa, Pinus strobiformis, Pinus chihuahuana, Quercus spp.

	$\left\{\begin{smallmatrix}68\% & 32\% \\ 0\% & 0\%\end{smallmatrix}\right\}^a$	$\left\{\begin{smallmatrix}99\% & 0\% \\ 0\% & 0\%\end{smallmatrix}\right\}^a$	$\left\{\begin{smallmatrix}95\% & 0\% \\ 0\% & 4\%\end{smallmatrix}\right\}^a$	$\left\{\begin{smallmatrix}10\% & 0\% \\ 45\% & 43\%\end{smallmatrix}\right\}^a$
	2740 m	2470 m	2180 m	2040 m
Age (years)	93^b	142^b	150^b	101^b
Trees/ha	2700	1100	1280	2780
Tree height (m)	12.8^b	18.4^b	15.2^b	7.5^b
Basal area (m²/ha)	39.4	46.3	34.9	26.0
Leaf area index	2.7^c	2.1^c	1.7^c	1.3^c
Stem volume (m³/ha)	253^d	425^d	265^d	98^d
Dry biomass (t/ha) Stem wood / Stem bark	}126	}213	}134	}79
Branches	28	30	23	28
Fruits etc.				
Foliage	7.3	6.8	5.4	6.6
Root estimate				
CAI (m³/ha/yr)	1.8^d	1.8^d	1.0^d	0.7^d
Net production (t/ha/yr) Stem wood	1.38^e	1.36^e	1.29^e	0.75^e
Stem bark	0.35^e	0.27^e	0.17^e	0.20^e
Branches	1.00^e	0.97^e	0.72^e	0.90^e
Fruits etc.	0.30	0.25	0.20	0.15
Foliage	3.09	2.90	2.55	2.35
Root estimate				

Ten to fifteen trees were sampled of each of the major species and stand values for
the above 0.1 ha plots were derived from regressions on D, wood volumes and surface
areas, and from other relationships. All trees and shrubs over 1 cm D were included.
a. Percentage of the total stem volume represented by *P. ponderosa, P. strobiformis,
P. chihuahuana* and *Quercus* spp., e.g. 68%, 32%, 0% and 0%, respectively, in the
far left column; *P. chichuahuana* is also known as *P. leiophylla* var. *chihuahuana.*
b. Weighted mean ages and heights.
c. All-sided LAI values were 7.6, 5.9, 4.7 and 3.7 in columns left to right.
d. Parabolic volumes.
e. Excluding woody litterfall and mortality.

Wilde, S.A. (1967). Production of energy material by forest stands as related to supply of soil water. *Silva fenn.* 1, 31-44.

Alban, D.H., Perala, D.A. and Schlaegel, B.E. (1978). Biomass and nutrient distribution in aspen, pine and spruce stands on the same soil type in Minnesota. *Can. J. For. Res.* 8, 290-299.

Singer, F.P. and Hutnik, R.J. (1966). Accumulation of organic matter in red pine and Norway spruce plantations of various spacings. *Penn. State Univ. Res. Briefs* 1, 22-28.

U.S.A. Plantations.	44°54'N 88°37'W ca.250 m Wisconsin, Keshena *Pinus resinosa* Sandy non-podzolic soil. (Wilde 1967)	47°20'N 94°30'W 400 m Minnesota, Pike Bay Expt. For. *P. resinosa* Fine sandy loam. (Alban *et al.* 1978)	ca.41°N 78°W — Central Pennsylvania *P. resinosa* (Singer and Hutnik 1966)		
Age (years)	32	40	42	42	42
Trees/ha	2175	1780	1076	2242	2990
Tree height (m)	14.4	17.6			
Basal area (m²/ha)	39	51.9			
Leaf area index					
Stem volume (m³/ha)	243	408[c]			
Dry biomass (t/ha) Stem wood	} 102.2	147.3 }	} 147.8	} 179.7	} 144.7
Stem bark		13.5			
Branches	} 29.1	25.4 } + 4.4[d]	31.5	36.7	29.3
Fruits etc.					
Foliage		14.1 }	13.7	16.7	12.7
Root estimate	35.8[a]	44			
CAI (m³/ha/yr)	10.2				
Net production (t/ha/yr) Stem wood	} 4.3				
Stem bark					
Branches	} 0.9 + 3.3[b]				
Fruits etc.					
Foliage					
Root estimate	1.6				

Wilde (1967) sampled several average-sized trees, including the root systems, and multiplied mean tree values by the numbers of trees per hectare.

Alban *et al.* (1978) sampled 10 trees in spring, excavated two root systems in August, and derived stand biomass values for ten 80 m² plots from regressions on D²H; nutrient contents were determined.

Singer and Hutnik (1966) sampled 9 trees and derived stand biomass values for the above three 400 m² plots from regressions on D and H; there was 1.7, 2.9 and 2.8 t/ha of dead branches in columns left to right.

a. Including stumps, weighting 7.8 t/ha.

b. Total litterfall.   *c.* Volume inside bark.

d. Understorey shrubs.

Leaf, A.L., Leonard, R.E., Wittwer, R.F. and Bickelhaupt, D.H. (1975). Four-year growth responses of plantation red pine to potash fertilization and irrigation in New York. *Forest Sci.* 21, 88-96.

Wittwer, R.F., Leaf, A.L. and Bickelhaupt, D.H. (1975). Biomass and chemical composition of fertilized and/or irrigated *Pinus resinosa* Ait. plantations. *Pl. Soil* 42, 629-651.

43°28'N 73°47'W   260 m   U.S.A., New York, Warrensburg.

Pinus resinosa

Plantations. Glacial outwash sandy soils.	Untreated	One application of 448 kg/ha K given 4 years previously. (A)	35 cm of irrigation water applied in each of the previous 4 years. (B)	Received treatments A and B.
Age (years)	35-40	35-40	35-40	35-40
Trees/ha	2638	2103	2106	2287
Tree height (m)	12.7^a	13.2^a	13.8^a	13.8^a
Basal area (m²/ha)				
Leaf area index				
Stem volume (m³/ha)				
Dry biomass (t/ha) — Stem wood	47.7	61.6	54.9	58.1
Dry biomass (t/ha) — Stem bark	8.0	9.3	8.7	8.8
Dry biomass (t/ha) — Branches	14.7	15.0	14.3	18.8
Dry biomass (t/ha) — Fruits etc.				
Dry biomass (t/ha) — Foliage	6.8	11.8	8.8	13.8
Dry biomass (t/ha) — Root estimate				
CAI (m³/ha/yr)				
Net production (t/ha/yr) — Stem wood				
Net production (t/ha/yr) — Stem bark				
Net production (t/ha/yr) — Branches				
Net production (t/ha/yr) — Fruits etc.				
Net production (t/ha/yr) — Foliage				
Net production (t/ha/yr) — Root estimate				

Five trees were sampled from each treatment in the autumn, and stand values for 800 m² plots were derived from regressions on individual tree basal area. Nutrient contents were determined.
a. Height of the dominant trees.

Madgwick, H.A.I. (1962). "Studies in the Growth and Nutrition of *Pinus resinosa* Ait."
Ph.D. thesis, State University College of Forestry, Syracuse University, Syracuse,
New York, U.S.A.

Madgwick, H.A.I., White, E.H., Xydias, G.K. and Leaf, A.L. (1970). Biomass of *Pinus
resinosa* in relation to potassium nutrition. *Forest Sci.* 16, 154-159.

43°28'N 73°47'W   260 m   U.S.A., New York, Warrensburg.

Plantations.
Brown, podzolic,
deep sand.

Pinus resinosa

	Untreated, thinned at age 16.	Slash[a] applied at age 16.	Fertile site	Slash[a] applied at age 14.	
Age (years)	30	30	32	30	30
Trees/ha	4990	7020	1530	6420	5730
Tree height (m)	10.4	9.2	15.2	10.2	9.9
Basal area (m²/ha)	30.6	37.3	36.4	39.6	44.0
Leaf area index	3.4[b]	3.1[b]	3.8[b]	4.2[b]	4.3[b]
Stem volume (m³/ha)					
Dry biomass (t/ha) — Stem wood	49.5	54.0	80.3	65.6	69.8
Stem bark	9.7	10.4	8.4	12.7	12.4
Branches	10.0	9.7	16.1	11.0	10.6
Fruits etc.					
Foliage	8.4	7.8	10.4	10.7	10.4
Root estimate					
CAI (m³/ha/yr)					
Net production (t/ha/yr) — Stem wood					
Stem bark					
Branches					
Fruits etc.					
Foliage	3.5[c]	3.2[c]	2.4[c]	3.5[c]	3.8[c]
Root estimate					

Five trees were sampled per plot during August-October (apart from the 'fertile'
site, where only 3 trees were sampled), and stand values were derived from regressions on D. There was 7.3, 9.5, 14.8, 10.4 and 9.9 t/ha of dead branches in columns
left to right. Nutrient contents were determined.
a. Thinnings and prunings from other plots.
b. All-sided LAI values were 9.6, 8.8, 10.7, 12.1 and 12.3 in columns left to right.
c. New foliage biomass.

Continued from p.314.

43°28'N 73°47'W   260 m   U.S.A., New York, Warrensburg.

Plantations.
Brown, deep,
podzolic sand,   *Pinus resinosa*
deficient in
potash.

	Unfertilized plots.			Unfertilized spacing experiment.			
Age (years)	29	30	31	32	32	32	32
Trees/ha	6520	6520	6520	1760	3830	6680	10720
Tree height (m)		7.5		10.2	10.1	8.6	7.6
Basal area (m² /ha)			26.5	24.6	30.4	26.9	25.5
Leaf area index	1.9^a	2.6^a	2.5^a	3.0^a	2.9^a	2.3^a	2.8^a

Stem volume (m³ /ha)

Dry biomass (t/ha)								
	Stem wood	27.8	32.5	33.3	43.8	49.7	35.1	32.0
	Stem bark	5.2	6.9	7.3	7.0	8.5	7.3	7.1
	Branches	7.9	9.9	8.4	14.5	11.7	8.8	9.1
	Fruits etc.							
	Foliage	5.3	6.2	5.9	7.9	7.4	6.0	6.9
	Root estimate							

CAI (m³ /ha/yr)

Net production (t/ha/yr)								
	Stem wood		2.8^b					
	Stem bark		1.1^b					
	Branches							
	Fruits etc.							
	Foliage	2.5^c	2.8^c	2.8^c	3.6^c	3.3^c	2.3^c	2.9^c
	Root estimate							

Five trees were sampled per plot during August-October, and stand values were
derived from regressions on D. There was 4.8, 6.3, 6.5, 7.3, 7.5, 6.1 and 5.2 t/ha
of dead branches in columns left to right. Nutrient contents were determined.
a. All-sided LAI values were 5.3, 7.3, 7.0, 8.5, 8.2, 6.6 and 8.1 in columns left
   to right.
b. Increments from ages 29 to 31, excluding any mortality.
c. New foliage biomass.

Olsvig-Whittaker, L. (1980). "A Comparative Study of Northeastern Pine Barrens Vegetation." Ph.D. thesis, Cornell University, Ithaca, N.Y., U.S.A.

U.S.A.	ca.40°N 74°30'W 10-100m New Jersey.		ca.41°25'N 74°40'W 200-400 m New York, Shawangunk Mountains.	
	Pinus rigida and *Quercus marilandica*		*Pinus rigida*	
	(48%)[a]	(70%)[a]		
Age (years)	10-20	12-24	20-24	20-23
Trees/ha	40000	45800	10600	19400
Tree height (m)	1.2	1.1	1.7	1.3
Basal area (m²/ha)	4.1	6.2	14.9	13.8
Leaf area index				
Stem volume (m³/ha)	2.1	3.1	12.7	9.1
Dry biomass (t/ha) Stem wood	0.94	1.25	4.48	3.21
Stem bark	0.44	0.63	2.22	1.66
Branches	1.34	1.37	2.79	2.15
Fruits etc.				
Foliage	0.82[b]	1.05[b]	1.88[b]	1.70[b]
Root estimate				
CAI (m³/ha/yr)				
Net production (t/ha/yr) Stem wood	0.17[c]	0.20[c]	0.30[c]	0.24[c]
Stem bark	0.09[c]	0.12[c]	0.14[c]	0.13[c]
Branches	0.42[c]	0.45[c]	0.48[c]	0.37[c]
Fruits etc.				
Foliage	0.55[d]	0.55[d]	0.57[d]	0.49[d]
Root estimate				

Stand values for the above 100 m² plots were derived from regressions on D, stem conic surface area, and stem volume, following the methods of Whittaker *et al.* (1974) (see p.259). Nutrient contents were determined.
a. Percentage of the total basal area accounted for by *P. rigida*.
b. Including the current year's twigs.
c. Excluding woody litterfall and mortality.
d. New foliage biomass, corrected for leaf fall and consumption, and including the current year's twigs.

Continued from p.316.

ca.40°45'N 73°W   10-100 m   U.S.A., New York, Long Island.

	Pinus rigida (83%)[a] *Quercus alba*, *Quercus velutina*, and *Quercus ilicifolia*.	*P. rigida* (92%)[a] *Quercus stellata* and *Q. velutina*.	*P. rigida* (86-88%)[a] and *Q. ilicifolia*. Shrublands on sandy coastal plain.	
Age (years)	60-70	65-80	30-40	30-45
Trees/ha	930 + 450	690 + 340	7000+16400	7800+21400
Tree height (m)	6.2	7.1	1.7	1.6
Basal area (m²/ha)	13.0 + 2.7	17.3 + 1.5	11.5 + 1.5	10.9 + 1.8
Leaf area index				
Stem volume (m³/ha)	73.4 + 13.3	118.0 + 8.0	10.4 + 0.7	9.3 + 0.8
Dry biomass (t/ha)				
Stem wood	26.9 + 6.5	44.3 + 4.0	3.7 + 0.4	3.3 + 0.4
Stem bark	6.0 + 1.5	8.9 + 0.9	1.8 + 0.1	1.6 + 0.2
Branches	10.7 + 2.6	18.7 + 2.3	2.3 + 0.5	2.0 + 0.4
Fruits etc.				
Foliage	6.4[b]+ 0.6[b]	8.3[b]+ 0.5[b]	1.6[b]+ 0.1[b]	1.5[b]+ 0.2[b]
Root estimate			4.7 + 2.1	4.2 + 2.0
CAI (m³/ha/yr)				
Net production (t/ha/yr)				
Stem wood	1.42[c]+0.22[c]	1.75[c]+0.16[c]	0.25[c]+0.04[c]	0.24[c]+0.05[c]
Stem bark	0.29[c]+0.04[c]	0.31[c]+0.03[c]	0.12[c]+0.02[c]	0.12[c]+0.02[c]
Branches	1.86[c]+0.30[c]	2.32[c]+0.25[c]	0.40[c]+0.05[c]	0.38[c]+0.06[c]
Fruits etc.				
Foliage	1.79[d]+1.53[d]	3.43[d]+0.47[d]	0.56[d]+0.13[d]	0.52[d]+0.18[d]
Root estimate				

Stand values for plots of 0.10 ha (two left-hand columns) or 0.01 ha (two right-hand columns) were derived from regressions on D, stem conic surface area, and stem volume, following the methods of Whittaker *et al.* (1974) (see p.259). Roots were excavated in the two plots on the coastal plain. Values are given above for *P. rigida* plus *Quercus* spp. (left and right in each column). Nutrient contents were determined.
a. Percentage of the total basal area.
b. Including the current year's twigs.
c. Excluding woody litterfall and mortality.
d. New foliage biomass, corrected for leaf fall and consumption, and including the current year's twigs.

Whittaker, R.H. (1963). Net production of heath balds and forest heaths in the Great Smoky Mountains. *Ecology* <u>46</u>, 176–182.

Whittaker, R.H. (1966). Forest dimensions and production in the Great Smoky Mountains. *Ecology* <u>47</u>, 103–121.

ca.35°40'N 83°30'W   (alt. given below)   U.S.A., Tennessee, Great Smoky Mountains, Mount Leconte.

	Pinus rigida, Pinus strobus, Pinus virginiana (74–78%)[a] *Quercus coccinea, et al.*		*Pinus pungens* (96%)[a] *et al.*	*P. pungens* (73%)[a] *Quercus prinus, et al.*
	610 m	550 m	1070 m	1340 m
Age (years)	Mature			
Trees/ha	$346 + 480^b$	$228 + 390^b$	$2630 + 500^b$	$2310 + 2280^b$
Tree height (m)	$20 \;\; (15.5)^c$	$17 \;\; (14.1)^c$	$18 \;\; (15.0)^c$	$12 \;\; (5.9)^c$
Basal area (m²/ha)	$34.4 + 0.3^b$	$25.4 + 0.2^b$	$25.6 + 0.3^b$	$19.1 + 1.5^b$
Leaf area index	0.5^b	0.3^b	1.4^b	1.5^b
Stem volume (m³/ha)	$227^d + 0.5^{bd}$	$162^d + 0.2^{bd}$	$165^d + 0.2^{bd}$	$74.2^d + 1.8^{bd}$
Dry biomass (t/ha) — Stem wood / Stem bark / Branches / Fruits etc. / Foliage / Root estimate	$180 + 1.5^b$	$130 + 1.2^b$	$86 + 5.8^b$	$52.0 + 5.7^b$
CAI (m³/ha/yr)	3.9^d	4.5^d	2.0^d	$0.7^d + 0.1^{bd}$
Net production (t/ha/yr) — Stem wood / Stem bark / Branches / Fruits etc. / Foliage / Root estimate	$8.2 + 0.5^{be}$	$9.5 + 0.4^{be}$	$4.0 + 1.6^{be}$	$2.1 + 1.7^{be}$

Stand values were estimated for plots of at least 0.1 ha from the weight of clippings of current year's twigs, from published regressions, from stem volumes, branch/stem biomass ratios, and other relationships.

a. Percentage of the total volume increment; 74–78% refers to the sum of the three *Pinus* species.

b. Understorey shrubs.

c. Weighted mean height (in the brackets).

d. Parabolic volumes.

e. Excluding woody litterfall and mortality; total foliage production of trees plus shrubs was 2.6, 2.7, 1.8 and 1.6 t/ha/yr in columns left to right.

Swank, W.T. and Schreuder, H.T. (1973). Temporal changes in biomass, surface area and net production for a *Pinus strobus* L. forest. In: "IUFRO Biomass Studies", pp.173-182. College of Life Sciences and Agriculture, University of Maine, Orono, U.S.A.

Swank, W.T. and Schreuder, H.T. (1974). Comparison of three methods of estimating surface area and biomass for a forest of young eastern white pine. *Forest Sci.* 20, 91-100.

35°04'N 83°26'W   706-988 m   U.S.A., North Carolina, Franklin, Coweeta.

Plantations.

Pinus strobus

Age (years)	10	12	15
Trees/ha	1790	1790	1790
Tree height (m)	ca.6.5	ca.7.5	9.8
Basal area (m²/ha)	7.3	12.6	23.4
Leaf area index	3.5[a]	6.2[a]	6.4[a]
Stem volume (m³/ha)			

Dry biomass (t/ha)		10	12	15
	Stem wood	}7.9	}23.5	}42.1
	Stem bark			
	Branches	7.6	15.3	22.8
	Fruits etc.			
	Foliage	2.5	4.7	4.7
	Root estimate			

Net production (t/ha/yr)	CAI (m³/ha/yr)	
	Stem wood	}6.84[b]
	Stem bark	
	Branches	3.04 + 0.02[c]
	Fruits etc.	
	Foliage	0.42 + 3.21[c]
	Root estimate	

Twenty trees were sampled at age 10, six at age 12 and thirteen at age 15, all in February. Stand biomass values for twenty 800 m² plots were derived from regressions on basal area per tree.

a. All-sided LAI values in February were 5.4, 9.7 and 9.9 at ages 10, 12 and 15, respectively; these values were increased by 80% to give late summer values, and were then divided by 2.8 to give projected areas.

b. Including estimated mortality.

c. Litterfall.

Switzer, G.L., Nelson, L.E. and Smith, W.H. (1966). The characterization of dry matter and nitrogen accumulation by loblolly pine (*Pinus taeda* L.). *Proc. Soil. Sci. Soc. Am.* <u>30</u>, 114-119.

ca.33°N 89°W   --   U.S.A., Mississippi, near Louisville.

Red-yellow   *Pinus taeda*. Natural regeneration.
podzols.

	Poor upland sites.					Good lowland sites.				
Age (years)	20	30	40	50	60	20	30	40	50	60
Trees/ha	2260	1400	880	680	580	1260	810	500	370	300
Tree height (m)	10	15	17	19	21	12	22	26	29	30
Basal area (m²/ha)	21	29	31	31	31	24	31	33	34	33
Leaf area index										
Stem volume (m³/ha)	50^a	160^a	220^a	240^a	260^a	150^a	290^a	380^a	420^a	440^a

Dry biomass (t/ha)		Poor upland					Good lowland				
	Stem wood										
	Stem bark										
	Branches	45	101	135	157	174	95	168	219	241	252
	Fruits etc.										
	Foliage										
	Root estimate										

CAI (m³/ha/yr)

Net production (t/ha/yr)	
	Stem wood
	Stem bark
	Branches
	Fruits etc.
	Foliage
	Root estimate

Five trees were sampled in each of 5 size classes at each site in March-April, and stand biomass values were derived from regressions on stem volume per tree and from stand volume tables. Nitrogen contents were determined.
a. Volume inside the bark.

Baker, J.B., Switzer, G.L. and Nelson, L.E. (1974). Biomass production and nitrogen recovery after fertilization of young loblolly pines. *Proc. Soil Sci. Soc. Am.* <u>38</u>, 958-961.

ca.34°30'N 89°30'W   100 m   U.S.A., Mississippi, Interior Upland Flatwoods.

Plantations. Coarse and fine loams.

Pinus taeda

	No fertilizers applied.				112 kg N applied per hectare at age 3.	224 kg N applied per hectare at age 4.		*a*
Age (years)	3	4	5	6	5	6	5	6
Trees/ha	2421	2421	2376	2376	2408	2870	2964	2655
Tree height (m)	1.0	2.1	3.3	4.3	3.4	4.4	3.3	4.4
Basal area (m²/ha)				8.0				8.4
Leaf area index								
Stem volume (m³/ha)								
Dry biomass (t/ha) Stem wood	} 0.3	} 1.2	2.1	6.6	2.9	3.1	7.1	6.6
Stem bark			0.7	2.1	0.9	1.0	2.3	2.0
Branches	0.4	0.5	1.6	4.0	2.3	2.7	3.8	4.4
Fruits etc.	0.0	0.0	0.0	0.0	0.0	0.0	0.0	0.0
Foliage	0.6	1.7	3.0	4.7	4.3	4.7	4.7	5.2
Root estimate								
CAI (m³/ha/yr)								
Net production (t/ha/yr) Stem wood								
Stem bark								
Branches								
Fruits etc.								
Foliage								
Root estimate								

Three trees were sampled per treatment per year during the autumn, and stand values for 400 m² plots were derived from regressions on D and H. Nitrogen contents were determined.

a. Data in this column refer to a plot which received N-fertilizer in two applications.

Demott, T.E. (1979). "Response to and Recovery of Nitrogen Fertilizer in Young Loblolly Pine Plantations." M.Sc. thesis, Mississippi State University, U.S.A.

ca.32°46'N 88°33'W   100 m   U.S.A., Mississippi, Interior Upland Flatwoods, near
De Kalb.

Plantations.

| | *Pinus taeda* | All fertilizer treatments were applied at age 8. | | | |
| | | Urea | | Ammonium nitrate | |
	Untreated	112 kg/ha	224 kg/ha	112 kg/ha	224 kg/ha
Age (years)	10	10	10	10	10
Trees/ha	1870	1870	1860	1820	1850
Tree height (m)	9.2	9.1	9.2	9.0	8.9
Basal area (m²/ha)	21.8	22.9	23.1	22.3	24.3
Leaf area index					
Stem volume (m³/ha)	79.1	82.6	83.3	80.0	87.5
Dry biomass (t/ha) Stem wood	33.6	34.8	35.2	34.0	36.8
Stem bark	6.0	6.2	6.1	6.0	6.7
Branches	11.0	11.8	12.2	11.4	12.8
Fruits etc.					
Foliage	5.8	6.1	6.3	5.8	6.5
Root estimate					
CAI (m³/ha/yr)					
Net production (t/ha/yr) Stem wood	8.60^a	8.85^a	8.85^a	8.55^a	9.90^a
Stem bark	0.75^a	0.85^a	0.85^a	0.75^a	1.10^a
Branches	2.45^a	2.65^a	2.75^a	2.45^a	3.15^a
Fruits etc.					
Foliage	0.20^a	0.40^a	0.50^a	0.25^a	0.60^a
Root estimate					

Stand values were obtained using regression methods. Nitrogen contents were determined.
a. Increments since age 8, excluding all woody and foliage litterfall.

Larsen, H.S., Carter, M.C., Gooding, J.W. and Hyink, D.M. (1976). Biomass and nitrogen distribution in four 13-year-old loblolly pine plantations in the hilly coastal plain of Alabama. *Can. J. For. Res.* 6, 187–194.

Burkhart, H.E. (1977). Biomass and nitrogen distribution in four 13-year-old loblolly pine plantations in the hilly coastal plain of Alabama: discussion. *Can. J. For. Res.* 7, 545–546.

---

ca.31°N 88°W 50 m U.S.A., Alabama, Wilcox County, hilly coastal plain.

Plantations.　　　　　　　　*Pinus taeda*
Paleudult soils.

	Unthinned stand.
Age (years)	13
Trees/ha	1439
Tree height (m)	11.6
Basal area (m²/ha)	25.7
Leaf area index	
Stem volume (m³/ha)	

Dry biomass (t/ha)		
Stem wood	50.9	
Stem bark	8.1	
Branches	12.1	+ 1.2^a
Fruits etc.		
Foliage	9.5	
Root estimate	9.7^b	

CAI (m³/ha/yr)	
Net production (t/ha/yr)	
Stem wood	
Stem bark	
Branches	
Fruits etc.	
Foliage	
Root estimate	

Thirty-two trees were sampled in August, and stand biomass values for a 0.62 ha plot in each of four plantations were derived by proportional basal area allocation. There was 7.9 t/ha of dead branches. Nitrogen contents were determined.
a. Understorey shrubs.
b. Stumps only.

Nemeth, J.C. (1973a). Dry matter production in young loblolly (*Pinus taeda* L.) and slash pine (*Pinus elliottii* Engelm.) plantations. *Ecol. Monogr.* 43, 21-41.

Nemeth, J.C. (1973b). Forest biomass estimation: permanent plots and regression techniques. In: "IUFRO Biomass Studies", pp.79-88. College of Life Sciences and Agriculture, University of Maine, Orono, U.S.A.

35°20'N 76°45'W   8 m   U.S.A., North Carolina, Beaufort County, near Aurora.

Plantations. Sandy loams. Alfisols and Ultisols.

	Pinus taeda and *Pinus elliottii*		*Pinus taeda*			
	(63% 37%)[a]	(48% 52%)[a]				
Age (years)	4-5	6-7	8	9-10	11	12
Trees/ha	1450	2030	900	1220	1400	1400
Tree height (m)	3.4[b]	4.3[b]	5.2	9.4	10.5	11.6
Basal area (m²/ha)	ca.2.3	ca.7.5	ca.6.3	ca.22.0	ca.24.1	ca.28.0
Leaf area index						
Stem volume (m³/ha)						

Dry biomass (t/ha)

Stem wood	} 7.0[c]	} 9.2	} 7.3	} 40.0	} 53.0	} 65.5
Stem bark						
Branches		6.5[c]	4.6[c]	16.2[c]	17.4[c]	18.6[c]
Fruits etc.		0.0	0.0	0.0	0.0	0.0
Foliage		4.7	4.2	9.2	7.8	6.9
Root estimate	1.3	3.7	2.9	11.9	14.2	16.6

CAI (m³/ha/yr)

Net production (t/ha/yr)

Stem wood		2.1	4.9	4.3	12.4	11.5
Stem bark		0.6	1.3	0.8	1.3	1.2
Branches		1.3+0.0[d]	2.4+0.0[d]	2.6+0.2[d]	1.2+1.0[d]	0.1+1.6[d]
Fruits etc.		0.0	0.0	0.0	0.0	0.0
Foliage		2.2[e]	4.2[e]	4.0[e]	4.4[e]	3.0[e]
Root estimate		1.0	1.9	1.8	2.9	2.6

A total of 4 *P. taeda* and 15 *P. elliottii* trees were sampled over 2 years during August-September (biomass values given above are the values in the first sampling year). Five root systems were excavated, and fine roots were core sampled. Stand biomass values in a total of twenty-eight 100 m² plots were derived from regressions on D and H. The production values given above refer to years 4-6, 6-8, 8-9, 9-10 and 11-12 in columns left to right.

a. Percentage of the total tree number (e.g. *P. taeda* 63%, *P. elliottii* 37% in left column). b. Weighted by the proportion of each species. c. Including dead branches. d. Estimated branch mortality (0.0, 0.0, 2.7, 5.9 and 8.5% of the branch biomass in columns left to right). e. New foliage biomass; foliage litterfall was 0.8, 1.4, 2.5, 3.5 and 3.5 t/ha/yr in columns left to right.

Pope, P.E. (1979). The effect of genotype on biomass and nutrient content of 11-year-old loblolly pine plantation. *Can. J. For. Res.* **9**, 224-230.

35°45'N   91°39'W   50-200 m   U.S.A., Arkansas, Batesville.

Plantations.   *Pinus taeda*
Loamy, skeletal,
siliceous soils.

Unthinned plots of four open-pollinated progenies
from parent trees in southern Arkansas.

Age (years)	11	11	11	11
Trees/ha	2990	2990	2990	2990
Tree height (m)	9.9	10.0	9.5	9.9
Basal area (m²/ha)	52.7	49.2	39.4	51.0
Leaf area index				
Stem volume (m³/ha)				
Dry biomass (t/ha) — Stem wood	61.1	47.8	41.1	73.3
Stem bark	7.8	6.9	6.2	9.1
Branches	20.7	18.9	11.8	25.3
Fruits etc.				
Foliage	6.7	3.8	7.5	7.8
Root estimate				
CAI (m³/ha/yr)				
Net production (t/ha/yr) — Stem wood				
Stem bark				
Branches				
Fruits etc.				
Foliage				
Root estimate				

Twelve trees were sampled of each progeny and stand biomass values for two 0.2 ha plots per progeny were derived from regressions on D and D²H. Nutrient contents were determined.

Ralston, C.W. (1973). Annual primary productivity in a loblolly pine plantation.
   In: "IUFRO Biomass Studies", pp.107-117. College of Life Sciences and Agriculture, University of Maine, Orono, U.S.A.
Kinerson, R.S., Ralston, C.W. and Wells, C.G. (1977). Carbon cycling in a loblolly
   pine plantation. *Oecologia* <u>29</u>, 1-10.
Harris, W.F., Kinerson, R.S. and Edwards, N.T. (1977). Comparison of belowground
   biomass of natural deciduous forest and loblolly pine plantations. *Pedobiologia*
   <u>17</u>, 369-381.

ca.36°N 79°W   135 m   U.S.A., North Carolina, near Saxapahaw, Triangle Site.

Plantation.   *Pinus taeda*
Sandy loam,
pH 4.5.
                    The same stand was measured in three successive years.

Age (years)	13	14	15
Trees/ha	1445	1445	1445
Tree height (m)	11.9^a	12.6^a	13.9^a
Basal area (m²/ha)	31.8	33.5	34.5
Leaf area index			
Stem volume (m³/ha)		3.9-6.6	

Dry biomass (t/ha)				
	Stem wood	} 54.4	} 65.3	} 74.8
	Stem bark			
	Branches	7.8	11.0	11.8
	Fruits etc.			
	Foliage	4.3	5.7	6.0
	Root estimate	18.0	20.2	21.8 *19%*

CAI (m³/ha/yr)		
Net production (t/ha/yr)	Stem wood	} 9.54
	Stem bark	
	Branches	$0.79^b + 0.37^c$
	Fruits etc.	
	Foliage	$0.30 + 5.07^c$
	Root estimate	ca.9.0^d

Twenty-six trees were sampled over the three years, and 7 root systems were excavated. Stand biomass values were derived from regressions on D, H and stem diameter
at the base of the crowns. There was at least 6.0 t/ha of dead branches at age 15.
a. Mean height of the dominant trees.
b. Branch increment, excluding the increase in biomass of standing dead branches,
   which Kinerson *et al.* 1977 estimated to be 5 to 6 t/ha/yr.
c. Litterfall.
d. Root increment and turnover, estimated by Harris *et al.* (1977).

Wells, C.G., Jorgensen, J.R. and Burnette, C.E. (1975). "Biomass and Mineral Elements in a Thinned Loblolly Pine Plantation at Age 16." U.S.D.A. Forest Service Res. Paper SE-126. Southeast For. Exp. Stn, Ashville, N.C., U.S.A.

ca.36°N　79°W　　149 m　　U.S.A., North Carolina, Duke Forest.

Plantation.
Coarse sandy loam,　　　　　*Pinus taeda*
of average fertility,
pH 4.5.

Age (years)	16
Trees/ha	2243
Tree height (m)	15.0
Basal area (m²/ha)	49
Leaf area index	
Stem volume (m³/ha)	

Dry biomass (t/ha)

Stem wood	109.6
Stem bark	15.2
Branches	14.6
Fruits etc.	
Foliage	8.0
Root estimate	36.3　*26%*

CAI (m³/ha/yr)

Net production (t/ha/yr)

Stem wood	} 5.6[a]
Stem bark	
Branches	1.9[a]
Fruits etc.	
Foliage	3.8[b] (or 4.8)[c]
Root estimate	

Sixteen trees were sampled in September, and two root systems were excavated. Stand biomass values for a 0.3 ha plot were derived from regressions on basal area per tree. Increments were estimated between ages 14 and 16. There was 8.6 t/ha of dead branches. Nutrient contents were determined.
a. Excluding woody litterfall and any mortality.
b. Determined by regression analysis.
c. The new foliage biomass.

Madgwick, H.A.I. (1968). Seasonal changes in biomass and annual production of an old field *Pinus virginiana* stand. *Ecology* <u>49</u>, 149-152.

37°15'N 80°25'W   700 m   U.S.A., Virginia, Blacksburg, Fishburn Tract.

Shallow   *Pinus virginiana*
Calvin
silt-loam.
              Natural regeneration on an abandoned field.

Age (years)		17
Trees/ha		5750
Tree height (m)		8.6[a]
Basal area (m²/ha)		25.3
Leaf area index		
Stem volume (m³/ha)		
Dry biomass (t/ha)	Stem wood	} 47.2
	Stem bark	
	Branches	19.3
	Fruits etc.	0.4
	Foliage	8.8
	Root estimate	
CAI (m³/ha/yr)		
Net production (t/ha/yr)	Stem wood	} 5.8
	Stem bark	
	Branches	3.6[b]
	Fruits etc.	0.6
	Foliage	5.3[c]
	Root estimate	

Five trees were sampled on each of 9 occasions during one year and stand biomass values were derived from regressions on D²H. The stem and branch biomass values given above are the means of the 9 samples; the leaf biomass value is the mean during June to September.

a. Height of the dominant trees.

b. Branch biomass divided by its mean age of 5.5 years; excluding woody litterfall.

c. New foliage biomass of 4.3 t/ha plus estimated growth of old foliage and loss in weight of the new foliage before sampling.

Dice, S.F. (1970). "The Biomass and Nutrient Flux in a Second Growth Douglas-fir
  Ecosystem." Ph.D. thesis. University of Washington, Seattle, Washington, U.S.A.
Cole, D.W., Gessel, S.P. and Dice, S.F. (1968). In:"Primary Productivity and Mineral
  Cycling in Natural Ecosystems" (H.E. Young, ed.), pp.197-233. Univ. Maine, Orono.
Cole, D.W. and Rapp, M. (1981). In: "Dynamic Properties of Forest Ecosystems"
  (D.E. Reichle, ed.), pp.341-409. Cambridge University Press, Cambridge, London.
Grier, C.C., Cole, D.W., Dyrness, C.T. and Fredriksen, R.L. (1974). In: "Integrated
  Research in the Coniferous Forest Biome". (see p.340)

---

ca.47°23'N   122°15'W   212 m   U.S.A., Washington, Cedar River, Thompson Research
                                  Center.

Plantation.
Infertile, well-
drained, gravelly
sandy loam.

Pseudotsuga menziesii

	a	*b*
Age (years)	36	36
Trees/ha	2223	2223
Tree height (m)	20	18
Basal area (m²/ha)	37.7	37.7
Leaf area index		
Stem volume (m³/ha)		

Dry biomass (t/ha)

	a	*b*
Stem wood	134.4	121.7
Stem bark	18.5	18.7
Branches	11.2	13.9
Fruits etc.		
Foliage	7.8	9.1
Root estimate	31.4	33.1

CAI (m³/ha/yr)

Net production (t/ha/yr)

	a	*b*	
Stem wood	9.55[c]	6.05	
Stem bark	1.32[c]	0.69	} + 0.70[d]
Branches	0.80	0.42	
Fruits etc.			
Foliage	2.20[e]	1.99[e]	
Root estimate	3.63	1.19	

Many trees were sampled in the autumn, and roots were excavated. Stand biomass
values were derived from regressions on D. Nutrient contents were determined.
a. Values from Dice (1970). He estimated there to be 5.8 t/ha of dead branches,
  and measured foliage litterfall of 1.9 t/ha/yr.
b. Values from Cole *et al.* (1968) and Cole and Rapp (1981). They estimated there
  to be 8.1 t/ha of dead branches, and measured foliage litterfall of 2.2 t/ha/yr.
c. Including estimated woody litterfall and mortality.
d. Woody litterfall, excluding any mortality.
e. New foliage biomass.

Heilman, P.E. (1961). "Effects of Nitrogen Fertilization on the Growth and Nitrogen Nutrition of Low-site Douglas-fir Stands." Ph.D. thesis, University of Washington, Seattle, Washington, U.S.A.

Heilman, P.E. and Gessel, S.P. (1963). The effect of nitrogen fertilization on the concentration and weight of nitrogen, phosphorus, and potassium in Douglas fir trees. *Proc. Soil Sci. Soc. Am.* 27, 102–105.

ca.46°40'N   122°20'W   (alt. given below)   U.S.A., Washington, La Grande, Charles Lathrop Pack Forest.

Gravelly infertile acid loams, fertilized during previous 2-9 years.

Pseudotsuga menziesii

	488 m		235 m	
	Untreated	560 kg/ha N	Untreated	740 kg/ha N
Age (years)	28–32	28–32	38	38
Trees/ha	2840	3520	2480	2000
Tree height (m)	6–7	7–9	11–17	11–17
Basal area (m²/ha)	13.2	22.5	39.9	43.4
Leaf area index				
Stem volume (m³/ha)	46	70	226	248
Dry biomass (t/ha) Stem wood	18.9	29.6	51.7	63.9
Stem bark	4.0	5.4	18.1	21.4
Branches	6.3	10.0	8.5	17.1
Fruits etc.				
Foliage	8.0	13.2	8.0	14.2
Root estimate	25.1	20.1	10.0	9.8
CAI (m³/ha/yr)	4.3	8.1	9.7	17.1
Net production (t/ha/yr) Stem wood	1.8^a	3.1^a	4.2^a	6.7^a
Stem bark				
Branches	0.0^b	0.0^b	0.3^b	0.3^b
Fruits etc.				
Foliage	1.3^b	1.7^b	2.1^b	2.5^b
Root estimate				

Eight trees were sampled per plot in winter, and stand biomass values were derived from regressions on D. Root biomass was estimated from soil core samples. There was 1.3, 0.9, 10.2 and 13.4 t/ha of dead branches in columns left to right. Nutrient contents were determined.

a. Excluding the bark, and assuming wood density to be 0.41, 0.38, 0.43 and 0.39 g/cm³ in columns left to right.

b. Litterfall only.

Continued from p.330.

U.S.A., Washington	48°15'N 121°37'W 152 m Near Darrington.		47°13'N 123°05'W 137 m Matlock, Shelton.	
Infertile, acid, sandy loam.	*Pseudotsuga menziesii*			
	Untreated	560 kg N applied per hectare at ages 25-29.	Untreated	336 kg N applied per hectare at age 37.
Age (years)	32	32	38	38
Trees/ha	4040	3520	1600	1440
Tree height (m)	9-11	9-12	9-12	9-12
Basal area (m²/ha)	14.5	23.4	35.8	32.1
Leaf area index				
Stem volume (m³/ha)	46	80	280	237
Dry biomass (t/ha) Stem wood	19.2	34.4	115.5	91.9
Stem bark	4.1	6.6	14.4	13.1
Branches	4.6	9.4	13.9	19.0
Fruits etc.				
Foliage	5.3	9.6	9.0	16.2
Root estimate	20.7	19.7	16.8	15.6
CAI (m³/ha/yr)	3.4	8.1	17.8	28.5
Net production (t/ha/yr) Stem wood	1.4^a	3.5^a	8.4^a	13.1^a
Stem bark				
Branches	0.1^b	0.0^b	0.4^b	0.3^b
Fruits etc.				
Foliage	1.0^b	1.4^b	1.8^b	2.4^b
Root estimate				

At Darrington, 10 trees were sampled per plot in winter; at Shelton 4 trees were sampled per plot in March. Stand biomass values were derived from regressions on D. Root biomass values were estimated from soil core samples. There was 2.6, 3.2, 6.7 and 6.1 t/ha of dead branches in columns left to right. Nutrient contents were determined.
a. Excluding the bark.
b. Litterfall only.

Continued from p.331.

48°17'N 122°40'W   91 m   U.S.A., Washington, near Oak Harbour, Whidbey Island.

	Untreated	224 kg N applied per hectare at age 50.
Sandy, acid, infertile loam.	*Pseudotsuga menziesii* with some *Tsuga heterophylla* and *Thuja plicata*.	
Age (years)	52	52
Trees/ha	2480	2160
Tree height (m)	16	17
Basal area (m²/ha)	46.6	52.1
Leaf area index		
Stem volume (m³/ha)	339	337

		Untreated	224 kg N applied
Dry biomass (t/ha)	Stem wood	147.5	142.5
	Stem bark	27.3	30.9
	Branches	17.9	21.3
	Fruits etc.		
	Foliage	12.0	13.9
	Root estimate	12.3	11.4
CAI (m³/ha/yr)		11.2	12.6
Net production (t/ha/yr)	Stem wood	5.3[a]	7.1[a]
	Stem bark		
	Branches	0.9[b]	0.8[b]
	Fruits etc.		
	Foliage	1.5[b]	2.1[b]
	Root estimate		

Four trees were sampled per plot in winter, and stand biomass values were derived from regressions on D. Root biomass was estimated from soil core samples. There was 11.2 and 9.0 t/ha of dead branches in the left and right columns, respectively. Nutrient contents were determined.
a. Excluding the bark.
b. Litterfall only.

Keyes, M.R. and Grier, C.C. (1981). Above- and below-ground net production in 40-
year-old Douglas-fir stands on low and high productivity sites. *Can. J. For. Res.*
11, 599–605.

46°40'N   122°20'W   320 m   U.S.A., Washington, 90 km SE of Seattle, Charles
Lathrop Pack Experimental Forest.

Pseudotsuga menziesii

		Poor site: gravelly loam sand, low in N and low base saturation.	Good site: colluvial soil with 33% base saturation.
Age (years)		40	40
Trees/ha			
Tree height (m)		23	33
Basal area (m²/ha)			
Leaf area index			
Stem volume (m³/ha)			
Dry biomass (t/ha)	Stem wood	188.5	368.8
	Stem bark	33.0	55.2
	Branches	17.1	27.7
	Fruits etc.		
	Foliage	10.0	16.0
	Root estimate	$49.3 + 8.3^a$	$85.4 + 2.7^a$
CAI (m³/ha/yr)			
Net production (t/ha/yr)	Stem wood	4.2^b	8.2^b
	Stem bark	0.9^b	1.7^b
	Branches	0.2^b	0.6^b
	Fruits etc.		
	Foliage	2.0^c	3.2^c
	Root estimate	$2.5 + {>}5.6^a$	$2.7 + {>}1.4^a$

Stand biomass values for one 400 m² plot per stand were derived using published
regressions of woody biomass on D and foliage biomass on sapwood basal area. Production was estimated over the previous one year. Fine root biomass and increment
values were estimated by soil core sampling and root observations at underground
windows.
a. Fine roots, not more than 2 mm in diameter.
b. Including estimated mortality, but excluding woody litterfall.
c. Assumed to be 20% of the foliage biomass.

Whittaker, R.H. and Niering, W.A. (1975). Vegetation of the Santa Catalina Mountains, Arizona. V Biomass, production and diversity along the elevation gradient. *Ecology* 56, 771-790.

Whittaker, R.H. and Niering, W.A. (1968). Vegetation of the Santa Catalina Mountains, Arizona. IV Limestone and acid soils. *J. Ecol.* 56, 523-544.

ca.32°20'N 110°50'W   2645 m   U.S.A., Arizona, Santa Catalina Mountains, near Tucson.

Soils described by Whittaker and Niering (1968).	*Pseudotsuga menziesii* (88%)[a] *Abies concolor* (7%)[a] et al.	*P. menziesii* (79%)[a] *A. concolor* (17%)[a] et al.
Age (years)	252[b]	321[b]
Trees/ha	340 380[c]	400 870[c]
Tree height (m)	27.6[b]	27.9[b]
Basal area (m²/ha)	70.5 + 0.3[c]	118.1 + 1.4[c]
Leaf area index	6.7[d]	7.3[d]
Stem volume (m³/ha)	980[e]	1666[e]

Dry biomass (t/ha)

Stem wood	} 363 + 0.3[c]	} 681 + 4.0[c]
Stem bark		
Branches	57 + 0.2[c]	82 + 2.3[c]
Fruits etc.		
Foliage	17 + 0.0[c]	20 + 2.3[c]
Root estimate		

CAI (m³/ha/yr)	3.1[e]	5.9[e]

Net production (t/ha/yr)

Stem wood	2.27[f] } + 0.02[cf]	3.40[f] } + 0.21[cf]
Stem bark	0.67[f]	1.13[f]
Branches	1.39[f] + 0.02[cf]	1.77[f] + 0.23[cf]
Fruits etc.	0.40	0.50
Foliage	3.60 + 0.03[c]	3.95 + 0.46[c]
Root estimate		

Ten to fifteen trees were sampled of each of the major species, and stand biomass values for the above two 0.1 ha plots were derived from regressions on D, wood volumes and surface areas, and from other relationships. All trees and shrubs over 1 cm D were included.
a. Percentage of the total stem volume.
b. Weighted mean ages and heights.
c. Understorey shrubs.
d. All-sided LAI values were 15.5 and 16.7 in columns left and right, respectively.
e. Parabolic volumes.
f. Excluding woody litterfall and any mortality.

Gholz, H.L. (1981). Environmental limits on aboveground net primary production, leaf area, and biomass in vegetation zones of the Pacific Northwest. *Ecology* (in press).

Gholz, H.L., Grier, C.C., Campbell, A.G. and Brown, A.T. (1979). "Equations for Estimating Biomass and Leaf Area of Plants in the Pacific Northwest." Forest Research Laboratory, Oregon State University, Corvallis, U.S.A. Research Paper no.41.

Gholz, H.L., Fitz, F. and Waring, R.H. (1976). Leaf area differences associated with old-growth forest communities in the western Oregon Cascades. *Can. J. For. Res.* 6, 49-57.

44-45°N   122-124°W   (alt. given below)   U.S.A., Oregon.

	Interior Coast Range	Western Cascade Mountains
	Pseudotsuga menziesii and *Abies grandis*	*P. menziesii*
	365 m	410 m
Age (years)	150	125
Trees/ha	312[a]	488[a]
Tree height (m)	35-55	35-55
Basal area (m²/ha)	84.2	54.5
Leaf area index	7.8[b]	6.5[b]
Stem volume (m³/ha)		

Dry biomass (t/ha)

	Interior Coast Range	Western Cascade Mountains
Stem wood	} 789	} 407
Stem bark		
Branches	60	30
Fruits etc.		
Foliage	16	12
Root estimate		

CAI (m³/ha/yr)

Net production (t/ha/yr)

	Interior Coast Range	Western Cascade Mountains
Stem wood	} 5.0[c]	} 3.0[c]
Stem bark		
Branches	0.5[c]	0.2[c]
Fruits etc.		
Foliage	5.0[d]	3.0[d]
Root estimate		

Stand values for plots of at least 0.25 ha were derived from regressions on D.

a. Trees over 10 cm D; there were 110 and 12 trees/ha less than 10 cm D in columns left and right, respectively.

b. All-sided LAI values were 18 and 15 in columns left and right, respectively.

c. Excluding woody litterfall and any mortality.

d. Assumed to be 20-30% of the foliage biomass, depending on the species.

Fujimori, T., Kawanabe, S., Saito, H., Grier, C.C. and Shidei, T. (1976). Biomass and primary production in forests of three major vegetation zones of the north-western United States. *J. Jap. For. Soc.* 58, 360–373.

Gholz, H.L. (1981). Environmental limits on aboveground net primary production, leaf area, and biomass in vegetation zones of the Pacific Northwest. *Ecology* (in press).

U.S.A., Oregon	44°10'N 122°31'W 450 m Blue River	44–45°N 121–122°W 1500 m West Cascades
At Blue River: red-brown gravelly silt clay loams.	*Pseudotsuga menziesii* (91%)[a] *Acer macrophyllum* (6%)[a] *et al.*	*P. menziesii*, with *Tsuga heterophylla* and *Abies amabilis*
	(Fujimori *et al.* 1976)	(Gholz 1981)
Age (years)	90–110	150
Trees/ha	478	1005[b]
Tree height (m)	62.6[c]	35–55
Basal area (m²/ha)	63.3	72.4
Leaf area index		9.6[d]
Stem volume (m³/ha)	1406	

Dry biomass (t/ha)		
Stem wood	529.9	} 467
Stem bark	71.2	
Branches	49.0	43
Fruits etc.		
Foliage	11.1	18
Root estimate		

Net production (t/ha/yr)		
CAI (m³/ha/yr)	19.1	
Stem wood	7.2[e]	} 5.0[e]
Stem bark	1.0[e]	
Branches	1.7[e]	0.5[e]
Fruits etc.		
Foliage	2.8	4.0[f]
Root estimate		

Fujimori *et al.* (1976) derived stand biomass values for a 0.38 ha plot, using regressions on D^2H for the main species and proportional basal area allocation for the minor species; branches and bark were assumed to grow at the same relative rates as stem wood.
Gholz (1981) derived stand biomass values for a plot of at least 0.25 ha from regressions on D.
a. Percentage of the total basal area.
b. Trees over 10 cm D; there were 1250 trees/ha less than 10 cm D.
c. Mean height of the dominant trees.   *d.* All-sided LAI was 22.
e. Excluding woody litterfall and mortality.   *f.* Foliage production was assumed to be between 20 and 30% of the foliage biomass, depending on the species.

Fogel, R. and Hunt, G. (1979). Fungal and arboreal biomass in a western Oregon Douglas-fir ecosystem: distribution patterns and turnover. *Can. J. For. Res.* <u>9</u>, 245-256.

44°28'N 123°29'W   305 m   U.S.A., Oregon, 11.3 km SW of Philomath.

Well-drained
gravelly loam.
pH 5.2-5.7.

Pseudotsuga menziesii (94%)[a] with
Castanopsis chrysophylla and *Alnus rubra*.

Overstocked second generation forest.

Age (years)	40 (35-50)
Trees/ha	1626[b]
Tree height (m)	24.1
Basal area (m²/ha)	49.1
Leaf area index	
Stem volume (m³/ha)	423[b]

Dry biomass (t/ha)

Stem wood	} 212.9
Stem bark	
Branches	22.8
Fruits etc.	
Foliage	14.7
Root estimate	(49.3 + 15.0)[c]

CAI (m³/ha/yr)

Net production (t/ha/yr)

Stem wood	} 1.37[d]
Stem bark	
Branches	0.27[e]
Fruits etc.	
Foliage	2.41[e]
Root estimate	9.16[f]

Stand biomass values were derived from regressions on D taken from Dice (1970) (see p.329). There was 7.7 t/ha of dead branches.
a. Percentage of the total basal area.
b. Live *P. menziesii* trees over 10.2 cm D.
c. Non-mycorrhizal (49.3 t/ha) plus mycorrhizal (15.0 t/ha) roots.
d. Mortality only, taken as dead log biomass divided by stand age; stem and branch increments were not determined.
e. Litterfall, measured over one year, adjusted for pre-fall losses.
f. Estimated turnover of mycorrhizal roots.

Franklin, J.F. and Waring, R.H. (1980). Distinctive features of the northwestern coniferous forest: development, structure and function. In: "Forests: Fresh Perspectives from Ecosystem Analysis" (R.H. Waring, ed.), pp.59-86. Oregon State University Press, Corvallis, Oregon, U.S.A.

---

ca.46°45'N   122°W   below 1200 m   U.S.A., Washington, Mount Rainier National Park.

Pseudotsuga menziesii with *Tsuga heterophylla*.

Age (years)	500^a	500^a	500^a	500^a	500^a	500^a	500^a	500^a
Trees/ha								
Tree height (m)								
Basal area (m² /ha)	50	81	76	65	89	69	98	74
Leaf area index	7.1^b	11.7^b	13.2^b	9.7^b	12.2^b	10.7^b	14.7^b	12.2^b
Stem volume (m³ /ha)								
Dry biomass (t/ha) Stem wood, Stem bark, Branches	303	567	559	586	933	520	760	539
Fruits etc.								
Foliage	14	23	26	19	24	21	29	24
Root estimate								
CAI (m³ /ha/yr)								
Net production (t/ha/yr) Stem wood								
Stem bark								
Branches								
Fruits etc.								
Foliage								
Root estimate								

The biomass of one hectare reference stands or permanent sample plots were derived using published regressions on D and H (Dice, 1970; Heilman, 1961; Grier and Logan, 1977) (see pp. 329, 330 and 340, respectively).
a. Age of the oldest trees.
b. Approximate all-sided LAI values can be obtained by multiplying by 2.3.

Continued from p.338.

44°15'N 122°20'W   360–1200 m   U.S.A., Oregon, 70 km E of Eugene, Andrews
                                  Experimental Forest.

Gravelly
silty clay.   *Pseudotsuga menziesii* with *Tsuga heterophylla*.

Age (years)	450^a	450^a	450^a	450^a	450^a	450^a	450^a	450^a
Trees/ha								
Tree height (m)								
Basal area (m²/ha)	68	84	92	99	118	116	92	129
Leaf area index	7.1^b	9.2^b	12.7^b	10.2^b	14.7^b	15.2^b	10.7^b	15.2^b
Stem volume (m³/ha)								

Dry biomass (t/ha)									
Stem wood								
Stem bark								
Branches	701	893	801	1203	1208	1107	1018	1392
Fruits etc.								
Foliage	14	18	25	20	29	30	21	30
Root estimate								

CAI (m³/ha/yr)

Net production (t/ha/yr)								
Stem wood								
Stem bark								
Branches								
Fruits etc.								
Foliage								
Root estimate								

See p.338.

Grier, C.C. and Logan, R.S. (1977). Old-growth *Pseudotsuga menziesii* communities of a western Oregon watershed: biomass distribution and production budgets. *Ecol. Monogr.* <u>47</u>, 373-400.

Grier, C.C., Cole, D.W., Dyrness, C.T. and Fredriksen, R.L. (1974). Nutrient cycling in 37- and 450-year-old Douglas fir ecosystems. In: "Integrated Research in the Coniferous Forest Biome" (R.H. Waring and R.L. Edmonds, eds), pp.21-36. US/IBP Bull. No.5, University of Washington, Seattle, Washington, U.S.A.

44°15'N 122°20'W   430-470 m   U.S.A., Oregon, 70 km E of Eugene, Andrews Experimental Forest.

Gravelly silty clay. pH 6.2.

Pseudotsuga menziesii, with *Tsuga heterophylla*, *Thuja plicata et al.*

	Xeric	Xeric	Warm mesic	Mesic	Cool moist
	90%[a]	91%[a]	90%[a]	90%[a]	81%[a]
Age (years)	450[b]	450[b]	450[b]	450[b]	450[b]
Trees/ha	ca.290	ca.290	ca.290	ca.290	ca.290
Tree height (m)	70[b]	70[b]	70[b]	70[b]	70[b]
Basal area (m²/ha)	ca.62	ca.62	ca.62	ca.62	ca.62
Leaf area index	ca.12	ca.12	ca.12	ca.12	ca.12

Stem volume (m³/ha)

Dry biomass (t/ha)

	Xeric	Xeric	Warm mesic	Mesic	Cool moist
Stem wood	538.8 ⎫	794.7 ⎫	399.1 ⎫	655.2 ⎫	438.3 ⎫
Stem bark	68.3 ⎬ +4.6[c]	96.9 ⎬ +4.9[c]	48.3 ⎬ +6.8[c]	78.7 ⎬ +1.1[c]	52.6 ⎬ +8.8[c]
Branches	41.7 ⎭	64.4 ⎭	32.7 ⎭	54.2 ⎭	47.9 ⎭
Fruits etc.					
Foliage	11.6 + 1.7[c]	14.5 + 1.8[c]	8.6 + 1.8[c]	14.1 + 4.5[c]	11.4 + 1.3[c]
Root estimate	143.8	204.0	104.9	172.8	122.6

CAI (m³/ha/yr)

Net production (t/ha/yr)

	Xeric	Xeric	Warm mesic	Mesic	Cool moist
Stem wood	⎫	⎫	⎫	⎫	⎫
Stem bark	⎬ -4.1 + 1.1[c] + 6.6[e]	⎬ -6.2 + 1.2[c]	⎬ -3.1 + 0.3[c]	⎬ -5.0 + 0.2[c]	⎬ -2.9 + 0.4[c]
Branches	⎬	⎬	⎬	⎬	⎬
Fruits etc.	⎬ 4.3[d] + 0.4[f]	⎬ 5.0[d] + 0.4[f]	⎬ 3.8[d] + 0.4[f]	⎬ 4.4[d] + 0.4[f]	⎬ 4.2[d] + 0.4[f]
Foliage	⎭	⎭	⎭	⎭	⎭
Root estimate	2.8[g]	3.0[g]	2.0[g]	3.3[g]	2.7[g]

Sixty-one trees were sampled and stand biomass values were derived using calculated regressions on D and regressions published by Dice (1970) and Heilman (1961) (see pp. 329 and 330). Fine roots were core sampled. There was 4.7, 5.3, 3.2, 5.3 and 4.3 t/ha of dead branches in columns left to right. Values given here are from Grier and Logan (1977) which updated those published earlier.

a. Percentage of the total biomass accounted for by *P. menziesii*. b. Age and height of the oldest trees. c. Understorey shrubs. d. Total litterfall, measured over 2 years. e. Mortality, estimated in this stand only. f. Through-fall and consumption losses. g. Coarse roots were assumed to grow at the same relative rate as above-ground woody parts and fine root production was assumed to be 20% of the fine root biomass; but see updated values on p.341.

Grier, C.C. and Logan, R.S. (1977). Old-growth *Pseudotsuga menziesii* communities of a western Oregon watershed: biomass distribution and production budgets. *Ecol. Monogr.* <u>47</u>, 373-400.

Santantonio, D., Hermann, R.K. and Overton, W.S. (1977). Root biomass studies in forest ecosystems. *Pedobiologia* <u>17</u>, 1-31.

Santantonio, D. (1979). Seasonal dynamics of fine roots in mature stands of Douglas fir of different water regimes - a preliminary report. In: "Root Physiology and Symbiosis" (A. Riedacker & J. Gagnaire-Michard, eds), pp.190-193. Nancy, France.

44°15'N   122°20'W   430-670 m   U.S.A., Oregon, 70 km E of Eugene, Andrews
                                   Experimental Forest.

Gravelly
silty clay,   *Pseudotsuga menziesii* (80%)^a, *Tsuga heterophylla*,
pH 6.2.
                          Thuja plicata with understorey shrubs.

Age (years)	up to 450
Trees/ha	290
Tree height (m)	5 to 70
Basal area (m²/ha)	62.7
Leaf area index	12.5
Stem volume (m³/ha)	

Dry biomass (t/ha)

Stem wood	575.8 ⎫
Stem bark	70.4 ⎬ + 5.3^b
Branches	47.8 ⎭
Fruits etc.	
Foliage	12.4 + 1.5^b
Root estimate	209^{cg}

CAI (m³/ha/yr)

Net production (t/ha/yr)

Stem wood	⎫
Stem bark	⎬ -4.3 + 0.6^b + 7.0^d
Branches	⎭
Fruits etc.	⎫ 4.3^e + 0.4^f
Foliage	⎭
Root estimate	8.5 to 10.2^g

These are average values for the whole watershed, and update values published earlier. Sixty-one trees were sampled and stand values were derived from regressions on D. There was an average of 4.8 t/ha of dead branches. Roots of three large trees were excavated and fine roots were core sampled.

a. Percentage of the total basal area.
b. Understorey shrubs.
c. Comprised of 198 t/ha of roots over 1 mm diameter and 11 t/ha of fine roots.
d. Mortality.
e. Total litterfall.
f. Throughfall and consumption losses.
g. From Santantonio *et al.* (1977) and Santantonio (1979).

Turner, J. and Long, J.N. (1975). Accumulation of organic matter in a series of Douglas fir stands. *Can. J. For. Res.* 5, 681–690.

Turner, J. (1981). Nutrient cycling in an age sequence of western Washington Douglas fir stands. *Ann. Bot.* 48, 159–169.

---

ca.47°50'N   123°00'W   210 m   U.S.A., Washington, Cedar River, Thompson Research Center.

Gravelly loams derived from glacial till.

Pseudotsuga menziesii with understorey shrubs.

	Natural regeneration		Plantation	Natural regeneration
Age (years)	22	30	30	42
Trees/ha	2756	2346	1800	822
Tree height (m)				
Basal area (m²/ha)	42.4	32.4	34.4	35.7
Leaf area index				
Stem volume (m³/ha)				

Dry biomass (t/ha)

Stem wood	99.4	121.3	128.5	157.5
Stem bark	13.9	16.1	17.4	19.5
Branches	13.2^a }$+7.6^b$	15.6^a }$+5.1^b$	16.7^a }$+10.9^b$	19.5^a }$+4.2^b$
Fruits etc.				
Foliage	5.0	6.2	6.5	8.3
Root estimate				

CAI (m³/ha/yr)

Net production (t/ha/yr)

Stem wood	}6.13^c	}4.76^c	}4.98^c	}6.39^c
Stem bark				
Branches	0.54^c }$+1.99^b$	0.53^c }$+0.95^b$	0.54^c	0.67^c }$+0.63^b$
Fruits etc.				
Foliage	2.10^d	3.14^d	2.10^d	2.23^d
Root estimate				

Stand values for the above 450 m² plots were derived from regressions on D calculated by Dice (1970) (see p.329). Nutrient contents were determined.
a. Including stem biomass above the base of the crowns.
b. Understorey shrubs.
c. Including woody litterfall and mortality, measured over one year.
d. New foliage biomass; total litterfall was 2.67, 3.53, 2.50 and 2.57 t/ha/yr in columns left to right.

Continued from p.342.

	Plantation 100%[a]	Plantation 73%[a]	Natural regeneration 95%[a]

ca. 47°50'N 123°00'W   210 m   U.S.A., Washington, Cedar River, Thompson Research Center.

Gravelly loams derived from glacial till.

Pseudotsuga menziesii with *Tsuga heterophylla*, *Thuja plicata*, and understorey shrubs.

	Plantation 100%[a]	Plantation 73%[a]	Natural regeneration 95%[a]
Age (years)	42	49	73
Trees/ha	1289	1067	1889
Tree height (m)			
Basal area (m²/ha)	44.5	41.6	57.2
Leaf area index			
Stem volume (m³/ha)			

Dry biomass (t/ha)

	Plantation 100%[a]	Plantation 73%[a]	Natural regeneration 95%[a]
Stem wood	182.6	178.4	237.0
Stem bark	23.6	22.8	30.4
Branches	23.2[b] } +3.4[c]	23.4[b] } +3.4[c]	26.2[b] } +2.8[c]
Fruits etc.			
Foliage	9.4	9.4	10.8
Root estimate			

CAI (m³/ha/yr)

Net production (t/ha/yr)

	Plantation 100%[a]	Plantation 73%[a]	Natural regeneration 95%[a]
Stem wood	} 3.65[d]	} 3.30[d]	} 2.50[d]
Stem bark			
Branches	0.48[d] } +0.54[c]	0.42[d] } +0.53[c]	0.33[d] } +0.57[c]
Fruits etc.			
Foliage	2.44[e]	2.20[e]	2.28[e]
Root estimate			

Stand values for the above 450 m² plots were derived from regressions on D calculated by Dice (1970) (see p.329). Nutrient contents were determined.

a. Percentage of the total tree number that were *P. menziesii*.
b. Including stem biomass above the base of the crowns.
c. Understorey shrubs.
d. Including woody litterfall and mortality, measured over one year.
e. New foliage biomass; total litterfall was 3.12, 2.28 and 3.73 t/ha/yr in columns left to right.

Long, J.N. and Turner, J. (1975). Above-ground biomass of understorey and over-storey in an age sequence of four Douglas fir stands. *J. appl. Ecol.* <u>12</u>, 178-188.

ca.47°50'N 123°00'W 210 m U.S.A., Washington, near Seattle, Thompson Research Center.			
Coarse gravelly glacial outwash soils.	*Pseudotsuga menziesii* with *Tsuga heterophylla* with an understorey of broadleaved species. Natural regeneration with incomplete canopy cover.		
	99%[a]	97%[a]	96%[a]	68%[a]

	99%[a]	97%[a]	96%[a]	68%[a]
Age (years)	22	30	42	73
Trees/ha	1664	1941	540	1137
Tree height (m)				
Basal area (m²/ha)	18.0	22.1	25.6	44.8
Leaf area index				
Stem volume (m³/ha)				

Dry biomass (t/ha)

	99%	97%	96%	68%
Stem wood	} 56.6	} 69.6	} 123.8	} 206.8
Stem bark				
Branches	4.0[b] } +7.6[c]	5.9[b] } +5.3[c]	7.3[b] } +4.2[c]	12.6[b] } +2.8[c]
Fruits etc.				
Foliage	2.5	3.1	5.6	9.3
Root estimate				

CAI (m³/ha/yr)

Net production (t/ha/yr)

Stem wood				
Stem bark				
Branches				
Fruits etc.				
Foliage				
Root estimate				

Stand biomass values for five 375 m² plots per stand were derived from regressions on D taken from Dice (1970) and Zavitkovski and Stevens (1972) (see pp. 329 and 252).

a. Percentage of the total basal area accounted for by *P. menziesii*.
b. Including the stems within the crowns.
c. Understorey shrubs.

Westman, W.E. and Whittaker, R.H. (1975). The pygmy forest region of northern California: studies on biomass and primary productivity. *J. Ecol.* <u>63</u>, 493-520.

Whittaker, R.H. (1966). Forest dimensions and production in the Great Smoky Mountains. *Ecology* <u>47</u>, 103-121.

ca.39°20'N   123°45'W   (alt. given below)   U.S.A., California, Mendocino, near Fort Bragg.

Sequoia sempervirens with *Pseudotsuga menziesii et al.*

	83%[a] Hugo soils on 25-50° slopes.			99%[a] Flat alluvial terraces.		
	270 m	270 m	50 m	30 m	60 m	240 m
Age (years)	Mature	Mature	Mature	Mature	Mature	Mature
Trees/ha	$240^b + 230^c$	$170^b + 810^c$	$570^b + 720^c$	$234^b + 78^c$	$400^b + 30^c$	$270^b + 0^c$
Tree height (m)	43^d	64^d	30^d	79^d	81^d	79^d
Basal area (m²/ha)	96	164	144	250	243	247
Leaf area index						
Stem volume (m³/ha)	2055^e	5188^e	2184^e	8980^e	9732^e	
Dry biomass (t/ha) — Stem wood, Stem bark, Branches, Fruits etc., Foliage, Root estimate	732	1799	934	2980	3280	3300
CAI (m³/ha/yr)	4.6^e	12.9^e	15.0^e	13.3^e	16.0^e	
Net production (t/ha/yr) — Stem wood, Stem bark, Branches, Fruits etc., Foliage, Root estimate	5.3^f	13.0^f	18.9^f	11.1^f	13.1^f	18.8^f

Stand values for the above plots of 0.1 to 0.5 ha were derived from regressions on various dimensions following Whittaker *et al.* (1974) (see p.259).

a. Percentage of the total basal area accounted for by *S. sempervirens*.

b. Number of *S. sempervirens*.

c. Number of *P. menziesii*.

d. Weighted mean height.

e. Parabolic volume.

f. Excluding all litterfall, estimated to be about 2.1 t/ha/yr, and excluding mortality.

Fujimori, T. (1977). Stem biomass and structure of a mature *Sequoia sempervirens* stand on the Pacific coast of northern California. *J. Jap. For. Soc.* <u>59</u>, 435-441.

40°20'N 124°00'W   80 m   U.S.A., California, Humboldt State Park, Bull Creek.

Alluvial soils.

Sequoia sempervirens

Age (years)	to over 1000
Trees/ha	167
Tree height (m)	88
Basal area (m²/ha)	338[a]
Leaf area index	
Stem volume (m³/ha)	10817

Dry biomass (t/ha)

Stem wood	⎫
Stem bark	⎬ 3461
Branches	⎭
Fruits etc.	
Foliage	
Root estimate	

CAI (m³/ha/yr)

Net production (t/ha/yr)

Stem wood	
Stem bark	
Branches	
Fruits etc.	
Foliage	
Root estimate	

Eight fallen old trees and eleven felled young trees were used to calculate stand values for a 1.44 ha plot from regressions on D²H.
a. Ninety-six per cent of the basal area was accounted for by 66 trees.

Reiners, W.A. (1972). Structure and energetics of three Minnesota forests. *Ecol. Monogr.* 42, 71–94.

Reiners, W.A. (1974). Foliage production by *Thuja occidentalis* L. from biomass and litter fall estimates. *Am. Midl. Nat.* 92, 340–345.

Reiners, W.A. and Reiners, N.M. (1970). Energy and nutrient dynamics of forest floors in three Minnesota forests. *J. Ecol.* 58, 497–519.

45°30'N   193°20'W   400 m   U.S.A., Minnesota, north of Minneapolis.

	Thuja occidentalis (29%)[a] *Fraxinus nigra* (32%)[a] et al. Marginal fen.	*T. occidentalis* (65%)[a] *Betula papyrifera* (16%)[a] et al. Cedar swamp.
Age (years)	45–50	70–100
Trees/ha	3348	2755
Tree height (m)	ca. 15	ca. 15
Basal area (m²/ha)	25.1	42.2
Leaf area index		
Stem volume (m³/ha)		
Dry biomass (t/ha)		
Stem wood	64.9	104.5
Stem bark	7.9	12.1
Branches	21.4	34.6
Fruits etc.	0.0	0.5
Foliage	3.9	7.8
Root estimate		
CAI (m³/ha/yr)		
Stem wood	2.17	3.19
Stem bark	0.28	0.37
Net production (t/ha/yr) Branches	1.19^{b} + 1.24^{c}	1.37^{b} + 1.28^{c}
Fruits etc.	0.01^{d}	1.10^{d}
Foliage	2.87^{e}	4.10^{e}
Root estimate		

Stand biomass values for sixteen 100 m² plots in each stand were derived from regressions on D and wood volume. Production values were estimated for the previous one year.

a. Percentage of the total biomass.
b. Increment of old branches plus new twigs.
c. Approximate values for woody litterfall; total litterfall was 4.12 and 4.88 t/ha/yr in columns left and right, respectively, according to Reiners and Reiners (1970).
d. Fruits etc. litterfall.
e. Broadleaved foliage biomass plus 35% of the foliage biomass of *T. occidentalis*.

Whittaker, R.H. (1963). Net production of heath balds and forest heaths in the Great Smoky Mountains. *Ecology* <u>46</u>, 176–182.

Whittaker, R.H. (1966). Forest dimensions and production in the Great Smoky Mountains. *Ecology* <u>47</u>, 103–121.

ca.35°40'N 83°30'W (alt. given below)   U.S.A., Tennessee, Great Smoky Mountains, Mount LeConte.

	Tsuga canadensis (40%)[a] *Fagus grandifolia* (28%)[a] et al.	*T. canadensis* (87%)[a] *Betula alleghaniensis* (4%)[a] with understorey of *Rhododendron maximum* (8%)[a]
	430 m	1280 m
Age (years)		
Trees/ha	1300 + 3810[b]	230 + 2730[b]
Tree height (m)	29 7.6[b] (21.1)[c]	34 5.2[b] (30.3)[c]
Basal area (m²/ha)	24.6 + 4.5[b]	56.3 + 4.4[b]
Leaf area index	3.6[b]	2.1[b]
Stem volume (m³/ha)	252[d] + 14.3[bd]	805[d] + 19.1[bd]
Dry biomass (t/ha) — Stem wood, Stem bark, Branches, Fruits etc., Foliage, Root estimate	170 + 23.0[b]	490.0 + 20.5[b]
CAI (m³/ha/yr)	5.4[d] + 0.1[bd]	5.3[d] + 0.5[bd]
Net production (t/ha/yr) — Stem wood, Stem bark, Branches, Fruits etc., Foliage, Root estimate	11.0[e] + 2.3[be]	8.5[e] + 1.7[be]

Stand values were estimated for plots of at least 0.1 ha from the weight of clippings of current year's twigs, from published regressions, from stem volumes, branch/stem biomass ratios, and other relationships.
a. Percentage of the total volume increment; *B. alleghaniensis* syn. *lutea*.
b. Understorey shrubs.
c. Weighted mean height (in brackets).
d. Parabolic volume.
e. Excluding woody litterfall and mortality; total foliage production of trees plus shrubs was 3.7 and 3.2 t/ha/yr in columns left and right, respectively.

Fujimori, T. (1971). "Primary Production of a Young *Tsuga heterophylla* Stand and
some Speculation about Biomass of Forest Communities on the Oregon Coast."
U.S.D.A. Forest Service Research Paper PNW-123, 1-11.
Grier, C.C. (1976). Biomass, productivity and nitrogen-phosphorus cycles in hemlock-
spruce stands of the central Oregon coast. In: "Western Hemlock Management", pp.
71-81. Univ. of Washington, Coll. of Forest Resources. Contribution no.34.
Fujimori, T., Kawanabe, S., Saito, H., Grier, C.C. and Shidei, T. (1976).(see p.336)
J. Jap. For. Soc. 58, 360-373.

45°02'N   123°56'W   50-100 m   U.S.A., Oregon, Cascade Head, near Otis.

Deep fertile red-brown porous loams. pH 3.7-4.4.	*Tsuga heterophylla* Thinned between ages 7 and 8.	*T. heterophylla* (76%)[a] and *Picea sitchensis*, with understorey shrubs.
Age (years)	26 (19 to 32)	121
Trees/ha	6627	373
Tree height (m)	10.0	47.7
Basal area (m²/ha)	49.4	98.2
Leaf area index	16.1[b]	20.2[b]
Stem volume (m³/ha)		1979

Dry biomass (t/ha)

	Tsuga heterophylla	*T. heterophylla* and *Picea sitchensis*
Stem wood	} 150.9	751.5
Stem bark		63.2 } + 2.1[c]
Branches	20.7	48.7
Fruits etc.		
Foliage	21.1	8.1 + 2.1[c]
Root estimate	38.4[d]	186.7[d]

CAI (m³/ha/yr)   16.0

Net production (t/ha/yr)

	Tsuga heterophylla	*T. heterophylla* and *Picea sitchensis*
Stem wood	} 20.4 } + 0.3[e]	6.1 } + 5.8[e] } + 0.4[c]
Stem bark	} + 1.0[f]	0.5 } + 3.4[f]
Branches	4.3	0.9
Fruits etc.		
Foliage	6.0[g] + 0.2[h]	2.9[g] + 0.1[h]
Root estimate	5.5[d]	2.7[d]

Ten trees were sampled from the 26-year-old stand, and stand values were derived by
proportional basal area allocation. Stand values for a 0.41 ha plot of the 121-year-
old trees were derived from regressions on D. Roots were assumed to grow at the same
relative rates as above-ground woody parts. Nutrient contents were determined.
a. Percentage of the total basal area.
b. All-sided LAI values were 37.0 and 46.5 in columns left and right, respectively.
c. Understorey shrubs.   *d.* 'Coarse' roots only.
e. Mortality.   *f.* Woody litterfall.
g. New foliage biomass; foliage litterfall was 6.0 and 2.8 t/ha/yr in columns left
and right, respectively.
h. Consumption.

Greene, S. (1981). Forest Science Laboratories, 3200 Jefferson Way, Corvallis, Oregon 97331, U.S.A. Personal communication.

45°05'N 124°00'W   46 m   U.S.A., Oregon, Cascade Head, Otis.

Tsuga heterophylla (61%)[a]

and *Picea sitchensis*.

Age (years)		up to 250
Trees/ha		367 + 49
Tree height (m)		
Basal area (m²/ha)		35.3 + 22.4
Leaf area index		
Stem volume (m³/ha)		570 + 475
Dry biomass (t/ha)	Stem wood	} 197.7 + 253.2
	Stem bark	
	Branches	70.7 + 21.4
	Fruits etc.	
	Foliage	13.2 + 1.9
	Root estimate	
CAI (m³/ha/yr)		
Net production (t/ha/yr)	Stem wood	
	Stem bark	
	Branches	
	Fruits etc.	
	Foliage	
	Root estimate	

Stand values for circular plots of up to 0.1 ha along 4 transects were derived from regressions on D, based on data from various tree samplings in the Pacific Northwest and British Columbia.

Values are given for *T. heterophylla* plus *P. sitchensis*, left and right, respectively.

a. Percentage of the total basal area.

Gholz, H.L. (1981). Environmental limits on aboveground net primary production, leaf area and biomass in vegetation zones of the Pacific Northwest. *Ecology* (in press).

Gholz, H.L., Grier, C.C., Campbell, A.G. and Brown, A.T. (1979). "Equations for Estimating Biomass and Leaf Area of Plants in the Pacific Northwest." Forest Research Laboratory, Oregon State University, Corvallis, U.S.A. Research Paper no.41.

Gholz, H.L., Fitz, F. and Waring, R.H. (1976). Leaf area differences associated with old-growth forest communities in the western Oregon Cascades. *Can. J. For. Res.* 6, 49–57.

44–45°N   122–124°W   (alt. given below)   U.S.A., Oregon.

	Western Coast Range		Cascade Mountains
	Tsuga heterophylla		*Tsuga mertensiana*
	200 m	200 m	1590 m
Age (years)	130	130	Mature
Trees/ha	294[a]	499[a]	804[a]
Tree height (m)	5	35–55	35–55
Basal area ( )		111.2	57.2
Leaf a		19.1[b]	4.3[b]
Stem v			
Dry biomass (t/ha) — Ste / S		} 1316	} 228
Dry biomass (t/ha) — B		144	35
Dry biomass (t/ha) — Fru			
Dry biomass (t/ha) — Fo		32	15
Dry biomass (t/ha) — Root			
CAI (m³/h)			
Net production (t/ha/yr) — Stem / Stem bark	} 7.0[c]	} 4.0[c]	} 1.0[c]
Net production (t/ha/yr) — Branches	2.0[c]	1.0[c]	0.2[c]
Net production (t/ha/yr) — Fruits etc.			
Net production (t/ha/yr) — Foliage	6.0[d]	8.0[d]	3.0[d]
Net production (t/ha/yr) — Root estimate			

Stand values for plots of 0.25 to 0.41 ha were derived from regressions on D.

a. Trees over 10 cm D; there were 2500, 1500 and 1700 trees/ha less than 10 cm D in columns left to right.

b. All-sided LAI values were 31, 44 and 10 in columns left to right.

c. Excluding woody litterfall and any mortality.

d. Assumed to be 20–30% of the foliage biomass, depending on the species.

Merzoev, O.G. (1981). In: "Dynamic Properties of Forest Ecosystems" (D.E. Reichle, ed.), pp.622-623. Cambridge University Press, Cambridge, London, New York, and Melbourne.

---

ca.41°N 48°E   2000 m   U.S.S.R., Azerbaijan, Caucasus.

Betula pendula syn. *verrucosa*

Age (years)	20	20
Trees/ha	9480	3560
Tree height (m)	4.5	6.5
Basal area (m²/ha)		
Leaf area index		
Stem volume (m³/ha)		
Dry biomass (t/ha) Stem wood	} 18.0	} 16.9
Stem bark		
Branches	5.3	4.1
Fruits etc.		
Foliage	2.3	1.3
Root estimate		
CAI (m³/ha/yr)		
Net production (t/ha/yr) Stem wood		
Stem bark		
Branches		0.07[a]
Fruits etc.		
Foliage	2.34[a]	1.27[a]
Root estimate		

a. Litterfall only.

Rodin, L.E. and Bazilevich, N.I. (1967). "Production and Mineral Cycling in Terrestrial Vegetation." Oliver and Boyd, Edinburgh and London. (Quoting Russian authors in their Table 32).

U.S.S.R. Podzols	ca.56°N 36°E -- Moscow Province *Betula verrucosa*[a]	ca.55°N 83°E -- Novosibirsk	
		B. verrucosa[a] *Populus* sp. *et al.*	*Betula pubescens* *Populus* sp. *et al.*
Age (years)	42	35	20
Trees/ha			
Tree height (m)	19.4	15.0	8.0
Basal area (m²/ha)			
Leaf area index			
Stem volume (m³/ha)			
Dry biomass (t/ha)			
Stem wood			
Stem bark	} 203.4	} 164.2	} 57.0
Branches			
Fruits etc.			
Foliage	3.6[b]	5.2[b]	2.7[b]
Root estimate	42.8	43.9	30.1
CAI (m³/ha/yr)			
Net production (t/ha/yr)			
Stem wood			
Stem bark	} 5.6 + 1.6[c]	} 5.5 + 1.2[c]	} 4.2 + 0.5[c]
Branches			
Fruits etc.			
Foliage	3.6[b]	5.2[b]	2.7[b]
Root estimate	1.9	1.9	2.9

Values given above include understorey shrubs. Nutrient contents were determined.
a. *B. verrucosa* syn. *pendula*.
b. Foliage litterfall.
c. Woody litterfall.

Rodin, L.E. and Bazilevich, N.I. (1967). "Production and Mineral Cycling in Terrestrial Vegetation." Oliver and Boyd, Edinburgh and London. (Quoting Russian authors in their Table 32.)

---

ca.45°N 36°E   --   U.S.S.R., Crimea.

Brown leached   *Carpinus betulus*, with a few *Fagus sylvatica*.
forest soils.   *Quercus* spp. *et al.* with understorey shrubs.

Age (years)	46	
Trees/ha		
Tree height (m)	17.0	
Basal area (m²/ha)		
Leaf area index		
Stem volume (m³/ha)		

Dry biomass (t/ha)	Stem wood	⎫
	Stem bark	⎬ 216.3
	Branches	⎭
	Fruits etc.	
	Foliage	6.4
	Root estimate	57.6

	CAI (m³/ha/yr)	
Net production (t/ha/yr)	Stem wood	
	Stem bark	
	Branches	
	Fruits etc.	
	Foliage	
	Root estimate	

Values given above include understorey shrubs. Nutrient contents were determined.

Rodin, L.E. and Bazilevich, N.I. (1967). "Production and Mineral Cycling in Terrestrial Vegetation." Oliver and Boyd, Edinburgh and London. (Quoting Russian authors in their Tables 9 and 32.)

U.S.S.R.	ca.52°N 39°E -- Voronezh Province *Populus tremula,* *Quercus* spp., *Aegopodium* sp. *et al.* Grey sandy loams		ca.56°N 30°E -- Velikiye Luki Province *P. tremula* (70%)[a], *Betula* spp., *Picea abies* *et al.* Peaty podzol
Age (years)	25	50	45–55
Trees/ha			
Tree height (m)	17.0	28.0	(6.6 26.3)[b]
Basal area (m²/ha)			
Leaf area index			
Stem volume (m³/ha)			
Dry biomass (t/ha) Stem wood	147.0	253.0	191.9
Stem bark			
Branches			
Fruits etc.			
Foliage	3.4	4.8	6.1
Root estimate	35.9	47.1	32.9
CAI (m³/ha/yr)			
Net production (t/ha/yr) Stem wood	13.5 + 2.0[c]	10.6 + 7.1[c]	5.6 + 2.9[c]
Stem bark			
Branches			
Fruits etc.			
Foliage	3.5[c]	4.7[c]	3.9[c]
Root estimate	3.7	2.1	1.2

Values given above include understorey shrubs. Nutrient contents were determined.
a. Percentage of the total tree number.
b. Heights of the lower and upper storeys.
c. Litterfall.

Safarov, I.S. and Djhalilov, K.G. (1973). Biological productivity of the *Quercus castaneifolia* forests of the Talysh region (Soviet Azerbaijan). *Lesovedenie* 3, 40-46.

Djhalilov, K.G. and Safarov, I.S. (1981). In: "Dynamics of Forest Ecosystems" (D.E. Reichle, ed.), p.625. Cambridge University Press, Cambridge, London, New York, Melbourne.

38°40-50'N 48°30'E   (alt. given below)   U.S.S.R., Azerbaijan Region.

Subtropical yellow soils.	*Quercus castaneifolia, Zelkova carpinifolia, Parrotia persica*	*Q. castaneifolia* (50%)[a], *Alnus glutinosa* var. *barbata* (30%)[a], *Ulmus carpinifolia* syn. *nitens* (10%)[a] et al.
	450 m	minus 22 m
Age (years)	75-80	50-60
Trees/ha	490	420
Tree height (m)	27.5	22.6
Basal area (m²/ha)	30.9	
Leaf area index		
Stem volume (m³/ha)		

		450 m	minus 22 m
Dry biomass (t/ha)	Stem wood	} 292.8	} 250.0
	Stem bark		
	Branches	49.7	79.0
	Fruits etc.		
	Foliage	7.2	8.2
	Root estimate		

CAI (m³/ha/yr)

		450 m	minus 22 m
Net production (t/ha/yr)	Stem wood	} 0.08[b]	} 0.07[b]
	Stem bark		
	Branches	0.91[b]	0.78[b]
	Fruits etc.	0.32[b]	0.15[b]
	Foliage	3.65[b]	5.67[b]
	Root estimate		

---

a. Percentage of the total tree number.
b. Litterfall only.

Rodin, L.E. and Bazilevich, N.I. (1967). "Production and Mineral Cycling in Terrestrial Vegetation." Oliver and Boyd, Edinburgh and London. (Quoting Russian authors in their Table 32.)

U.S.S.R.	ca.52°N 39°E -- Voronezh Province *Quercus robur* (60-90%)[a], *Fraxinus excelsior*, *Acer platanoides*, *et al.* Grey sandy loams.				ca.45°N 36°E -- Crimea *Q. robur* with a few *F. excelsior*, *Cornus* sp. *et al.* Brown leached soils.
Age (years)	12	43	48	220	40
Trees/ha					
Tree height (m)	5.0	17.5	23.0	30.0	11.0
Basal area (m²/ha)					
Leaf area index					
Stem volume (m³/ha)					
Dry biomass (t/ha) Stem wood	} 43.0	} 104.9	} 187.0	} 402.8	} 118.9
Stem bark					
Branches					
Fruits etc.					
Foliage	3.3	3.8	3.6	3.7	4.1
Root estimate	22.7	45.9	70.2	97.3	31.8
CAI (m³/ha/yr)					
Net production (t/ha/yr) Stem wood	} 1.9+1.1[b]	} 3.7+2.0[b]	} 3.6+1.1[b]	} 2.5+1.8[b]	
Stem bark					
Branches					
Fruits etc.					
Foliage	3.3[b]	3.7[b]	3.6[b]	3.7[b]	
Root estimate	2.2	0.6	1.6	0.3	

Values given above include understorey shrubs. Nutrient contents were determined.
a. Percentage of the total tree number.
b. Litterfall.

Goryshina, T.K. (1981). In: "Dynamic Properties of Forest Ecosystems" (D.E. Reichle, ed.), pp.626-627. Cambridge University Press, Cambridge, London, New York.

Goryshina, T.K. (1974a). Investigations of biological productivity and factors affecting it in the Les na Vorske forest-steppe oak wood. *Ekologija* 3, 5-10.

Goryshina, T.K. (ed.) (1974b). "Biological Productivity and its Factors in the Oaks of the Forest Steppe." Scientific Notes, Series of Biological Science No.367, 2. Leningrad University Press, Leningrad.

50°38'N 35°58'E   200 m   U.S.S.R., Belgorod Region, Les na Vorske.

Grey soils.
pH 4.9-6.4

Quercus robur, Tilia cordata syn. *parvifolia,*
Acer platanoides and *Ulmus scabra,*
with understorey shrubs.

Age (years)	80	250
Trees/ha	446	557
Tree height (m)	25	31
Basal area (m²/ha)	25	37
Leaf area index		
Stem volume (m³/ha)		

Dry biomass (t/ha)

Stem wood	$161.2 + 3.4^a$	$191.1 + 2.5^a$
Stem bark	$21.7 + 1.5^a$	$36.9 + 0.3^a$
Branches	$55.3 + 3.0^a$	$67.9 + 3.2^a$
Fruits etc.		
Foliage	$2.7 + 0.1^a$	$3.6 + 0.1^a$
Root estimate	45.9	91.7

CAI (m³/ha/yr)

Net production (t/ha/yr)

Stem wood	$\Big\} 2.52 + 0.43^b$	$\Big\} 2.37 + 2.20^b$
Stem bark		
Branches	$1.64 + 0.99^a + 1.40^c$	$1.27 + 0.58^a + 2.30^c$
Fruits etc.		
Foliage	3.56^c	3.76^c
Root estimate		

a. Understorey shrubs.
b. Mortality.
c. Litterfall.

Rodin, L.E. and Bazilevich, N.I. (1967). "Production and Mineral Cycling in Terrestrial Vegetation." Oliver and Boyd, Edinburgh and London. (Quoting Russian authors in their Table 32.)

---

ca.54°N 45°E   --   U.S.S.R., Mordovskaya.

Grey forest   *Tilia cordata* syn. *parvifolia*
sandy loams.   with a few
                     Picea abies, *Betula* spp. *et al.*

Age (years)	40	74
Trees/ha		
Tree height (m)	18.0	22.6
Basal area (m²/ha)		
Leaf area index		
Stem volume (m³/ha)		

Dry biomass (t/ha)			
	Stem wood		
	Stem bark	} 116.8	} 165.1
	Branches		
	Fruits etc.		
	Foliage	2.8	4.5
	Root estimate	38.9	55.4

CAI (m³/ha/yr)			
Net production (t/ha/yr)	Stem wood		
	Stem bark	} $4.9 + 1.6^a$	} $1.8 + 0.7^a$
	Branches		
	Fruits etc.		
	Foliage	2.9^a	4.5^a
	Root estimate	1.8	0.9

Values given above include understorey shrubs. Nutrient contents were determined.
a. Litterfall.

Karpov, V.G. (ed.) (1973). "Structure and Productivity of Spruce Forests of the Southern Taiga." Nauka, Leningrad Branch, Academy of Sciences, Leningrad, USSR.

56°30'N 32°40'E   200 m   U.S.S.R., Central Forest Reserve.

Clayed weak
podzol,   *Picea abies*
pH 4.0-4.5.

Age (years)	110
Trees/ha	678
Tree height (m)	26.5
Basal area (m²/ha)	
Leaf area index	9.7
Stem volume (m³/ha)	

Dry biomass (t/ha)		
	Stem wood	155.9
	Stem bark	12.6
	Branches	23.5
	Fruits etc.	
	Foliage	12.5
	Root estimate	68.1

CAI (m³/ha/yr)

Net production (t/ha/yr)	
Stem wood	
Stem bark	
Branches	
Fruits etc.	
Foliage	
Root estimate	

Kazimirov, N.I. and Morozova, R.M. (1981). In: "Dynamic Properties of Forest Eco-systems" (D.E. Reichle, ed.), pp.629-645. Cambridge University Press, Cambridge, London, New York and Melbourne.

Kazimirov, N.I. and Morozova, R.M. (1973). "Biological Cycling of Matter in Spruce Forests of Karelia." Nauka, Leningrad Branch, Academy of Sciences, Leningrad, U.S.S.R.

---

ca.62°N 34°E   80-200 m   U.S.S.R., Karelia.

Humus iron
podzols.   *Picea abies*
pH 4.1-4.4

Age (years)	22	37	45	54
Trees/ha	34800	13750	9240	4820
Tree height (m)	2.6	6.8	8.8	11.1
Basal area (m²/ha)	10.6	21.9	23.5	24.8
Leaf area index	1.8	3.0	3.2	3.6
Stem volume (m³/ha)				

Dry biomass (t/ha)

	22	37	45	54
Stem wood	12.4	38.1	51.2	67.2
Stem bark	1.5	4.2	5.1	5.8
Branches	6.5	10.6	12.1	14.2
Fruits etc.				
Foliage	5.5	9.1	9.8	10.9
Root estimate	6.2	14.1	15.8	21.6

CAI (m³/ha/yr)

Net production (t/ha/yr)

Stem wood	1.05	$+0.42^a$	1.67	$+0.84^a$	1.88	$+0.93^a$	1.98	$+1.08^a$
Stem bark	0.13	$+0.24^b$	0.18	$+0.36^b$	0.18	$+0.40^b$	0.17	$+0.42^b$
Branches	0.29		0.24		0.19		0.15	
Fruits etc.								
Foliage	0.30	$+1.43^b$	0.21	$+2.31^b$	0.14	$+2.51^b$	0.07	$+2.73^b$
Root estimate	0.52		0.48		0.52		0.53	

There was 1.0, 3.4, 4.2 and 5.6 t/ha of standing dead wood in columns left to right.
a. Mortality.
b. Litterfall, omitting consumption.

Continued from p.361.

---

ca.62°N   34°E   80-200 m   U.S.S.R., Karelia.

Picea abies

	Peat pH 6.0	Peat pH 3.6	Sandy podzol pH 4.4	Sandy podzol pH 4.6
Age (years)	41	42	43	38
Trees/ha	9930	9410	6310	4480
Tree height (m)	6.7	5.8	9.8	12.2
Basal area (m²/ha)	17.7	14.8	23.2	25.4
Leaf area index	2.4	2.0	4.3	4.4
Stem volume (m³/ha)				

Dry biomass (t/ha)

	Peat pH 6.0	Peat pH 3.6	Sandy podzol pH 4.4	Sandy podzol pH 4.6
Stem wood	23.4	18.7	53.2	60.1
Stem bark	2.6	2.3	5.3	5.2
Branches	8.2	7.0	12.5	12.2
Fruits etc.				
Foliage	7.5	6.5	9.5	9.9
Root estimate	9.5	8.0	16.8	18.3

CAI (m³/ha/yr)

Net production (t/ha/yr)

	Peat pH 6.0	Peat pH 3.6	Sandy podzol pH 4.4	Sandy podzol pH 4.6
Stem wood	1.22	1.07	2.11	2.27
Stem bark	0.15 $\Big\}$ + 0.75[a]	0.13 $\Big\}$ + 0.59[a]	0.23 $\Big\}$ + 1.26[a]	0.23 $\Big\}$ + 1.51[a]
Branches	0.16 $\Big\}$ + 0.26[b]	0.14 $\Big\}$ + 0.21[b]	0.25 $\Big\}$ + 0.40[b]	0.27 $\Big\}$ + 0.45[b]
Fruits etc.				
Foliage	0.11 + 1.81[b]	0.10 + 1.51[b]	0.15 + 2.85[b]	0.16 + 3.12[b]
Root estimate	0.53	0.29	0.53	0.61

There was 3.4, 2.7, 6.2 and 6.9 t/ha of standing dead wood in columns left to right.
a. Mortality.
b. Litterfall, omitting consumption.

Continued from p.362.

ca.62°N   34°E   80-200 m   U.S.S.R., Karelia.

Picea abies

	Eluvium debris pH 3.3	Humus iron podzols		
		pH 4.2	pH 3.8	pH 4.3
Age (years)	37	39	45	68
Trees/ha	9010	9980	9620	2336
Tree height (m)	4.2	7.8	6.9	14.2
Basal area (m²/ha)	13.3	20.6	17.9	29.9
Leaf area index	1.8	3.4	2.6	3.8
Stem volume (m³/ha)				

Dry biomass (t/ha)

	Eluvium debris pH 3.3	pH 4.2	pH 3.8	pH 4.3
Stem wood	14.5	43.4	25.8	98.6
Stem bark	2.0	4.8	3.2	7.4
Branches	6.1	11.2	8.7	15.1
Fruits etc.				0.1
Foliage	5.7	10.2	8.2	11.5
Root estimate	6.6	14.6	10.1	29.1

CAI (m³/ha/yr)

Net production (t/ha/yr)

	Eluvium debris pH 3.3	pH 4.2	pH 3.8	pH 4.3
Stem wood	0.83	1.84	1.34	1.89
Stem bark	0.11 $\Big\}$ + 0.42[a] + 0.16[b]	0.21 $\Big\}$ + 1.16[a] + 0.38[b]	0.17 $\Big\}$ + 0.81[a] + 0.29[b]	0.15 $\Big\}$ + 1.12[a] + 0.41[b]
Branches	0.12	0.22	0.17	0.09
Fruits etc.				0.01
Foliage	0.08 + 1.40[b]	0.17 + 2.61[b]	0.13 + 2.20[b]	0.00 + 2.86[b]
Root estimate	0.27	0.48	0.36	0.50

There was 2.0, 5.3, 3.7 and 6.9 t/ha of standing dead wood in columns left to right.
a. Mortality.
b. Litterfall, omitting consumption.

Continued from p.363.

ca.62°N 34°E   110-140 m   U.S.S.R., Karelia.

Humus iron
podzols.   *Picea abies*
pH 3.8-4.1

Age (years)	82	98	109	126	138
Trees/ha	1898	1319	1080	856	1087
Tree height (m)	17.1	19.6	20.0	22.6	22.8
Basal area (m²/ha)	32.3	33.1	38.9	40.5	38.0
Leaf area index	3.8	3.6	3.2	2.7	2.4
Stem volume (m³/ha)					

Dry biomass (t/ha)

	82	98	109	126	138
Stem wood	109.0	149.0	156.4	174.9	167.9
Stem bark	7.0	8.9	8.6	9.0	8.1
Branches	16.8	16.5	17.6	16.6	17.1
Fruits etc.	0.3	0.4	0.4	0.5	0.4
Foliage	11.4	10.8	9.7	8.1	7.4
Root estimate	33.2	41.0	45.0	46.0	47.5

CAI (m³/ha/yr)

Net production (t/ha/yr)

Stem wood	1.63	$+1.17^a$	1.29	$+1.06^a$	0.92	$+0.89^a$	0.39	$+0.87^a$	0.05	$+0.91^a$
Stem bark	0.11	$+0.36^b$	0.07	$+0.26^b$	0.05	$+0.25^b$	0.03	$+0.14^b$	0.00	$+0.10^b$
Branches	0.06		0.04		0.02		0.01		0.01	
Fruits etc.	0.00		0.01		0.00		0.00		0.00	
Foliage	$-0.02+2.88^b$		$-0.04+2.63^b$		$-0.05+2.48^b$		$-0.06+2.25^b$		$-0.08+2.17^b$	
Root estimate	0.42		0.25		0.15		0.06		0.02	

There was 8.1, 8.0, 7.1, 7.1 and 7.4 t/ha of standing dead wood in columns left to right.
a. Mortality.
b. Litterfall, omitting consumption.

Rodin, L.E. and Bazilevich, N.I. (1967). "Production and Mineral Cycling in Terrestrial Vegetation." Oliver and Boyd, Edinburgh and London. (Quoting Russian authors in their Table 9).

ca.56°N 30°E — U.S.S.R., Velikiye Luki Province.	*Picea abies*, with *Populus* spp. and *Betula* spp. Podzolized loams.		*P. abies* with understorey shrubs. Peaty podzol.
Age (years)	110	72	83
Trees/ha			
Tree height (m)	(9.2　19.2)[b]	22.0	26.9
Basal area (m²/ha)			
Leaf area index			
Stem volume (m³/ha)			
Dry biomass (t/ha) Stem wood / Stem bark / Branches / Fruits etc.	182.3	214.7	260.9
Foliage	14.9	11.5	19.4
Root estimate	76.6	64.6	77.6
CAI (m³/ha/yr)			
Net production (t/ha/yr) Stem wood / Stem bark / Branches / Fruits etc.	$2.3 + 1.1^c$	$7.2 + 2.0^c$	$3.3 + 2.3^c$
Foliage	3.2^c	1.3^c	2.8^c
Root estimate	1.2	0.4	1.3

Values given above include understorey shrubs. Nutrient contents were determined.
a. Percentage of the total tree number.
b. Heights of lower and upper storeys.
c. Litterfall.

Rodin, L.E. and Bazilevich, N.I. (1967). "Production and Mineral Cycling in Terrestrial Vegetation." Oliver and Boyd, Edinburgh and London. (Quoting Russian authors in their Table 9).

Marchenko, A.N. and Karlov, Ye.M. (1962). Mineral exchange in spruce forests of the northern taiga and forest tundra in Arkhangelsk oblast. *Soviet Soil Sci.* 7, 722-734.

U.S.S.R.	ca.68°N 34°E -- Kola Peninsula	ca.65°N 47°E -- Arkhangelsk Region		ca.59°N 40°E -- Vologda Province
	Picea abies (70%)[a] *Betula* spp. (20%)[a] *et al.* Humus-iron podzol.	*P.* *abies* Poor site.	*P. abies* *Betula* spp. *et al.* Sandy, silt podzol.	*P. abies* (90%)[a] *Betula* spp. Poor peaty gley.
Age (years)	120	125	200	130
Trees/ha		1050	700	
Tree height (m)	10.9	15.0	30–32	17.6
Basal area (m²/ha)				
Leaf area index				
Stem volume (m³/ha)				
Dry biomass (t/ha)				
Stem wood	} 35.5	} 97.8+1.4[b]	188.6	} 118.9
Stem bark			12.4	
Branches		21.3+0.7[b]	33.6	
Fruits etc.		0.7		
Foliage	4.5	7.3+0.3[b]	16.3+4.8[b]	12.3
Root estimate	11.3	40.7	85.1	65.7
CAI (m³/ha/yr)				
Net production (t/ha/yr)				
Stem wood		} 1.1+0.9[c]	} 3.2+1.8[c]	} 1.8+1.3[c]
Stem bark	0.5[c]			
Branches				
Fruits etc.				
Foliage	1.3[d]	4.3	3.3	2.2
Root estimate		0.6	1.4	1.4

Values in the third column from the left are from Marchenko and Karlov (1962). Nutrient contents were determined.
a. Percentage of the total tree number.
b. Understorey shrubs; other biomass and production values in this table include the understorey shrubs.
c. Woody litterfall.
d. Foliage litterfall; foliage litterfall in the other columns was 2.7, 2.8 and 2.0 t/ha/yr left to right.

Kolli, R. and Kahrik, R. (1970). Phytomass and net primary production in the forests of the *Fragaria - Hepatica* type. *Trans. Estonian Agric. Acad. Soil Regimes and Processes (Sbornik nauctnykh trudov Estonskoi Selskokhozyaistvenov)* <u>65</u>, 69-91.

ca.58°N   25-30°E   --   U.S.S.R., Estonia.

	Mihkli.	Kaarma.
	Picea abies (88%)[a] *Quercus robur* (7%)[a] *Betula* spp. *et al.*	*P. abies* (66%)[a] *Alnus incana* (23%)[a] *Betula* spp. *et al.*
Age (years)	84	51
Trees/ha	538 + 70	724 + 376
Tree height (m)	23.7	17.5
Basal area (m²/ha)		
Leaf area index		
Stem volume (m³/ha)	249 + 38	201 + 39

Dry biomass (t/ha)

	Mihkli	Kaarma
Stem wood	104.6 + 15.4 ⎫	78.4 + 16.5 ⎫
Stem bark	8.1 + 2.0 ⎪	7.1 + 3.5 ⎪
Branches	23.2 + 3.4 ⎬ + 4.6[b]	20.7 + 3.9 ⎬ + 3.6[b]
Fruits etc.	0.2 ⎪	0.3 ⎪
Foliage	13.5 + 0.3 ⎭	11.0 + 0.7 ⎭
Root estimate	64.6	45.8

CAI (m³/ha/yr)

Net production (t/ha/yr)

	Mihkli	Kaarma
Stem wood	2.45 + 0.35 ⎫	3.01 + 1.10 ⎫
Stem bark	0.18 + 0.05 ⎬ c	0.29 + 0.23 ⎬ c
Branches	2.11 + 0.18 ⎭	2.31 + 0.66 ⎭
Fruits etc.	0.23	0.26
Foliage	2.82 + 0.28	2.70 + 0.72
Root estimate	5.06	4.49

Values are given above for *P. abies* plus broadleaved trees (left and right, respectively, in each column). There was 5.1 and 3.8 t/ha of dead branches in columns left and right, respectively.

a. Percentage of the total tree number.
b. Understorey shrubs.
c. Including estimated woody litterfall.

Dylis, N. (1971). Primary production of mixed forests. In: "Productivity of Forest Ecosystems" (P. Duvigneaud, ed.), pp.227-230. UNESCO, Paris.

---

ca.55°00'N 37°30'E   150 m   U.S.S.R., S. of Moscow, Pakhra River.

Well-drained
podzolic soils.
        Picea abies (75%)[a], *Betula pubescens*,
                Quercus spp. and *Populus* spp.

Thinned at various ages.

Age (years)		85
Trees/ha		
Tree height (m)		24
Basal area (m²/ha)		
Leaf area index		
Stem volume (m³/ha)		

Dry biomass (t/ha)	Stem wood	} 176.1
	Stem bark	
	Branches	23.9[b]
	Fruits etc.	} 17.4[c]
	Foliage	
	Root estimate	63.4

CAI (m³/ha/yr)

Net production (t/ha/yr)	Stem wood	} 6.19[d]
	Stem bark	
	Branches	1.47[d]
	Fruits etc.	0.23
	Foliage	3.08[e]
	Root estimate	

Three trees of average dimensions were sampled per species and their roots were excavated. Stand values were obtained by multiplying mean tree values by the numbers of trees per hectare.
a. Percentage of the total stem biomass.
b. Including dead branches.
c. Including scales, fruits and other green parts as well as the foliage.
d. Excluding woody litterfall and any mortality.
e. New foliage biomass.

Rodin, L.E. and Bazilevich, N.I. (1967). "Production and Mineral Cycling in Terrestrial Vegetation." Oliver and Boyd, Edinburgh and London. (Quoting Russian authors in their Table 9).

U.S.S.R.	ca.68°N 34°E -- Kola Peninsula. *Pinus sylvestris* (60%)[a] *Betula* spp. and *Picea abies*	ca.54°N 45°E -- Mordovskaya ASSR. *P. sylvestris*	59°N 77°E -- Vasyuganye swamp, W. Siberia. *P. sylvestris*
	Peaty humus podzol	Good site	Upland moss bog
Age (years)	100	71	100
Trees/ha			
Tree height (m)	7.8	24.1	5.0
Basal area (m²/ha)			
Leaf area index			
Stem volume (m³/ha)			
Dry biomass (t/ha) Stem wood			
Stem bark	56.7	202.4	17.9
Branches			
Fruits etc.			
Foliage	6.2	13.9	15.1
Root estimate	17.8	63.6	4.0
CAI (m³/ha/yr)			
Net production (t/ha/yr) Stem wood			
Stem bark	0.7[b]	2.8 + 2.2[b]	0.2 + 0.1[b]
Branches			
Fruits etc.			
Foliage	2.3[c]	2.4	3.2
Root estimate		0.9	0.1

Values given above include understorey shrubs. Nutrient contents were determined.
a. Percentage of the total tree number.
b. Woody litterfall.
c. Foliage litterfall; foliage litterfall in the other two columns was 2.1 and 2.3 t/ha/yr left and right, respectively.

Klinge, H. (1978). Studies on the ecology of Amazon caatinga forest in southern Venezuela. *Acta cient. venez.* <u>29</u>, 258-262.

Herrera, R. and Klinge, H. (1981). Phytomass of tall Amazon caatinga forest near San Carlos de Rio Negro, south Venezuela. *Vegetatio* (in press).

Klinge, H. and Herrera, R. (1978). Biomass studies in Amazon caatinga forest in southern Venezuela. I Standing crop of composite root mass in selected stands. *Trop. Ecol.* <u>19</u>, 93-110.

1°54'N 67°06'W   50-100 m   Venezuela, near San Carlos de Rio Negro.

Infertile podzols.

Micranda spruceana, Eperua leucantha and about 130 other species.

Blackwater Creek

	$72\%^a$	$43\%^a$	$54\%^a$	$66\%^a$	$19\%^a$	$16\%^a$
Age (years)						
Trees/ha	5600^b	6200^b	9400^b	9700^b	14300^b	13400^b
Tree height (m)	20.1	18.6	14.1	13.6	14.8	16.1
Basal area (m²/ha)	36.7^b	72.9^b	34.3^b	34.1^b	16.1^b	25.5^b
Leaf area index	4.4	9.5	5.8	5.4	3.1	5.7
Stem volume (m³/ha)	403^b	677^b	241^b	263^b	84^b	200^b

Dry biomass (t/ha)

	$72\%^a$	$43\%^a$	$54\%^a$	$66\%^a$	$19\%^a$	$16\%^a$
Stem wood	}178	}379	}107	}154	}52	}184
Stem bark						
Branches	112	215	74	80	52	74
Fruits etc.						
Foliage	4.8 (7)^c	14.7 (15)^c	5.3 (9)^c	8.6 (8)^c	4.5 (5)^c	5
Root estimate	71	141	42	101	91	115

CAI (m³/ha/yr)

Net production (t/ha/yr)

Stem wood						
Stem bark						
Branches						
Fruits etc.						
Foliage						
Root estimate						

The fresh weights of all vegetation, including 'extractable' roots, were determined in each of the above 100 m² plots. Six hundred and fifty-one individuals were sampled for dry weight and nutrient determinations. Stand dry biomass values were derived from regressions on D. There was 8.2, 6.5, 1.9, 14.8, 12.6 and 37.0 t/ha of standing dead wood in columns left to right. Nutrient contents were determined.
a. Percentage of the total biomass accounted for by *M. spruceana*.
b. Stems over 1 cm D, which accounted for over 95% of the total biomass.
c. Values in brackets are the total leaf biomass values of woody plus herbaceous vegetation; the unbracketed values refer to the trees only.

Continued from p.370.

Same as p.370.

	57%[a]	13%[a]	20%[a]	46%[a]	10%[a]	33%[a]	Bana woodland 0.1%[a]
Age (years)							
Trees/ha	15800[b]	11700[b]	10300[b]	10700[b]	10600[b]	13300[b]	17500[b]
Tree height (m)	16.0	16.9	16.4	15.9	19.2	15.5	10.1
Basal area (m²/ha)	22.0[b]	50.7[b]	33.0[b]	40.0[b]	59.5[b]	33.6[b]	21.8[b]
Leaf area index	2.8	9.8	3.9	4.7	4.6	3.2	3.0
Stem volume (m³/ha)	181[b]	419[b]	221[b]	292[b]	221[b]	200[b]	70[b]
Dry biomass (t/ha)							
Stem wood	}108	}263	}96	}183	}276	}107	}58
Stem bark							
Branches	72	256	81	91	166	56	32
Fruits etc.							
Foliage	4.6(5)[c]	14.7(17)[c]	7.3(7)[c]	6.5(10)[c]	7.5(8)[c]	5.5(5)[c]	5.0(5)[c]
Root estimate	173	336	172	83	195	92	142
CAI (m³/ha/yr)							
Net production (t/ha/yr) Stem wood							
Stem bark							
Branches							
Fruits etc.							
Foliage							
Root estimate							

Same as p.370, except

There was 10.3, 16.1, 4.9, 5.3, 19.9, 22.2 and 10.9 t/ha of standing dead wood in columns left to right.

Jordan, C.F. and Uhl, C. (1978). Biomass of a 'terra firma' forest of the Amazon basin. *Oecologia Plant.* 13, 387-400.

Stark, N. and Spratt, M. (1977). Root biomass and nutrient storage in rainforest oxisols near San Carlos de Rio Negro. *Trop. Ecol.* 18, 1-9.

Jordan, C.F. and Escalante, G. (1980). Root productivity in an Amazonian rainforest. *Ecology* 61, 14-18.

1°54'N 67°06'W   50-100 m   Venezuela, Amazonas, near San Carlos de Rio Negro.

Infertile, leached
unflooded, sandy   Tropical rainforest.
lateritic soil.

Age (years)	Mature
Trees/ha	1760 + 9457[a]
Tree height (m)	ca.14
Basal area (m²/ha)	34.3
Leaf area index	5.2
Stem volume (m³/ha)	

Dry biomass (t/ha)

Stem wood	
Stem bark	
Branches	} 315.7[b]
Fruits etc.	
Foliage	8.0
Root estimate	56.0

CAI (m³/ha/yr)

Net production (t/ha/yr)

Stem wood	
Stem bark	
Branches	} 6.00 + 3.40[c]
Fruits etc.	
Foliage	6.00[c]
Root estimate	>2.01[d] (16.80)[e]

Forty-two trees of 28 species were sampled, ignoring a few lianes, and stand biomass values for four plots of 0.5 to 1.0 ha were derived from regressions on D²H. Values given above are for the authors' plot 1. Roots were excavated in 18 pits in two plots. There was 2 to 8 t/ha of standing dead wood (8.3 t/ha in plot 1).
a. Trees under 5 cm D.
b. Including 30.6 t/ha of bark.
c. Litterfall, measured over 3 years.
d. Increment of surface roots only.
e. Estimated total root increment plus root death.

Bartholomew, W.V., Meyer, J. and Laudelout, H. (1953). "Mineral Nutrient Immobilization under Forest and Grass Fallow in the Yangambi (Belgian Congo) Region, with
some Preliminary Results on the Decomposition of Plant Material on the Forest
Floor." Publs Inst. natn. Etude agron. Congo belge Sér. Sci. No.57.

ca.0°50'N　24°25'E　　200-500 m　　Zaire, near Kisangani, Yangambi region.

Musanga cecropioides, et al.

Regeneration after clear-felling and burning.

Age (years)	5	8	17 to 18
Trees/ha			
Tree height (m)			
Basal area (m²/ha)			
Leaf area index			
Stem volume (m³/ha)			

Dry biomass (t/ha)

	5	8	17 to 18
Stem wood			
Stem bark	71.1	116.3	114.6
Branches			
Fruits etc.			
Foliage	5.6	5.4	6.4
Root estimate	25.8	22.8	31.2

CAI (m³/ha/yr)

Net production (t/ha/yr)

	5	8	17 to 18
Stem wood			
Stem bark	14.5^a	15.1^a	
Branches			$+ 14.2^b$
Fruits etc.			
Foliage			
Root estimate			

All vegetation was harvested and weighed in an area of 300 m² at each of the three
sites, and root samples were taken at random. Values given above include understorey shrubs, but exclude dead wood which was not weighed.
a. Wood increment, excluding litterfall.
b. Estimated total litterfall.

Malaisse, F., Alexandre, J., Freson, R., Goffinet, G. and Malaisse-Mousset, M. (1972). The miombo ecosystem; a preliminary survey. In: "Tropical Ecology with an Emphasis on Organic Productivity" (P.M. Golley and F.B. Golley, eds), pp.363-403. Institute of Ecology, University of Georgia, Athens, Georgia, U.S.A.

Malaisse, F. (1981). In: "Dynamic Properties of Forest Ecosystems" (D.E. Reichle, ed.), p.672. Cambridge University Press, Cambridge, London, New York, Melbourne.

11°37'S   27°29'E   1244 m   Zaire, Luanza, Kasapa.

Latosol,   *Brachystegia boehmii, Pterocarpus angolensis,*
pH 5.0.   *Marquesia macroura, et al.*

Miombo woodland

	Miombo woodland
Age (years)	120
Trees/ha	446
Tree height (m)	14 to 18
Basal area (m²/ha)	22.0^a
Leaf area index	3.5
Stem volume (m³/ha)	

		Miombo woodland
Dry biomass (t/ha)	Stem wood	52.8
	Stem bark	10.7
	Branches	78.2
	Fruits etc.	0.5
	Foliage	2.6
	Root estimate	25.5

CAI (m³/ha/yr)

Net production (t/ha/yr)	Stem wood		(or 0.53)^c
	Stem bark	0.89^b	
	Branches		(or 0.65)^c
	Fruits etc.	0.19^b	(or 0.54)^c
	Foliage	2.98^b	(or 4.26)^c
	Root estimate		

The biomass of all vegetation was measured in one 625 m² clear-felled plot. There was 24.7 t/ha of standing dead wood.

a. The mean basal area of the Kasapa miombo as a whole was 12.7 m²/ha.
b. Litterfall only, measured over 3 years, from Malaisse *et al.* (1972).
c. Litterfall plus decomposition and other losses given by Malaisse (1981).

Guy, P.R. (1981). Changes in the biomass and productivity of woodlands in the
Sengwa wildlife research area, Zimbabwe. *J. appl. Ecol.* <u>18</u>, 507-519.

18°10'S 28°14'E ca.1200 m Zimbabwe, Sengwa Wildlife Research Area.				
Sandy soils.	Miombo woodland.	Riverine woodlands.		Mopane woodland.
	Julbernardia globiflora, Brachystegia boehmii, et al.	*Acacia, Combretum, Diospyros, et al.*	*Acacia albida*	*Colophospermum mopane, et al.*
Age (years)				
Trees/ha	321	49	30	542
Tree height (m)				
Basal area (m²/ha)	9.2	4.9	9.8	23.3
Leaf area index				
Stem volume (m³/ha)				
Dry biomass (t/ha) — Stem wood, Stem bark, Branches, Fruits etc., Foliage, Root estimate	$21.9 + 1.2^{a}$	$19.8 + 3.7^{a}$	$52.2 + 0.0^{a}$	$64.5 + 1.3^{a}$
CAI (m³/ha/yr)				
Net production (t/ha/yr) — Stem wood, Stem bark, Branches, Fruits etc., Foliage, Root estimate	$0.45 + 0.07^{a}$	$0.26 + 0.19^{a}$	$0.47 + 0.05^{a}$	$1.21 + 0.07^{a}$

Biomass values were derived from regression on D for trees measured along 68 transects in 1972. Tree numbers refer to those over 6 cm D or 3 m height. Basal areas
were measured above the basal swelling.
These stands were grazed by elephants, and smaller biomass and production values
were recorded for 1974 and 1976; values for undisturbed miombo woodlands are given
by Malaisse *et al.* (1972) (see p.374).
a. Understorey shrubs.

AUTHOR INDEX

Abbott, W. 216
Adams, W.R. 299
Aeberli, B.C. 13
Akai, T. 175, 177, 178, 179, 182
Akhtar, M.A. 285
Alban, D.H. 270, 296, 299, 312
Albrektson, A. 225
Alexandre, J. 374
Alvera, B. 221
Ambroes, P. 25, 26, 28
Andersson, F. 223
Ando, T. 94, 125, 142, 145, 155, 171, 172, 181, 186
Arkley, R.J. 293
Arnold, L.E. 285
Ashton, D.H. 16
Attiwell, P.M. 14, 245
Auclair, A.N.D. 41
Auclair, D. 65, 67
Azzollini, I. 86

Baker, J.B. 267, 321
Bandhu, D. 81
Bandola-Ciolczyk, E. 213
Barney, R.J. 297
Bartholomew, W.V. 373
Baskerville, G.L. 40
Bazilevich, N.I. 353, 354, 355, 357, 359, 365, 366, 369
Beets, P. 203
Beets, P.N. 204
Berglund, J.V. 48
Bernhardt-Reversat, F. 91
Bickelhaupt, D.H. 249, 313
Bidkupsky, V. 56
Bindiu, C. 219, 220
Blackmon, B.G. 267
Bockheim, J.G. 273
Bormann, F.H. 259
Boysen Jensen, P. 60
Bradstock, R. 11
Bray, J.R. 36, 269
Brown, A.T. 289, 294, 310, 335, 351
Brunig, E. 30
Bullock, J.A. 195
Bunce, R.G.H. 240
Burger, H. 229, 230, 231, 232
Burkhart, H.E. 323
Burnette, C.E. 327

Cabanettes, A. 71
Cameron, J.N. 9
Campbell, A.G. 289, 294, 310, 335, 351
Campbell, N.A. 20
Cannell, M.G.R. 189
Cantiani, M. 87
Carey, M.L. 85
Carter, M.C. 266, 323
Chan, Y.H. 37
Chew, A.E. 262
Chew, R.M. 262
Chiba, K. 125, 186
Child, G.I. 209, 210, 211, 285
Christensen, B. 233
Cintrón, B. 218
Clarke, A.R.P. 9
Clements, R.G. 209, 210, 211
Cleve, K. van 251, 297
Coaldrake, J.E. 7
Cole, D.W. 253, 329, 340
Connor, W.H. 288
Cooper, A.W. 272
Cooper, D.T. 267
Cooper, J.M. 246, 247
Cottam, G. 276
Cragg, J.B. 37
Cromer, R.N. 9, 10
Crow, T.R. 217, 252, 274

Danaeyer-De Smet, S. 28
Day, F.P. 283
Day, J.W. 288
Deans, J.D. 242
Decei, I. 219
Demott, T.E. 322
Deselm, H.R. 256, 298
Dice, S.F. 329
Dimmock, G.M. 8, 12
Djhalilov, K.G. 356
Doi, K. 155, 173
Donita, N. 219
Doucet, R. 48
Drift, J. van der 201
Droste zu Hülshoff, B. von 73
Dudkiewicz, L.A. 36, 269
Duever, M.J. 209, 210, 211
Dugger, K. 218
Duvigneaud, P. 24, 25, 26, 28, 29
Dylis, N. 368
Dyrness, C.T. 329, 340

377

SPECIES INDEX

'Pure' means the species occurred in pure stands. 'Major' means the species was a major component of a mixed species stand, comprising between about 15% and 99% of the total basal area. 'Minor' means the species was a minor component of a mixed species stand, comprising less than 15% of the total basal area. 'Prod.' (short for 'production') means that some components of the current annual dry matter increment and/or litterfall were measured. '(Tropical)' means that the species was part of a mixed species tropical forest or woodland. Readers wishing to locate data on all tropical mixed species forests should consult the subject index.

The authorities of Latin names, which have been omitted from the tables, are given here.

SUBJECT INDEX

The following eight subject lists are provided to help readers locate data other than by species or country.

1. 'Age series' means that data are given for stands which differ in age, but are otherwise comparable.

2. 'Density series' means that data are given for stands with different stocking densities, but of the same age.

3. 'Fertilizer treatment effects' means that data were collected for stands or plots that received different fertilizer treatments.

4. 'Leaf area index' means that projected LAI values are given.

5. 'Litterfall' means that the weights of all or some components of the litterfall are given in the tables or footnotes.

6. 'Nutrient content data available' means that the mineral nutrient contents of the biomass were determined, and that these data can be found in the original publications.

7. 'Site differences' means that data are given for similar stands growing near each other on different site types.

8. 'Tropical forests' gives a complete listing of the tropical forest stands.